SELECTED LECTURES OF THE ROYAL SOCIETY

VOLUME 3

SELECTED LECTURES OF THE ROYAL SOCIETY

Reprinted from the
Philosophical Transactions or the
Proceedings of The Royal Society

VOLUME 3

1970
THE ROYAL SOCIETY
and
ACADEMIC PRESS · LONDON · NEW YORK

ACADEMIC PRESS INC. (LONDON) LTD.
Berkeley Square House
Berkeley Square
London, W1X 6BA

U.S. Edition published by
ACADEMIC PRESS INC.
111 Fifth Avenue
New York, New York 10003

SBN: 12–573653–3

MADE AND PRINTED BY OFFSET IN GREAT BRITAIN BY
WILLIAM CLOWES AND SONS, LIMITED, LONDON AND BECCLES

Contents

THE LEEUWENHOEK LECTURE, 1967

Teichoic acids and the molecular structure of bacterial walls

BY J. BADDILEY, F.R.S.

Microbiological Chemistry Research Laboratory, Department of Organic Chemistry, University of Newcastle upon Tyne

(*Delivered* 9 *November* 1967—*Received* 22 *December* 1967)

Proceedings of the Royal Society, B, *volume* 170, *pp*. 331–348, 1968

A glance at previous Leeuwenhoek Lectures reveals that in a number of cases the Lecturer has felt obliged to preface his account with an apology to the effect that the subject of his lecture would seem to bear little relationship to the distinguished work of Antony van Leeuwenhoek. In my case, however, such apology is unnecessary as a reasonably direct connexion can be established between those remarkable observations in the seventeenth century and the work described here. Leeuwenhoek's description of bacteria is based on their appearance under his primitive but remarkably effective microscopes. Such appearance is, of course, governed by shape and we see from the published record that cocci, bacilli and spirochaetes were observed and accurately described in terms of size and shape. These features are functions of the nature and rigidity of the wall surrounding the organisms and so are related to molecular structure; the purpose of my lecture is to describe certain aspects of the recently acquired knowledge of the molecular architecture of bacterial cell walls.

The great interest that has developed in the chemistry and biochemistry of bacterial cell walls during the last ten years or so has arisen for a number of reasons (cf. Salton 1964). In the first place, walls are of interest because they represent a substantial proportion of the metabolic products of the cell; they frequently comprise up to 20 % of the dry weight of cells and so must be regarded as important metabolites. Moreover, their chemical structure is interesting in view of their physical properties. Thus, besides being reasonably rigid structures with considerable strength they are nevertheless freely permeable towards cellular products and nutrients. Under certain conditions even large molecules such as antibodies, extracellular enzymes and nucleic acids can penetrate the wall, although the full extent and nature of this permeability, and the effect of negatively charged polymers (e.g. teichoic acids) in the wall, has not been clearly defined. The wall is frequently the site of important antigenic material; for example, in many cases it has been shown that group- or type-specific antigens are located in the outer structures of the cell, including the wall. Consequently, a better understanding of the immunological properties of bacteria, and particularly such features as pathogenicity, require a full understanding of the structure, function and biosynthesis of wall components.

In recent times a particularly important feature of cell-wall biochemistry has emerged through developments in our understanding of the mechanism of action of a number of antibiotics. It is now known that the penicillins, cephalosporins, bacitracin, vancomycin, ristocetin, cycloserine and possibly certain other antibiotics exert their characteristic action on bacteria by interfering with stages in the biosynthesis of wall macromolecules. A full understanding of the action of these antibiotics, with possible consequence in the development of synthetic and semi-synthetic antibacterial agents, requires a detailed knowledge of many aspects of the molecular structure, biochemistry and general biology of walls.

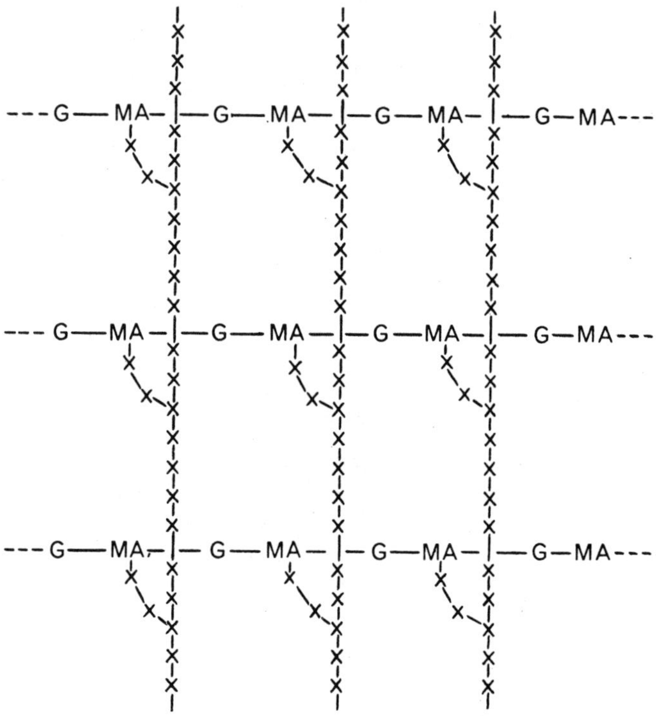

FIGURE 3. Schematic representation of the peptidoglycan in the walls of strains of *Staphylococcus aureus*. In the horizontal polysaccharide chains G = *N*-acetylglucosamine, MA = *N*-acetylmuramic acid. In the vertical polypeptide chains X = amino acid. At branch points on the polypeptide chains X = lysine. In some peptidoglycans from other bacteria the branch point is 2,4-diaminopimelic acid.

Many of the recent advances in this field have been made possible by the development of satisfactory procedures for the preparation of clean walls in useful amounts. Modern methods usually depend upon the mechanical rupture of bacterial cells followed by differential centrifugation in order to remove the broken walls from other cell components. The original procedure developed by Salton & Horne (1951) involves shaking suspensions of cells with glass powder (Ballotini beads); walls prepared by this method are shown in figure 1, plate 1. Usually such preparations, when examined by electron microscopy, show no regularity of surface structure; in rare cases, however, a regular pattern can be seen (figure 2, plate 2). This apparent

(*a*)

(*b*)

FIGURE 1. (*a*) Electron microphotograph of walls of *Lactobacillus lactis* 7278. (Magn. × 40 000.) (*b*) Electron microphotograph of walls of *Staphylococcus lactis* I3. The point of cell rupture and the thickened region of cell division are clearly seen. (Magn. × 28 000.)

PLATE 1

(*Facing p.* 2)

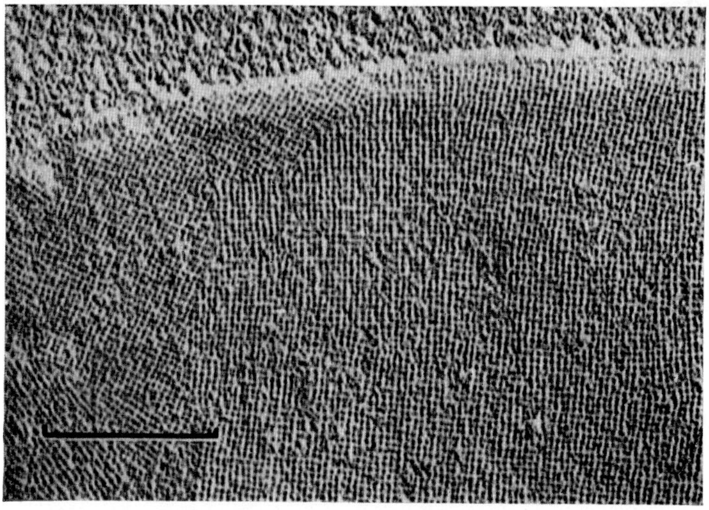

FIGURE 2. Electron microphotograph of part of a wall of *Bacillus polymyxa*. The black dots represent an ordered array of protein-polysaccharide granules. The scale mark represents 0·25 μm. From Goundry, Davison, Archibald & Baddiley (1967).

PLATE 2

regularity has been studied in detail in the case of *Bacillus polymyxa*, where it has been shown that the pattern represents an ordered array of granules containing protein and polysaccharide (Goundry, Davison, Archibald & Baddiley 1967; Nermut & Murray 1967).

The walls of all bacteria contain a substance called mucopeptide or peptidoglycan which is responsible for their strength and rigidity. In Gram-negative bacteria this is usually a relatively minor component, whereas in Gram-positive bacteria it may represent 50 to 90 % of the dry weight of the wall. The composition of this wall component varies in different organisms. The general structure in figure 3 is an approximation intended to describe the peptidoglycan from the walls of strains of *Staphylococcus aureus*. It is composed of polysaccharide chains (represented horizontally) in which *N*-acetylglucosamine and *N*-acetylmuramic acid alternate regularly. Attached to the carboxyl group of the muramic acid is a peptide composed of L-alanine, D-alanine, D-glutamic acid (as D-isoglutamine), L-lysine and about five glycine residues. These peptide chains cross-link to the ω-amino of lysine in a neighbouring peptide, and in this way they join parallel polysaccharide chains to form an open network with a certain amount of flexibility and great strength. The repeating unit in the peptidoglycan from *Staphylococcus aureus* is illustrated in more detail in structure (1).

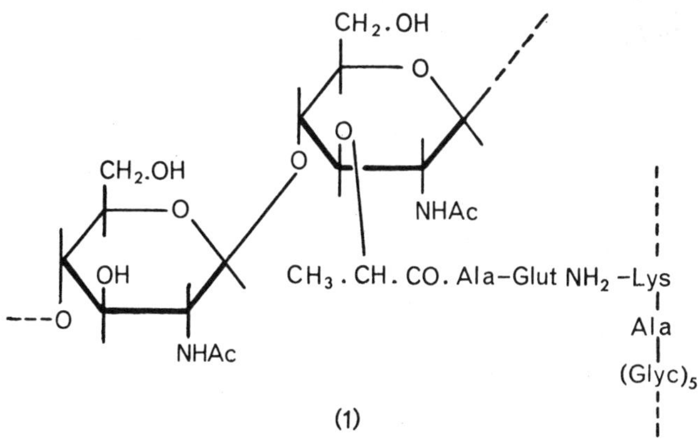

(1)

Whereas it is likely that the peptidoglycan in a number of staphylococci is represented approximately by the structure (1), it is now believed that in many bacteria the chemical structure of this wall component may differ markedly from the relatively regular network illustrated here. For example, cross-linking may be less complete, it may occur in part between units in the same polysaccharide chain, and variations in the nature and number of amino acids in the peptide chains have been observed. In other bacteria considerable variations in peptidoglycan structure have been noted; some lack the pentaglycine chains, whereas in others considerable numbers of muramic carboxyl groups are unsubstituted. It is noteworthy that the composition of the peptide chains can be varied in some cases in a given organism by variation of the amino acid composition of the medium. It is now known that the primary action of the penicillins on bacteria is to interfere with a step in the formation of the peptide chains. In cell-free systems from staphylococci this step is a late one involving the formation of cross-links between the terminal amino group of a pentaglycine chain and the carboxyl group of the subterminal alanine on a neighbouring peptide chain (Wise & Park 1965; Tipper & Strominger 1965; Izaki, Matsuhashi & Strominger 1966). In contrast to the variations in the composition of peptide chains, it appears that there is little variation in the nature of the polysaccharide chains; in all cases studied in detail there is believed to be a regular alternation of *N*-acetylglucosamine and *N*-acetylmuramic acid units along the chain, all glycosidic linkages possessing the β-configuration and involving the 4-hydroxyl group on amino sugar residues.

The walls of Gram-negative bacteria are complex. In addition to peptidoglycan they usually contain phospholipids, lipopolysaccharides and proteins. Although there has been much important work on them (cf. reviews by Lüderitz, Staub & Westphal (1966) and by Horecker (1966) on lipopolysaccharides and related compounds), this is not the place for their consideration in detail. On the other hand, the walls of Gram-positive bacteria, although of considerable structural complexity, are somewhat simpler in their general overall composition. In a number of

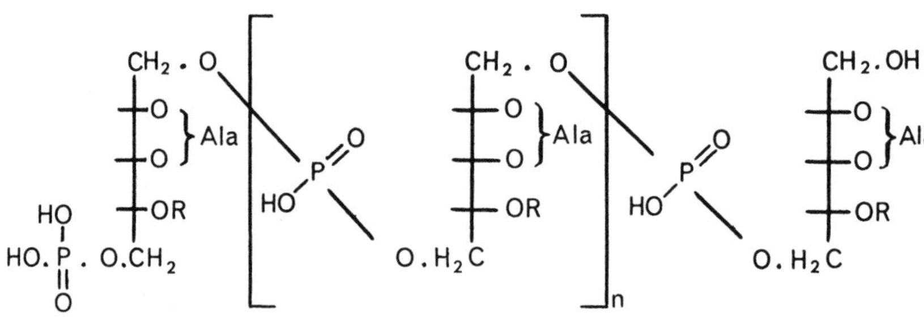

R = glycosyl

(2)

cases they contain polysaccharides of various degrees of complexity attached in an as yet unestablished manner to the peptidoglycan. In many cases teichoic acids are important wall components (Armstrong, Baddiley, Buchanan, Carss & Greenberg 1958). These components, which often comprise between 20 and 50 % of the weight of the wall, are polymeric substances containing either glycerol or ribitol, phosphate and usually sugar residues and D-alanine (for a review see Archibald & Baddiley 1966). The first examples were found to be composed of chains in which ribitol phosphate units are joined to each other through phosphodiester linkages (scheme 2). They can be regarded as polymers of D-ribitol 5-phosphate (L-ribitol 1-phosphate). Carbohydrate residues are usually attached to the D-4 (L-2) position and alanine is in labile ester linkage with a hydroxyl group at position 2 or 3 on ribitol. Such polymers have been found in the walls of a large range of Gram-positive bacteria, and in the early studies structural variation was thought to be largely confined to the number and nature of the carbohydrate residues. In many cases the substituent is D-glucose, but *N*-acetyl-D-glucosaminyl substituents are often found. The size of these polymers is not known accurately but it is generally thought that they vary from 7 to 15 or so units in length. Such teichoic acids have been observed in bacilli, (Armstrong, Baddiley & Buchanan 1960), staphylococci (Baddiley, Buchanan, Martin & RajBhandary 1962) and lactobacilli (Archibald, Baddiley & Buchanan 1961). Their composition is usually fairly constant but some variation has been noted in organisms grown in different media. Glycerol teichoic acids (scheme 3) (Ellwood, Kelemen & Baddiley 1963), in which the chain can be regarded as a polymer of glycerol phosphate with phosphodiester linkages joining positions 1 and 3 in adjacent glycerol residues, are also found in the walls of many bacteria. These usually possess both carbohydrate and D-alanine substituents. Their chain length is known with even less certainty than it is for the ribitol teichoic acids but it is likely that they are larger than the latter polymers.

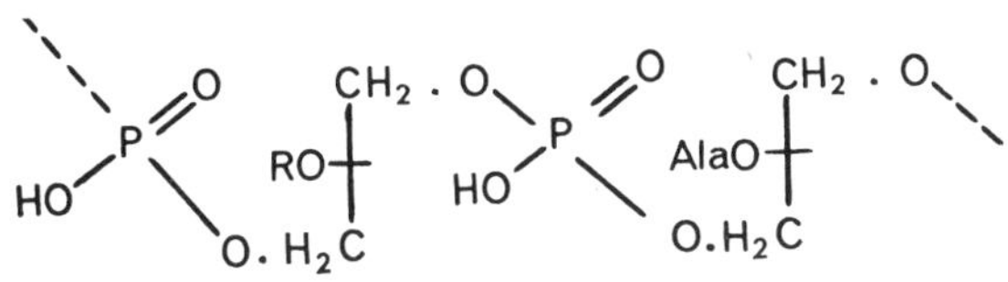

R = glycosyl

(3)

Although there have been no detailed studies on the effect of changes in medium or growth conditions on the composition of teichoic acids, general experience suggests that such effects would not be significant in the majority of cases. Nevertheless, the number of glucosyl substituents on the ribitol teichoic acid from the walls of *Lactobacillus arabinosus* 17-5 does depend partly upon the concentration of glucose in the growth medium (A. R. Archibald, unpublished). On the other hand, the amount of teichoic acid in walls, in contrast to their general composition, can be markedly influenced by the composition of the medium (Ellwood & Tempest 1967), and probably by other factors concerned with growth conditions.

In cases where the ratio of carbohydrate substituent to polyol is not an integer, there arises the question whether the preparation is a mixture of regularly substituted and unsubstituted polymers or whether all the molecules are incompletely substituted. A serological method has been used to show that in one case such a preparation appears to be a mixture containing unsubstituted and fully substituted molecules (Burger 1966), whereas analysis of products of alkali hydrolysis has shown that in other cases incompletely substituted chains are present (Ellwood *et al.* 1963; Shaw & Baddiley 1964). The ribitol teichoic acid from strains of *Staphylococcus aureus* is interesting; in most preparations *N*-acetylglucosamine is attached to ribitol by both α and β glycosidic linkages (Baddiley *et al.* 1962), and a serological study has shown that these preparations are in fact mixtures of polymers containing entirely α-linkages and others containing entirely β-linkages (Torii, Kabat & Bezer 1964).

Although polymers of glycerol phosphate and ribitol phosphate of the type described above occur in the walls of many bacteria, more extensive studies have shown that fundamental structural variations are common. For example, a teichoic acid (scheme 4) in the wall of *Streptomyces griseus* is a polymer of ribitol phosphate in which glucose residues are attached to some of the ribitol, whereas others bear glycerol phosphate substituents joined to the 3-position on ribitol through phosphodiester linkage (B. Bews, J. G. Buchanan & J. Baddiley, unpublished). Thus, this teichoic acid is a mixed ribitol-glycerol polymer in which the glycerol phosphate residues are appendages to the main chain. It is interesting that in this case, and in other teichoic acids from *Streptomyces* spp. (Naumova, Shaba-

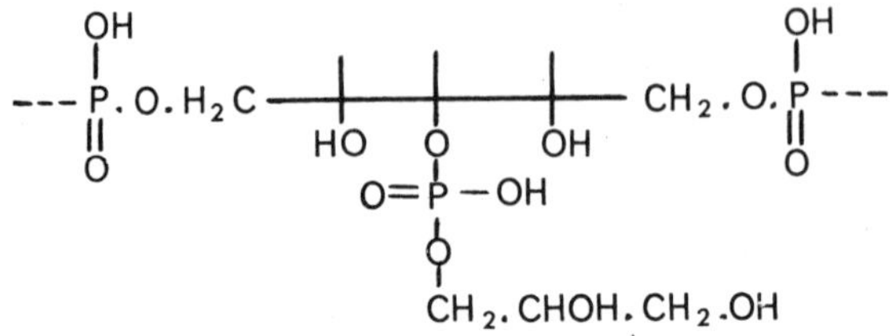

(4)

rova & Belozersky 1963), alanine is absent but succinate half esters are found. A common variation on the original structural type is one in which sugar residues form a part of the polymer chain. These were first encountered in strains of *Bacillus subtilis* (Glaser & Burger 1964) where the polymer chain is composed of repeating units (scheme 5) in which glucose or galactose is directly attached to

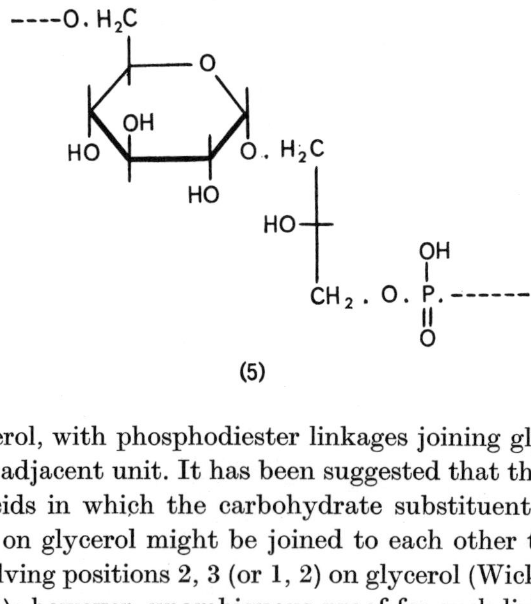

(5)

position 1 on glycerol, with phosphodiester linkages joining glycerol to position 6 of the sugar on an adjacent unit. It has been suggested that the repeating units of several teichoic acids in which the carbohydrate substituent is attached to the primary hydroxyl on glycerol might be joined to each other through phosphodiester linkages involving positions 2, 3 (or 1, 2) on glycerol (Wicken 1966; Naumova & Zeretskaya 1966); however, unambiguous proof for such linkages is lacking.

More complex examples in which carbohydrate substituents form a part of the chain are encountered in pneumococci. The long-known, serologically characteristic C-substance in the walls of a number of different *Pneumococcus* types is, in fact, a teichoic acid comprising ribitol phosphate to which is attached a disaccharide containing *N*-acetylgalactosamine and a diaminotrideoxyhexose (Brundish & Baddiley 1967); some of the details of this structure are not yet settled. Closely related are the serologically type-specific substances from the capsules of certain pneumococci; the one illustrated (scheme 6) is from Type 29 (M. J. Watson, E. V. Rao, J. G. Buchanan & J. Baddiley, unpublished). Examples of capsular substances from different pneumococcal types have been shown to possess similar structures. Although these are not normally regarded as wall components, they are so closely related to teichoic acids that it is convenient to classify them as such.

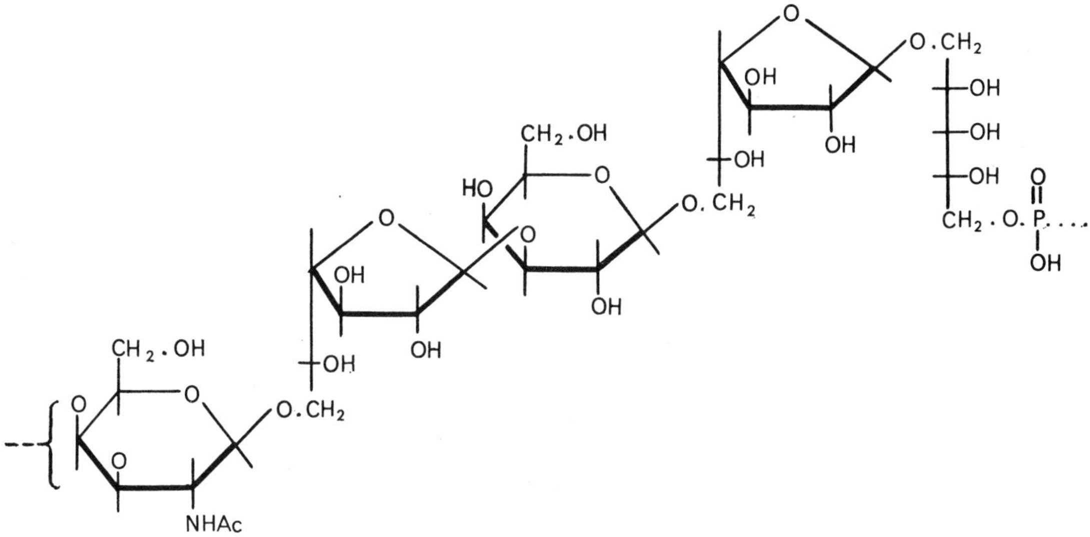

(6)

Recent work on wall teichoic acids from micrococci has shown that in these organisms structural variations are even more diverse than those already discussed. For example, *Staphylococcus lactis* I 3 contains a teichoic acid (scheme 7)

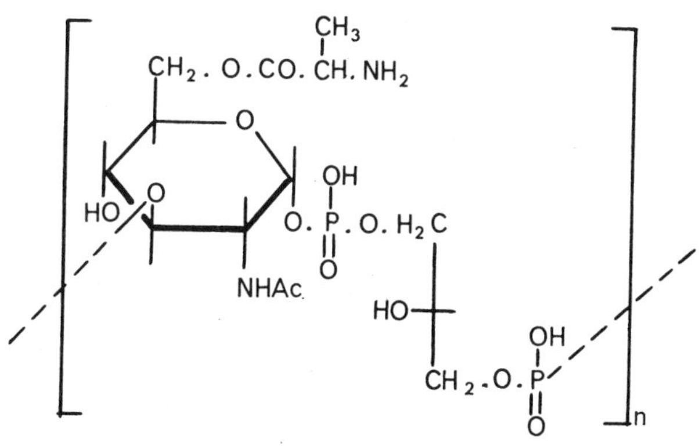

(7)

in which *N*-acetylglucosamine 1-phosphate is attached through its phosphate to glycerol phosphate, which in turn is attached to a 3-hydroxyl group in the amino sugar residue of an adjacent unit. Alanine is also present, but in this case it is attached through an ester linkage to the 6-hydroxyl of the amino sugar residue (Archibald *et al.* 1965). This is an example of an increasing number of polymers from micro-organisms in which sugar 1-phosphate linkages occur. Other examples include the extracellular phosphomannans and phosphogalactans from yeasts and a number of as yet incompletely characterized teichoic acids mainly from micrococci. Recent studies have shown that another strain of *Staphylococcus lactis* contains in its wall a polymer of *N*-acetylglucosamine 1-phosphate; linkage between units in this polymer involves the phosphate at position 1, which is attached to position 6 of a neighbouring amino sugar residue (unpublished). Details of the structure of this polymer have not been completely elucidated but there is evidence for a small amount of branching as shown in the provisional formula (scheme 8). Neither glycerol nor ribitol is present, and alanine is also apparently absent. Nevertheless, its location in the wall and its general chemical nature make it convenient to include it and similar wall polymers with the teichoic acids.

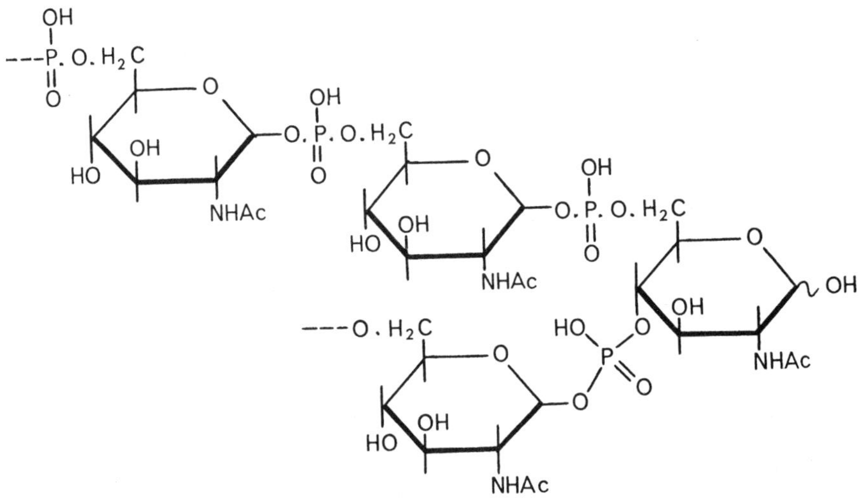

(8)

The distribution and location of teichoic acids are of considerable biological significance. A reasonably extensive survey (Baddiley & Davison 1961; Davison & Baddiley 1963, and unpublished work) of Gram-positive bacteria has shown that, whereas many contain either glycerol or ribitol teichoic acids in their walls, there are examples of organisms lacking these wall polymers; a small selection from this survey is shown in table 1. On the other hand, the survey reveals in all cases the presence of glycerol teichoic acids elsewhere in the cell. This so-called 'intracellular' teichoic acid is found in the supernatant from disrupted cells after removal of walls. In most disrupted cell preparations it can be sedimented by high-speed centrifugation, but occasionally it remains in the supernatant. Studies with protoplasts (Hay, Wicken & Baddiley 1963; Shockman & Slade 1964; Smith & Shattock 1964) have shown that 'intracellular' glycerol teichoic acids are located in the vicinity of the membrane, presumably attached to it in an undefined manner. For this reason they are now referred to as membrane teichoic acids. The structures of several membrane teichoic acids have been determined (Kelemen & Baddiley 1961; Critchley, Archibald & Baddiley 1962; Wicken & Baddiley 1963;

RajBhandary & Baddiley 1963; D. Button, unpublished) and it is found that all conform to the general formula (3), although the extent of carbohydrate substitution on glycerol shows considerable differences in material from different organisms. It is interesting that even in those organisms where the wall teichoic acid is atypical, e.g. in *Staphylococcus lactis* I3, the membrane teichoic acid is of the conventional type.

TABLE 1. OCCURRENCE AND DISTRIBUTION OF TEICHOIC ACIDS IN A SELECTION OF GRAM-POSITIVE BACTERIA

organism	wall	membrane
Lactobacillus arabinosus 17-5	R	G
L. plantarum NCIB 7220	R	G
L. helveticus NCIB 8025	G	G
L. jugurti NCIB 2889	G	G
L. buchneri NCIB 8001	G	G
L. brevis NCIB 8169	G	G
L. casei ATCC 7469	—	G
L. bulgaricus NCIB 76	—	G
L. fermentii NCIB 6691	—	G
Bacillus subtilis	R	G
B. megaterium NCIB 7581	—	G
B. cereus NCIB 2600	—	G
B. licheniformis ATCC 9945	G	G
B. subtilis NCTC 3610	G	G
Staphyloccocus aureus (many strains)	R	G
S. lactis NCTC 7944	G	G
S. lactis I3	G	G
S. saprophyticus (many strains)	G	G
S. epidermidis NCTC 7291	G	G
Streptococcus faecium NCIB 8191	R	G

R = ribitol teichoic acid. G = glycerol teichoic acid.

The widespread occurrence of wall teichoic acids, and the probably universal occurrence of membrane teichoic acids in Gram-positive bacteria, suggests that they may have an important function in the life processes of the cell. Evidence for function is still lacking, but it is possible that they are concerned in some way with the control of passage of ions through the wall-membrane region of the cell, or they may act as a barrier to the approach of harmful substances. In this connexion, it is noteworthy that Gram-negative bacteria, although they lack teichoic acids, nevertheless possess similar acidic polymers (lipopolysaccharides) in their outer regions.

An important property of bacteria during infection of animals is their behaviour towards antibodies in the serum and other body tissues. Although the role of teichoic acids in the processes of infection and the animal's defences against bacterial diseases is not yet known, many of them are powerfully antigenic. They exhibit serological specificity that is markedly dependent upon the nature and configuration of their glycosyl substituents. Thus, wall or membrane teichoic acids are group-specific components in most of the serological groups of lactobacilli (Baddiley & Davison 1961; Sharpe, Davison & Baddiley 1964). Similarly, their importance as specific components of staphylococci (Haukenes, Ellwood, Baddiley & Oeding 1961; Nathensen & Strominger 1962; Haukenes 1962; Morse 1962; Oeding, Myklestad & Davison 1967) and streptococci (McCarty 1959; Wicken, Elliott & Baddiley 1963) has been established.

Although teichoic acids are freely soluble in water, their extraction from walls is slow and usually requires acidic conditions. The probable existence of a covalent linkage between the macromolecular components of walls was indicated from studies of the action of autolytic enzymes. From enzymically hydrolyzed walls it has been possible to obtain soluble, macromolecular products, in which teichoic acid is attached to degraded peptidoglycan (Ghuysen 1964; Young, Tipper & Strominger 1964). Moreover, such preparations contain no phosphomonoester groups and it is concluded that, in some cases at least, the linkage between teichoic acid and the wall involves the terminal phosphate on teichoic acid (Ghuysen,

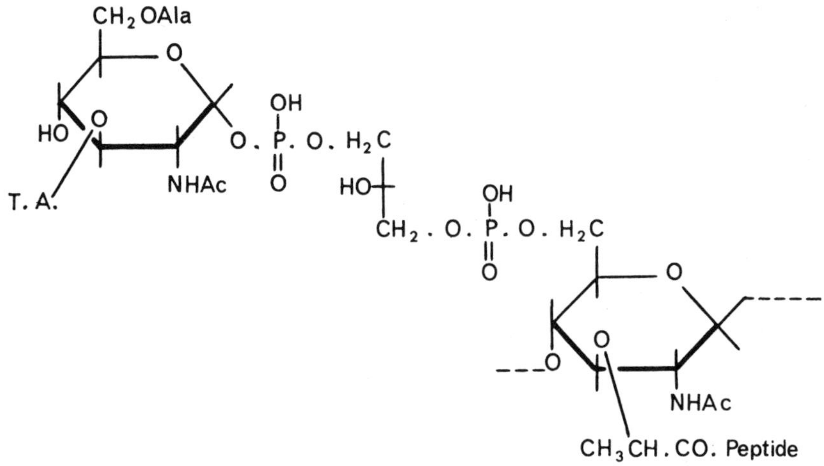

(9)

Tipper & Strominger 1965). In only one case, however, has a direct demonstration of the chemical nature of this linkage been obtained. Walls of *Staphylococcus lactis* I3 possess autolytic enzymes that degrade the peptidoglycan. From autolysates of washed walls a preparation has been obtained in which the teichoic acid is still chemically joined to a small fragment of peptidoglycan. In studies involving both chemical and enzymic hydrolysis (Button, Archibald & Baddiley 1966) it has been shown that a glycerol phosphate residue at one end of the teichoic acid is attached through phosphodiester linkage to a hydroxyl, presumably at position 6, on a muramic acid residue in the peptidoglycan (scheme 9). Hydrolysates of walls from other bacteria frequently contain a phosphate of muramic acid (Liu & Gotschlich 1967), and it is possible that linkage between teichoic acid and peptidoglycan in these cases is similar to that for the staphylococcus. It has also been suggested that the linkage between peptidoglycan and a wall polysaccharide in a strain of *Lactobacillus casei* might involve a phosphodiester between muramic acid and the 1-position of a sugar at the end of the polysaccharide chain (Hall & Knox 1965).

The discovery of teichoic acids followed the earlier discovery (Baddiley *et al.* 1956) of the nucleotides, cytidine diphosphate glycerol (CDP-glycerol) (scheme 10) and cytidine diphosphate ribitol (CDP-ribitol) (scheme 11), which are now known to be widely distributed in the Gram-positive bacteria. It was suggested (Armstrong *et al.* 1958) that these nucleotides might be involved in the biosynthesis of polymers containing polyol phosphate residues. After the discovery of the teichoic acids, it was assumed that the nucleotides would be precursors of these polymers and that the biosynthesis route would involve the successive transfer of polyol phosphate residues from the appropriate nucleotide to the extending polymer chain. The demonstration in a cell-free system of the transfer of polyol phosphates from nucleotides to teichoic acids confirmed and extended the earlier suggestions.

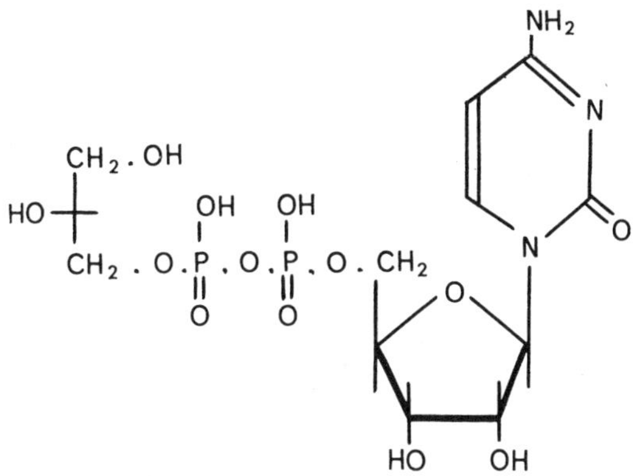

(10)

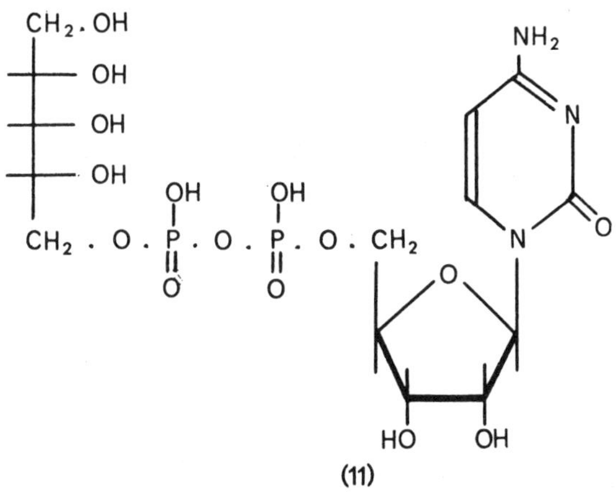

(11)

The generally accepted mechanism (Glaser 1964) for the biosynthesis of ribitol teichoic acids from D-ribulose 5-phosphate is outlined in figure 4. The ribitol phosphate is produced by reduction of D-ribulose 5-phosphate (Glaser 1963), and the nucleotide CDP-ribitol is formed by a reversible reaction between cytidine triphosphate and the ribitol phosphate (Shaw 1962). The enzyme responsible for the transfer of ribitol phosphate residues from the nucleotide to the polymer is located in the protoplast membrane. It is not known at what stage D-alanine is incorporated into the final product, but it is likely that an enzyme that catalyses a reaction between D-alanine and *ATP* is concerned in the process (Baddiley & Neuhaus 1960). Sugar residues are attached to the polymer by transfer from the appropriate nucleoside diphosphate sugar after the transfer of polyol phosphate (Glaser 1964; Ishimoto & Strominger 1966). A similar scheme for the biosynthesis of glycerol teichoic acid has been established (Burger & Glaser 1964; Glaser & Burger 1964).

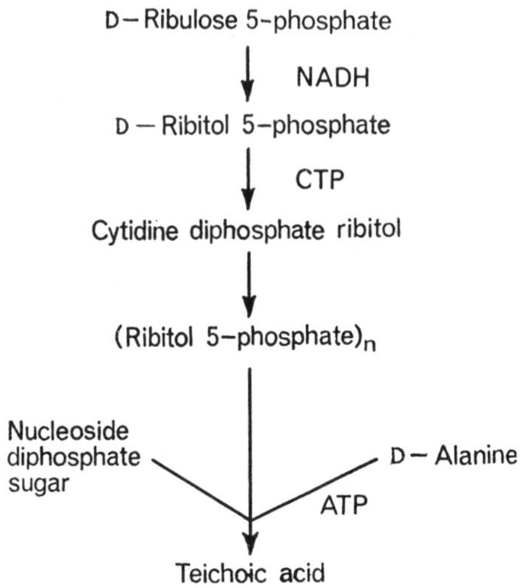

FIGURE 4. Biosynthesis of a ribitol teichoic acid.

Teichoic acids in which the sugar residues are a part of the polymer chain presumably do not conform to the biosynthetic route already described. In the polymer (scheme 5), and the corresponding one containing galactose, the chain is built up by alternate transfer from uridine diphosphate sugar (UDP-sugar) and CDP-glycerol (Burger & Glaser 1966). The teichoic acid (scheme 7) from *Staphylococcus lactis* I3 is particularly interesting in this respect; not only is the sugar a part of the polymer chain, but the repeating unit contains two phosphate groups. A particulate enzyme preparation representing fragmented membrane has been obtained from this organism, and the biosynthesis of the teichoic acid with this system requires the simultaneous presence of CDP-glycerol and UDP-*N*-acetylglucosamine (Blumsom, Douglas & Baddiley 1966). Moreover, the enzymes responsible for the biosynthesis of the two nucleotide precursors are themselves attached to the membrane preparation; the biosynthetic scheme is shown in figure 5. Using doubly labelled (^{14}C in acetyl and ^{32}P in phosphate) UDP-*N*-acetylglucosamine it was shown that CDP-glycerol contributes the part of the repeating unit enclosed on the right-hand side of figure 6, and UDP-*N*-acetylglucosamine contributes not only the amino sugar but the phosphate attached to its 1-position, as shown on the left-hand side of figure 6. This is the first example of a nucleoside diphosphate sugar donating a sugar 1-phosphate residue to a polymer. The mechanism of incorporation of the D-alanine is unknown, but as polymer synthesis occurs in the absence of amino acid it is likely that attachment of the latter occurs as a final step in the biosynthesis.

There have been interesting recent developments in our knowledge of the biosynthesis of other bacterial wall components, e.g. peptidoglycan and lipopolysaccharides. It is now known that the transfer of sugar residues from the appropriate nucleotides during the formation of polysaccharide chains in peptidoglycan

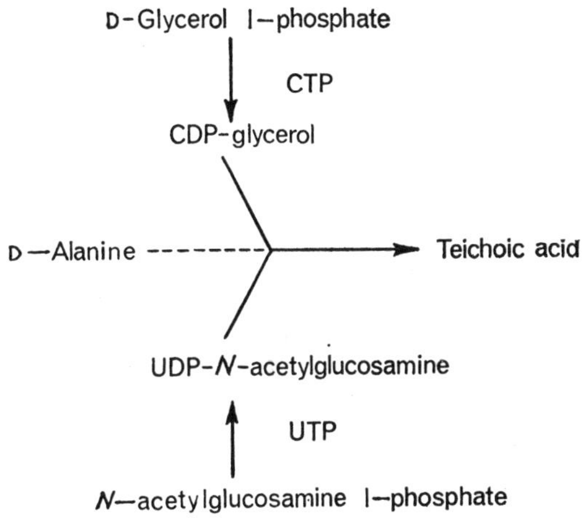

FIGURE 5. Biosynthesis of the glycerol teichoic acid (scheme 7) in the walls of *Staphylococcus lactis* I3.

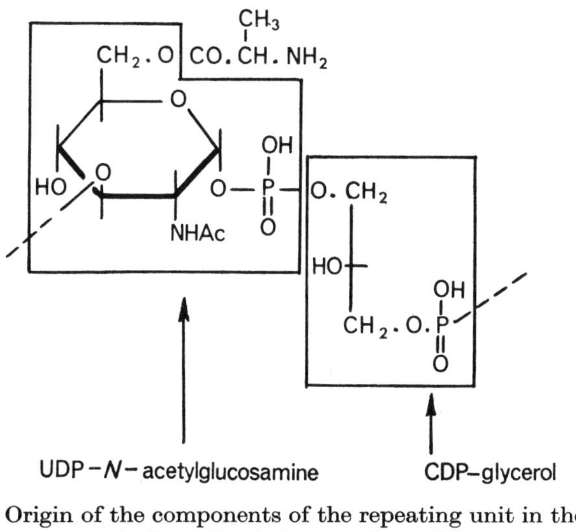

FIGURE 6. Origin of the components of the repeating unit in the teichoic acid from *Staphylococcus lactis* I3.

(Higashi, Strominger & Sweeley 1967) and in lipopolysaccharide (Wright, Dankert, Fennessey & Robbins 1967) is indirect. The sugars are transferred from nucleotides to a phospholipid in the particulate enzyme preparation, thereby producing a lipid intermediate containing a complete repeating unit; this unit then becomes attached to the extending chain of the polymer. The mediation of lipid intermediates in the biosynthesis of peptidoglycan is outlined in figure 7. The route starts at the top left-hand corner with the nucleotide, uridine diphosphate *N*-acetylmuramyl pentapeptide. This nucleotide, which was first isolated from bacteria grown in the presence of penicillin (Park 1952) and later recognized as a wall precursor (Park & Strominger 1957), possesses a peptide chain attached to the carboxyl of the muramic acid in the sequence L-alanyl-D-isoglutamyl-L-lysyl alanyl-D-alanine. The steps whereby the amide amino is introduced on the isoglutamyl, the five glycine residues are introduced on the ω-amino of the lysyl, and the cross-linkage is formed between the subterminal D-alanyl of a neighbouring

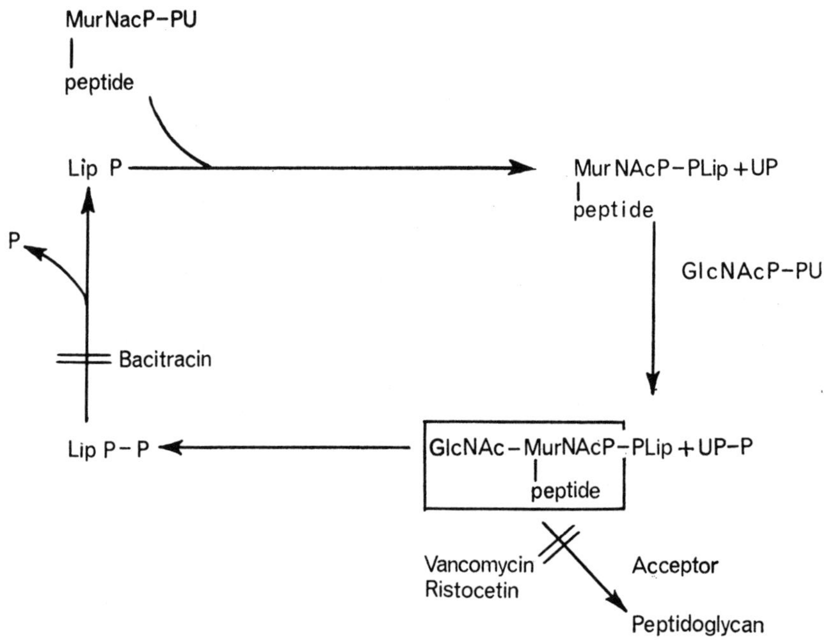

FIGURE 7. Lipid intermediates and inhibition by antibiotics during the formation of peptidoglycan.

MurNAcP-PU = uridine diphosphate *N*-acetylmuramyl-pentapeptide.
GlcNAcP-PU = uridine diphosphate *N*-acetylglucosamine.

chain and the amino of glycyl with production of free D-alanine have been omitted for the sake of clarity. For the extension of the polysaccharide chains reaction occurs between Park's nucleotide and a phospholipid to form a lipid pyrophosphate derivative which then accepts *N*-acetylglucosamine from UDP-*N*-acetylglucosamine. After the attachment of glycine residues the unit is transferred to an acceptor that can be regarded as the extending polysaccharide chain of the peptidoglycan; cross-linkage of the peptide chains occurs after transfer. The antibiotics vancomycin and ristocetin inhibit the transfer, whereas penicillin inhibits the cross-linkage process. The other product of transfer is a lipid pyrophosphate. The cycle is completed by the action of a pyrophosphatase that yields inorganic phosphate and the original phospholipid which is now available for repetition of the cycle. Bacitracin interferes with wall synthesis by inhibiting the pyrophosphatase. A similar cyclic sequence has been demonstrated independently for the biosynthesis of the oligosaccharide chains in the lipopolysaccharide in *Escherichia coli* (Wright *et al.* 1967), and in both cases the phospholipids (scheme 12) are derivatives of an isoprenoid alcohol containing eleven isoprene units.

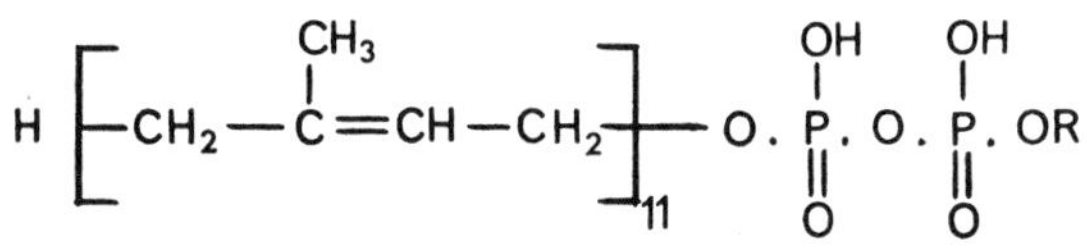

(12)

Preliminary experiments (L. J. Douglas, N. L. Blumsom & J. Baddiley, unpublished) have shown that a system involving lipid intermediates participates in the biosynthesis of the teichoic acid in *Staphylococcus lactis* I3, In the cell-free system UDP-*N*-acetylglucosamine containing ^{14}C in its acetyl group was readily incorporated into a lipid fraction in the enzyme preparation; the isotope appeared

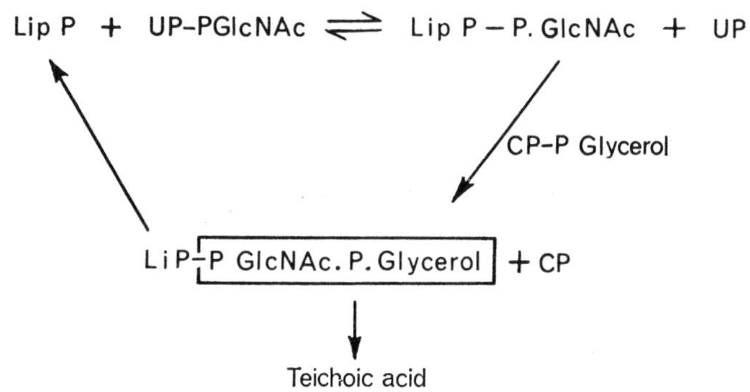

FIGURE 8. Possible route for the biosynthesis of a teichoic acid through lipid intermediates.

in teichoic acid after subsequent addition of unlabelled nucleotide and CDP-glycerol. Similarly, UDP-*N*-acetylglucosamine containing ^{32}P attached to the amino sugar was used to demonstrate that both sugar and phosphate groups are transferred to the lipid. Although studies on the detailed structure of the lipid intermediates are incomplete, it seems likely that the major component is a lipid pyrophosphate containing *N*-acetylglucosamine. A possible route for the biosynthesis of the teichoic acid through lipid intermediates is shown in figure 8. The incorporation of isotope from CDP-glycerol into lipid was much less effective than it was from UDP-*N*-acetylglucosamine. This suggests that either there is an independent mechanism for the introduction of glycerol phosphate, or the phosphate residue is attached to the repeating unit after the amino sugar, and that subsequent transfer of this unit to polymer is a rapid process. It is noteworthy that in the route outlined in figure 8 the transfer of repeating units would yield a lipid monophosphate, and thus the scheme differs at this point from the analogous one in the biosynthesis of lipopolysaccharides and peptidoglycans

Some features of the mechanism of biosynthesis of the teichoic acid in *Staphylococcus lactis* I 3 are inadequately explained at the present time. For example, it is not possible to exclude a scheme in which *N*-acetylglucosamine is transferred from a lipid to polymer, whereas glycerol phosphate is transferred directly from

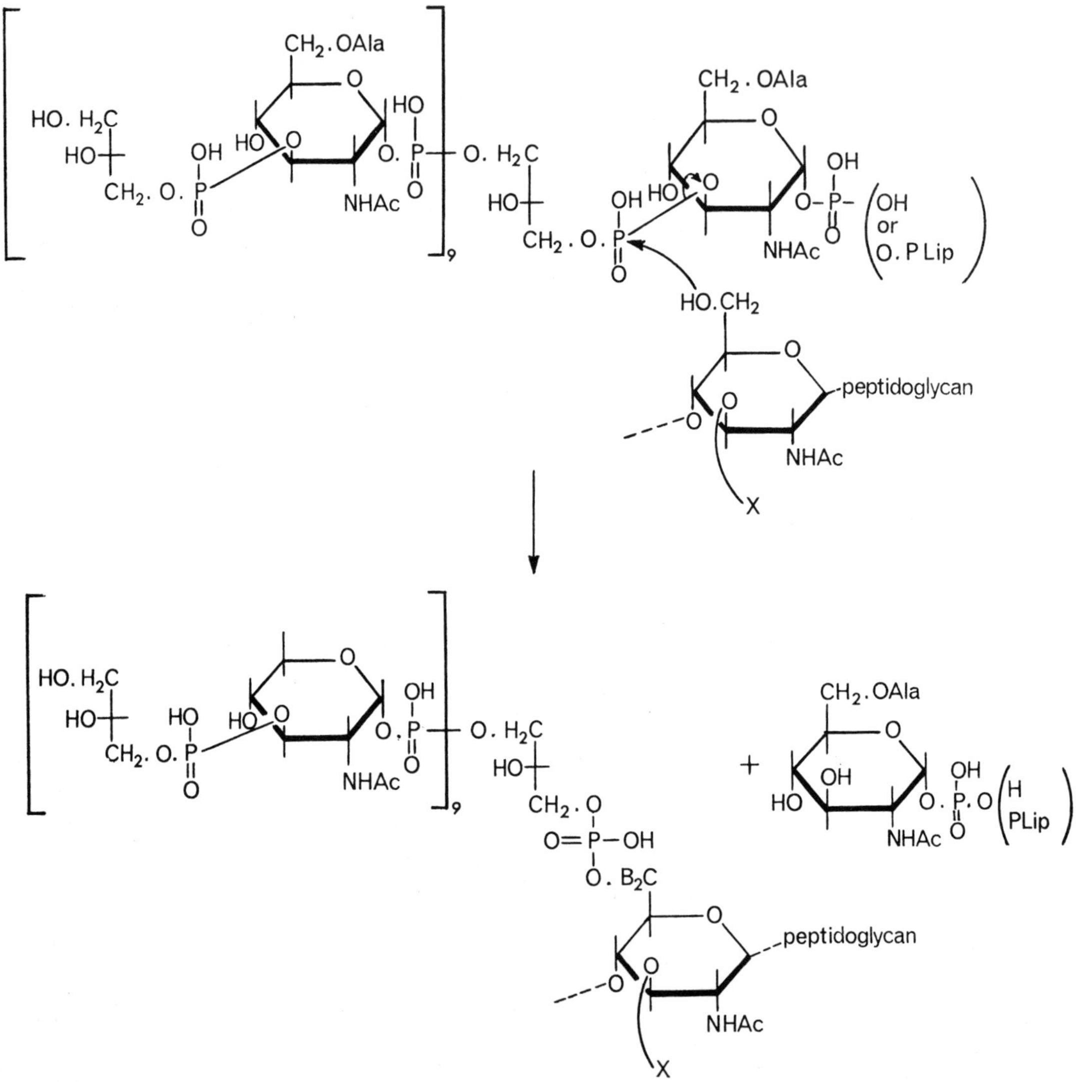

FIGURE 9. Suggested mechanism for the attachment of teichoic acid to peptidoglycan in *Staphylococcus lactis* I 3.

CDP-glycerol without preliminary attachment to a lipid intermediate. It is not known at which end of the polymer chain extension occurs, and the mechanism of attachment of teichoic acid to the muramyl residue in the peptidoglycan is not fully understood. A polymer comprising an integral number of repeating units, each with the structure of the unit on the lipid illustrated in figure 8, might be expected to be attached to the peptidoglycan through a phosphodiester linkage from a terminal *N*-acetylglucosamine. In fact, the attachment is through a phosphate to glycerol (see scheme 9); this interesting observation could be explained if the polymer is composed of a non-integral number of units. Recent experiments have shown that this is the case, as it appears that not only does the polymer terminate in glycerol phosphate at the point of attachment to muramic acid, but it terminates in glycerol at its other end. The process of attachment of teichoic acid to peptidoglycan would require first the formation of a polymer of about ten units, each of the structure shown on the lipid in figure 8, and then a phosphate ester exchange would occur between the 6-hydroxyl on a muramic acid residue and the phosphodiester group adjacent to the P-terminal end of the polymer chain; an *N*-acetylglucosamine 1-phosphate residue would be detached from the end of the teichoic acid during the transfer. This is illustrated in figure 9.

It is likely that work in the general field of bacterial cell walls will continue to develop in a number of directions in the future. Already it is clear that there are many important differences in the chemistry and biochemistry of walls from different organisms grown under varying conditions. Our knowledge of the nature and mechanism of formation of the linkage between macromolecular wall components will develop considerably. Immunological studies and the elucidation of details of the mechanisms of action of antibiotics will continue, and increasing attention may be given to the function of wall polymers and their relation to the function and properties of the cytoplasmic membrane. The difficult problems associated with attempts to correlate chemical structures with physical appearance of the outer regions of bacteria remain largely unsolved, and it is to be hoped that this important field of biology at the molecular level will stimulate much further work.

I am greatly indebted to my many colleagues who have contributed so much to the work described. They are too numerous to mention individually but I am particularly grateful to Drs A. R. Archibald and J. G. Buchanan, who have been associated with different aspects of this work over a number of years.

References

Archibald, A. R. & Baddiley, J. 1966 *Advanc. Carbohydrate Chem.* **21**, 323.

Archibald, A. R., Baddiley, J. & Buchanan, J. G. 1961 *Biochem. J.* **81**, 124.

Archibald, A. R., Baddiley, J. & Button, D. 1965 *Biochem. J.* **95**, 8c.

Armstrong, J. J., Baddiley, J. & Buchanan, J. G. 1960 *Biochem. J.* **76**, 610.

Armstrong, J. J., Baddiley, J., Buchanan, J. G., Carss, B. & Greenberg, G. R. 1958 *J. Chem. Soc.* p. 4344.

Baddiley, J., Buchanan, J. G., Carss, B., Mathias, A. P. & Sanderson, A. R. 1956 *Biochem. J.* **64**, 599.

Baddiley, J., Buchanan, J. G., Martin, R. O. & RajBhandary, U. L. 1962 *Biochem. J.* **85**, 49.

Baddiley, J. & Davison, A. L. 1961 *J. Gen. Microbiol.* **24**, 295.

Baddiley, J. & Neuhaus, F. C. 1960 *Biochem. J.* **75**, 579.

Blumsom, N. L., Douglas, L. J. & Baddiley, J. 1966 *Biochem. J.* **100**, 26c.

Brundish, D. E. & Baddiley, J. 1967 *Biochem. J.* **105**, 30c.

Burger, M. 1966 *Proc. Nat. Acad. Sci., Wash.* **56**, 910.

Burger, M. & Glaser, L. 1964 *J. Biol. Chem.* **239**, 3168.

Burger, M. & Glaser, L. 1966 *J. Biol. Chem.* **241**, 494.

Button, D., Archibald, A. R. & Baddiley, J. 1966 *Biochem. J.* **99**, 11c.

Critchley, P., Archibald, A. R. & Baddiley, J. 1962 *Biochem. J.* **85**, 420.
Davison, A. L. & Baddiley, J. 1963 *J. Gen. Microbiol.* **32**, 271.
Ellwood, D. C., Kelemen, M. V. & Baddiley, J. 1963 *Biochem. J.* **86**, 213.
Ellwood, D. C. & Tempest, D. W. 1967 *Biochem. J.* **104**, 69 P.
Ghuysen, J. M. 1964 *Biochim. biophys. Acta* **83**, 132.
Ghuysen, J. M., Tipper, D. J. & Strominger, J. L. 1965 *Biochemistry* **4**, 474.
Glaser, L. 1963 *Biochim. Biophys. Acta* **67**, 528.
Glaser, L. 1964 *J. Biol. Chem.* **239**, 3178.
Glaser, J. & Burger, M. 1964 *J. Biol. Chem.* **239**, 3187.
Goundry, J., Davison, A. L., Archibald, A. R. & Baddiley, J. 1967 *Biochem. J.* **104**, 1 C.
Hall, E. A. & Knox, K. W. 1965 *Biochem. J.* **96**, 310.
Haukenes, G. 1962 *Acta Path. Microbiol. Scand.* **55**, 463.
Haukenes, G., Ellwood, D. C., Baddiley, J. & Oeding, P. 1961 *Biochim. Biophys. Acta* **53**, 425.
Hay, J. B., Wicken, A. J. & Baddiley, J. 1963 *Biochim. Biophys. Acta* **71**, 188.
Higashi, T., Strominger, J. L. & Sweeley, C. C. 1967 *Proc. Nat. Acad. Sci., Wash.* **57**, 1798.
Horecker, B. L. 1966 *Ann. Rev. Microbiol.* **20**, 253.
Ishimoto, N. & Strominger, J. L. 1966 *J. Biol. Chem.* **241**, 639.
Izaki, K., Matsuhashi, M. & Strominger, J. L. 1966 *Proc. Nat. Acad. Sci. Wash.* **55**, 656.
Kelemen, M. V. & Baddiley, J. 1961 *Biochem. J.* **80**, 246.
Liu, T. & Gotschlich, E. C. 1967 *J. Biol. Chem.* **242**, 471.
Lüderitz, O., Staub, A. M. & Westphal, O. 1966 *Bacteriol. Rev.* **30**, 192.
McCarty, M. 1959 *J. Exp. Med.* **109**, 361.
Morse, S. I. 1962 *J. Exp. Med.* **116**, 229.
Nathenson, S. G. & Strominger, J. L. 1962 *J. Biol. Chem.* **237**, PC 3839.
Naumova, I. B., Shabarova, Z. A. & Belozersky, A. N. 1963 *Dokl. Akad. Nauk SSSR*, **152**, 1471.
Naumova, I. B. & Zaretskaya, M. Z. 1966 *Biochem. Biophys. Res. Comm.* **25**, 7.
Nermut, M. V. & Murray, R. G. E. 1967 *J. Bact.* **93**, 1949.
Oeding, P., Myklestad, B. & Davison, A. L. 1967 *Acta Path. Microbiol. Scand.* **69**, 458.
Park, J. T. 1952 *J. Biol. Chem.* **194**, 877, 885, 897.
Park, J. T. & Strominger, J. L. 1957 *Science, N.Y.* **125**, 99.
RajBhandary, U. L. & Baddiley, J. 1963 *Biochem. J.* **87**, 429.
Salton, M. R. J. 1964 *The bacterial cell wall.* Amsterdam: Elsevier Publishing Co.
Salton, M. R. J. & Horne, R. W. 1951 *Biochim. Biophys. Acta* **7**, 177.
Sharpe, M. E., Davison, A. L. & Baddiley, J. 1964 *J. Gen. Microbiol.* **34**, 353.
Shaw, D. R. D. 1962 *Biochem. J.* **82**, 297.
Shaw, N. & Baddiley, J. 1964 *Biochem. J.* **93**, 317.
Shockman, G. D. & Slade, H. D. 1964 *J. Gen. Microbiol.* **37**, 297.
Smith, D. G. & Shattock, P. M. F. 1964 *J. Gen. Microbiol.* **34**, 165.
Tipper, D. J. & Strominger, J. L. 1965 *Proc. Nat. Acad. Sci. Wash.* **54**, 1133.
Torii, M., Kabat, E. A. & Bezer, A. E. 1964 *J. Exp. Med.* **120**, 13.
Wicken, A. J. 1966 *Biochem. J.* **99**, 108.
Wicken, A. J. & Baddiley, J. 1963 *Biochem. J.* **87**, 84.
Wicken, A. J., Elliott, S. D. & Baddiley, J. 1963 *J. Gen. Microbiol.* **31**, 231.
Wise, E. M. & Park, J. T. 1965 *Proc. Nat. Acad. Sci. Wash.* **54**, 75.
Wright, A., Dankert, M., Fennessey, P. & Robbins, P. W. 1967 *Proc. Nat. Acad. Sci., Wash.* **57**, 1798.
Young, F. E., Tipper, D. J. & Strominger, J. L. 1964 *J. Biol. Chem.* **239**, PC 3600.

Review Lecture

Plant breeding for crop improvement in Britain: methods, achievements and objectives

By G. D. H. Bell, F.R.S.

Plant Breeding Institute, Cambridge

(*Delivered* 7 *March* 1968—*Received* 8 *April* 1968)

Proceedings of the Royal Society, B, *volume* 171, *pp*. 145–173, 1968

British agriculture and crop production

British agriculture possesses a number of interesting features which determine its place in the life of the country and which characterize its structure and economic contribution. Of the total United Kingdom land area, 90% is devoted to food production, there being 18 million acres of temporary and permanent grass, 18 million acres of arable, and nearly 18 million acres of rough and common grazing. On this land the farming community, which is less than 4% of the economically active population, produces two-thirds of the country's temperate food supplies, the current value of the gross output being approximately £1900 million, of which 70% is livestock and livestock products, 20% crops and 10% fruit, vegetables and flowers. However, the spread and versatility of home agriculture can best be appreciated by considering the break-down of the home contribution to temperate food supplies consumed in this country (table 1).

Table 1. Per cent contribution to Britain's temperate food supply from home agriculture

stock products		crop products	
beef and veal	70	wheat	48
mutton and lamb	40	barley	96
pork	98	potatoes	96
bacon	36	sugar	29
milk	100	fruit	65
butter	6	vegetables	80
cheese	42		
eggs and poultry	100		

Thus, in terms of cash value and of contribution to the country's food requirements, British agriculture is founded on livestock products and crop products, including grass, which contribute to the livestock industry and the important items such as potatoes, sugar beet, wheat and barley which are utilized directly, or as processed articles, for human consumption. It is not possible to measure in cash the value of the grass crop, but it is the most important single crop in terms of livestock products. The value of the arable crops is approximately £384 million, of which cereals contribute £240 million, potatoes £84 million and sugar beet £40 million. It is in this sector of agricultural production that it is considered that a major contribution can be made to the economy of the country by increasing production and thereby saving a portion of the £1500 million being spent on imported plant foods of temperate origin.

There is every reason to anticipate that the economy of Britain must look to agriculture for an ever increasing contribution and it is fortunate that the industry is in a position to respond by increasing its efficiency. Home food production has more than doubled since before World War II, and has increased by 40% in the last 10 years, while the per caput agricultural index increased from the 1952–56 base of 100 to 135 in 1965 compared with the world figure of 108. This figure for British agriculture is, with the exception of Greece and Cyprus, the highest of any country, and is a direct reflexion of the industry's rapid growth and development. In relation to other home industries, agriculture has in recent years been increasing its manpower productivity at nearly twice the national average, while its rising total production has been achieved in spite of an annual decrease in the labour force of 20000 and an annual loss of 40000 acres of land.

These achievements are the result of organizational improvements and greater technical efficiency which are reflected in an increase in net inputs which include maintenance and improvements, feeding stuffs, machinery, seeds, fertilizers and livestock. Moreover, there is justification for maintaining that British agriculture has remained equally competitive with alternative sources of food supply which is a vitally important consideration for the future if there is to be a greater reliance on home production not only to meet the anticipated rise in consumer expenditure on food of 1·3% annually, but also an increased saving in imports. Undoubtedly, one of the key factors to a successful accomplishment of the task will be raising the national yield averages for the major crops and to do this there must be varieties of crops with improved performance. This is recognized in all considerations of policy planning for the future of agriculture and targets for crop production are based on the likelihood of better varieties becoming available. Indeed, in a recent pronouncement dealing with the World Food Programme, it is stated that high hopes are being placed on technological breakthroughs, particularly new high-yielding varieties which are responsive to fertilizers.

There is a sound basis for this conclusion which places so much reliance for increased productivity on improved varieties, and plant breeding has an ever increasing role to play in British agriculture. The tempo of improvement in agricultural crop productivity has increased with technological advance; it required 500 years to double the yield of wheat in the period ending in the middle of the eighteenth century and 150 years to double again. After a period of stability, there was a further doubling in 30 years resulting in the unprecedented high yield in 1964 of 33 cwt per acre, and this period coincided with the greatest technological innovations and the most rapid advance in varietal improvement yet known. Although all the major crops do not show a similar pattern, and yields have been improved without any material change in the varieties available in some, there is no doubt that the combination of improved varieties and a more efficient agriculture is essential for the most rapid advances (table 2).

It is, of course, characteristic of the important crop plants of the world that they provide a wide range of genetic variation which has provided the basic plant material from which the cultivated forms and varieties have been derived. The evolution of the different species, or taxonomic groups, which provide the staple crops, has occurred through natural mutation and hybridization and it is only to be expected that man has used those species which expressed the valuable variation

TABLE 2. U.K. YIELDS

	1946/47	1956/57	1960/61	1966/67
	(cwt/acre)			
wheat	19·1	24·8	28·5	31·2
barley	17·8	24·1	25·2	28·7
oats	16·3	19·4	20·9	24·4
	(tons/acre)			
potatoes	7·1	8·2	8·6	9·7
sugar beet	10·5	12·2	16·7	14·6

he could exploit. The success of this exploitation has depended very largely on the richness of the available genotypes which could be selected, and it is generally accepted that for all practical purposes the scene was set before domestication of plants started. Nevertheless, it is also evident that much additional valuable variation occurred under domestication, which provided essential conditions for the production and survival of new cultivable forms. Artificial selection then became a potent factor in crop improvement, but it is only comparatively recently that artificially controlled hybridization or the experimental production of mutations on a rational and scientific basis have replaced the natural process in a conscious attempt to produce desirable genetic variation.

CEREALS

New varieties by hybridization

The greatest impact in plant breeding has occurred in the twentieth century since the establishment of genetic and cytological science, and although the contributions of the first 30 years of this century were considerable and remarkable, the pace was slow. But with the accumulating knowledge and the outstanding examples in the country of what could and had been achieved by the breeding of such wheat varieties as Little Joss and Yeoman, and barley varieties such as Spratt–Archer and Plumage–Archer, the stage was set for more ambitious research. The effectiveness of this new approach which had learned the lessons of a more scientific attitude made possible by the advances in relevant disciplines may be illustrated from some work at Cambridge on cereals.

In 1933 a decision was made to try and change the type of spring malting barley available to British farmers by combining the so-called Scandinavian mature plant model with the very high grain quality of the dominant varieties Spratt–Archer and Plumage–Archer which were characteristic of the type current for well over 100 years, though the varieties themselves had been released in 1914 and 1905 respectively. The Scandinavian types were characterized by short and stronger straw, a greater number of ears to each plant, rather earlier ripening and higher yield, but unsuitable malting quality. A scheme of hybridization was started whereby two Scandinavian varieties were crossed with a range of British varieties which represented virtually the full spectrum of improved varieties that had been cultivated in this country for something like 80 years. The purpose of this approach was to exploit as wide a range of adapted genotypes of proven worth ecologically, agriculturally, and industrially as was practicable, and although this comprehensive experiment was to be treated as a unit, the crosses were kept in groups according to the native parental type for greater convenience and ease of handling as well as for a more efficient elimination of the less useful parental combinations.

This experiment showed, apart from the new objective, several departures from the then accepted practice in devising and handling hybridization programmes. First, the wider range of parental combinations; secondly, the large total hybrid population to allow the most exceptional recombinants to appear; thirdly, a severe selection procedure based on a preconceived mature plant model which was applied on the basis of elementary biometrics; fourthly, the use of selection criteria for malting quality that had not previously been used and which were only evolved as the experiment proceeded; fifthly, those of a more protracted period of general selection over a considerably greater number of generations than is generally used; sixthly, the use of single grain selections; seventhly, final trials over 5 years instead of the more usual 3 years, in order to try and match the field performance with the higher yielding parent and to assure as far as possible a stability in the expression of malting quality, a character of notorious susceptibility to the effects of environment.

The results of small-scale yield trials of the most promising five selections carried out over 5 years showed that the yield of the higher yielding parent, Kenia, had been achieved, or even surpassed, the highest means being 7, 8 and 14% above Kenia. These levels of yield were associated with a close resemblance to the Kenia mature plant model, while malting tests indicated that the grain quality was equivalent to that of the standard variety Spratt–Archer and superior to that of Kenia (table 3). The five hybrids were therefore sent for national trials and after

TABLE 3. MALTING QUALITY OF HYBRIDS AND TWO PARENTS

	hot water extract		cold water extract		modification index	
	1946	1947	1946	1947	1946	1947
Spratt-Archer	101·9	101·3	22·8	22·7	36·7	40·6
Kenia	100·5	100·7	19·7	20·4	33·5	38·2
HB 34/2/3/33	101·6	101·1	20·4	21·4	34·8	39·2
HB 34/15/2/39	101·9	101·7	21·2	21·4	38·7	42·5
HB 35/99/3/3	101·1	100·7	21·9	22·6	38·9	39·3
HB 35/99/4/10	102·5	101·1	20·6	22·2	38·2	39·5
HB 40/66/5	100·9	101·4	20·2	20·7	35·3	41·0

three years the hybrid HB 35/99/4/10, which had proved outstanding, was finally chosen, multiplied and marketed as Proctor (figure 1). The results obtained by farmers fully justified the trials and Proctor quickly usurped the place occupied in this country by all other spring varieties of malting barley. The rigorous selection on mature plant characters, particularly on the criterion of ability to stand for long periods in the field without damage to straw or loss of grain, was well rewarded as Proctor proved highly successful for combine harvesting. In addition, after a difficult initiation, Proctor showed a more acceptable behaviour on the

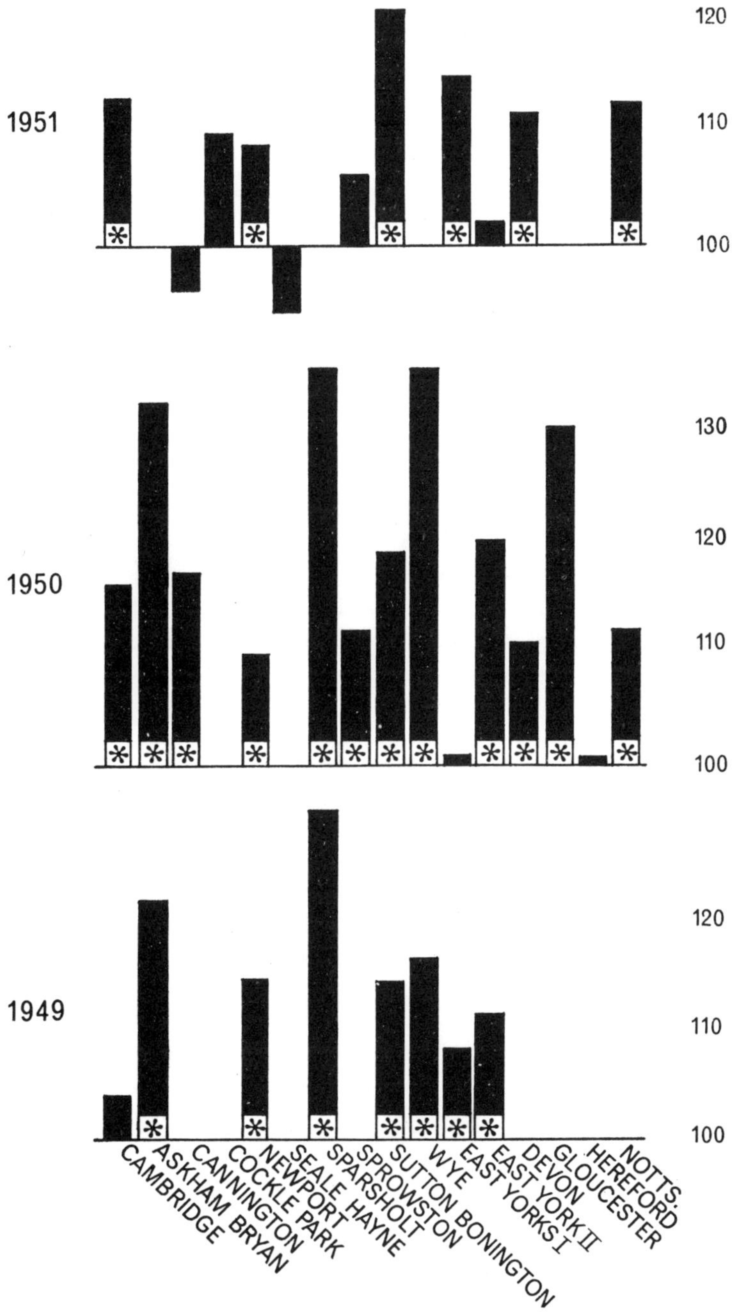

FIGURE 1. Yield of Procter given as percentage of Kenia during three years of national trials (N.I.A.B.). * shows yield significantly different from control.

malting floor and higher extracts in brewing were obtained commercially than had hitherto been customary. In short, Proctor exceeded either parent in field performance and processing behaviour, while it also became a widely acceptable general purpose variety that could be grown for feeding grain or malting quality. Proctor's wide adaptability was responsible for raising the national yield level by approximately 12 %. It is still the most widely grown variety of any cereal, after 15 years in cultivation, largely because of the vast increase in the barley acreage which is now 6000000.

A similar, but considerably more limited, experiment was conducted to breed a winter-hardy malting barley of the two-row type generally acceptable to maltsters. Two winter-hardy continental varieties—Tschermaks and Stadlers—were used as the source of winter hardiness, and a range of native varieties as the other parents. The objective was, in effect, to breed a winter-hardy Spratt–Archer variety which at the time was the most widely grown malting barley but lacking in winter hardiness. The winter-hardy parents were entirely unacceptable as commercial varieties on all characters—length of straw, yield and malting quality —so that the objective was essentially the transference of winter hardiness into a desirable genotype. No artificial hardiness testing equipment was available, and consequently all testing had to be conducted in the field, relying on selection over a number of winters to isolate the most acceptable combinations of field characters, malting quality and hardiness. This was achieved through a succession of hard winters, which were particularly effective for early generation selection and later for selection in field trials. During this experiment, new statistical layouts were used which allowed larger numbers of hybrids to be included in one trial. In particular, an incomplete randomized block was used for the first time in this country on breeding material, and as a result the variety Pioneer was produced which had been grown for 21 years as a recommended variety. Pioneer has now been superseded by the variety Maris Otter, which is a hybrid between Proctor and Pioneer, and combines the field and grain characters of the former with much of the winter hardiness of the latter. Maris Otter is, however, interesting in that it gives a 5% higher yield than Pioneer and is an 'alternative' variety, possessing winter hardiness, but no cold requirement for coming into ear. It can, therefore, be used for sowing in the autumn or the spring, thereby providing a greater elasticity in the management of the barley crop by the farmer without proliferating the number of varieties.

Further improvements in barley have subsequently been achieved by building on the successful Proctor and Pioneer genotypes through hybridization schemes and selection methods similar to those previously used. By the transference of a single major gene carrying resistance to a race of the fungus (*Erysiphe graminis* f.sp. *hordei*) derived from the related species *Hordeum spontaneum*, two varieties Maris Badger and Maris Concord were obtained. The effectiveness of this gene in preventing yield losses from the powdery mildew disease is shown in table 4, which indicates clearly the association between the severity of the attack and the level of loss. While these two resistant varieties are distinct phenotypically from Proctor, a third variety Maris Druid is identical, but the limited protection conferred by the single gene is no longer adequate. It is significant, however, that Maris Concord is classed as a higher yielding variety than Proctor, its assessment on a national basis being 104% of that variety. However, by exploiting a mildew resistant genotype allied to Proctor by hybridization with the continental variety Wisa

TABLE 4. MILDEW RESISTANT HYBRIDS IN THE 1958 TRIALS

	Yields as a percentage of Proctor				
	Seale Hayne	Headley Hall	Sprowston	Cambridge	mean
Proctor yield (cwt/acre)	37	21	17	27	
HB 246/5/5	104	121	160	131	129
HB 279/5/1/1	103	102	144	111	115
HB 279/5/1/3	96	96	119	105	104
HB 281/6/3/10	97	119	149	118	121
HB 281/6/6/5	102	119	164	113	125
L.S.D. 5%	8	14	14·5	11·1	
estimated mildew infection of controls	trace	10–25%	50%	10–25%	

(Hybrid 380/22) it has been possible to advance the level of yield by 12%. Similarly, with winter barley, the new hybrid HB 429/57, which has a single mildew resistant gene from *H. spontaneum* is indicating yields of 12% over Maris Otter, or 20% above Pioneer. These results from the newest hybrids are from preliminary national trials.

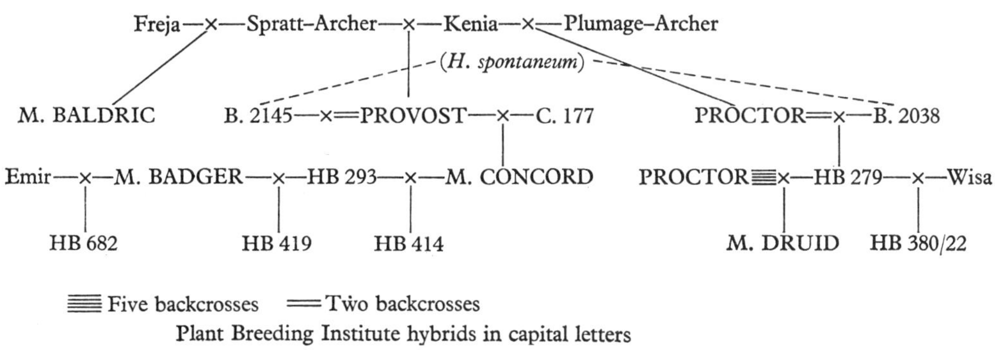

FIGURE 2. Pedigree of Cambridge hybrids: spring barley

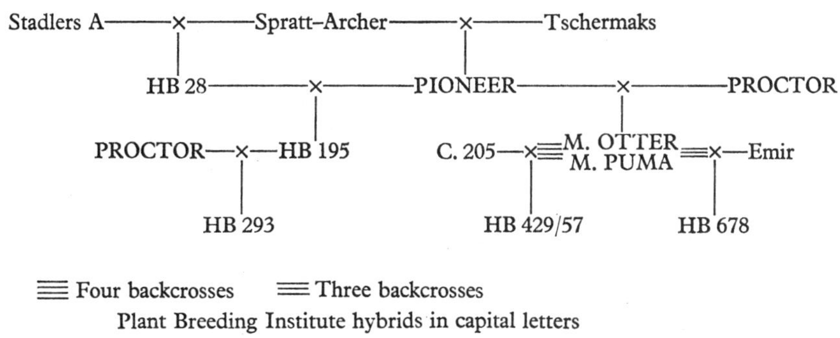

FIGURE 3. Pedigree of Cambridge hybrids: winter barley

The newer hybrids of barley have resulted from significant changes in selection techniques which have improved the efficiency of handling hybrid populations. First, by experimenting with the possibilities of introducing quantitative selection criteria in the earlier generations, two modified methods—the F_2 progeny method and the pedigree trial method—were introduced. These modifications are designed to introduce a greater measure of accuracy in selection in the earlier generations by the use of statistical analysis for quantitative characters, particularly yield. This approach was used in the winter wheat programme directed especially to the breeding of a new variety combining high yield and good baking quality, and also to winter oat improvement for raising yield levels while maintaining an adequate level of winter hardiness. The problem in winter wheat is essentially to try and break the inverse correlation between yield and protein percentage while exploiting the physico-chemical characters of the protein on which bread making characters largely depend. The outcome of the work was the breeding and marketing of the variety Maris Widgeon which is a hybrid of the varieties Cappelle Desprez and Holdfast and the effective result is that Maris Widgeon combines the yield of the former with the baking quality of the latter. This is a combination in advance of anything previously achieved and is particularly significant in view of the need to save on import costs, a feasible proposition with a baking wheat equivalent in yield to that of the most widely grown variety (figure 4).

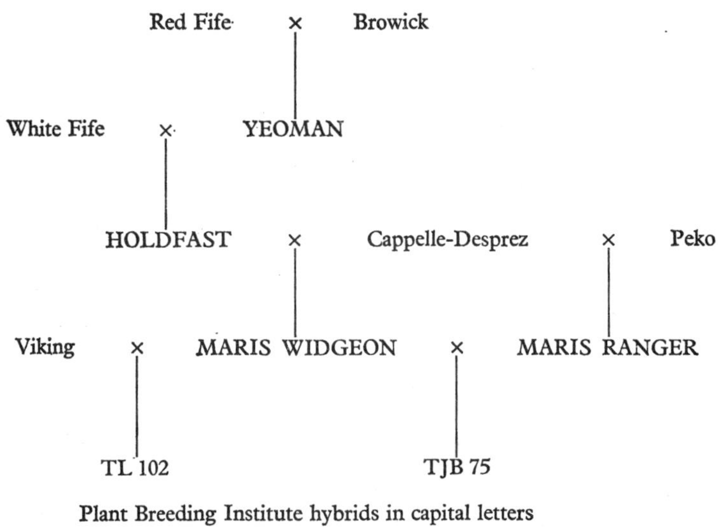

FIGURE 4. Pedigree of Cambridge hybrids: winter wheat

It is, of course, a more straightforward problem, but still requiring the use of carefully associated techniques of statistical analysis and phenotypic selection, to raise yields without reference to a specialized grain requirement such as baking quality. Nevertheless, breeding for increased yield is still the most difficult single objective in plant breeding because it depends on the expression of as yet unpredictable genetic transgressive segregation. Numerous attempts have been made to choose parental combinations that are most likely to give segregants showing yield improvements over either parents. The only available means of doing this is by suitable statistical analysis of yield, and yield parameters, of parents and their F_1 or F_2 generations when studied in crosses involving all combinations of parents when several such parents are concerned. When such a method is used it does have limited application, but accurate cross-prediction for yield is still not available when it is most needed, i.e. for yield improvement. Until more accurate methods are available for genetic analysis of parents, and there is a greater promise of this in wheat, the breeder has to rely on detailed screening of large populations derived from hybridizing carefully chosen parental combinations based on field performance. Two successful applications of this widely used method have recently been accomplished in the breeding of the winter wheat variety Maris Ranger—a hybrid of Cappelle Desprez and Peko—which marks a new high level on yield, showing an improvement of 11% over Cappelle Desprez, and the other is the winter oat variety Maris Quest—a hybrid of Blenda and S. 172—which with new varieties from other breeders is setting new target levels for yield. Still higher yields in winter wheat compared with Maris Ranger have been obtained by a hybrid now in national trials: TJB 32/219 has given a mean of 22% above the control Hybrid 46 and Cappelle Desprez.

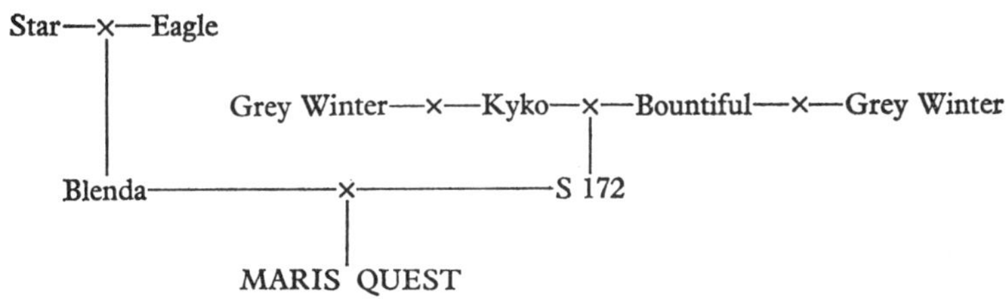

Plant Breeding Institute hybrid in capital letters

FIGURE. 5. Pedigree of Cambridge hybrids: winter oats

There is no doubt, however, that breeding for high yield should become a more precise and efficient research procedure if more knowledge was available on the physiology of yield in terms of metabolic processes and developmental characters throughout the life of the plant. Yield may be regarded as a plant breeder's character measured in terms of the end-product, and no selection pressure can be exerted on the components which have the greatest significance if these are not known. A great deal of work has been done on the morphological components in terms of the mature plant model, and the significance of such characters as ear number, grain number and grain size is reasonably well understood. This is true in so far as work with cereals has indicated that in hybrid populations selection for these yield components taken singly can have very limited positive effect because of compensating interactions of the components. This limitation is set by the simple additive genetic systems which determine the expression of each component, whereas the inheritance of yield is complicated by non-allele interactions and the effect of environment on the interaction of yield components. This has been shown clearly by a study of the inheritance of yield and some of its components in winter wheat where yield behaved as a dominant character at high seed rates and as a recessive at low seed rates due to high tillering capacity being a recessive character and exercising a decisive function when plants are widely spaced.

Physiology of yield

Such information is, of course, extremely useful in demonstrating the limitations of such selection criteria, but it has emphasized the necessity to turn to the physiological characters which are basic to plant productivity. It is difficult to draw the line between such physiological characters and the morphological characters associated with them, as, for example, in the case of the genotypic differences in net assimilation rate and leaf area index. Is it photosynthetic efficiency or leaf area index which are primarily determining differences in growth patterns? Each is claimed to be the key factor according to different schools of thought. There is good experimental evidence that certain growth parameters, as, for example, shoot weight before anthesis, are most significant in affecting yield while leaf area is only of indirect importance in determining photosynthetic capacity. On the other hand it is of the greatest importance that genotypic differences in growth parameters have been established, and that there is no doubt that a high expression of certain parameters are significant in determining yield. The evidence from hybrid population analysis shows, however, that the influence exerted by these parameters differs with individual parental combinations, thus supporting the well-established view that suitable parents for hybridization can be chosen which are most likely to give high-yielding progeny if the parental characters are known.

Patterns of growth in cereals in terms of developmental, morphological and physiological characters are now showing themselves with some clarity with reference to their influence on yield. The importance of mature plant characters such as short and strong straw in increasing yield potential is probably pre-eminent in their contribution to the higher yielding capacity of modern varieties, but this is not sufficient in itself. It is necessary to envisage a life cycle comprising vigorous young growth to provide a large photosynthetic surface after anthesis or ear emergence with the framework for ample storage of carbohydrates in the ears. It is suggested that this may be achieved in wheat by selecting for vigorous young growth, early initiation of the ears and a mature plant structure capable of providing the maximum sink for deposition of the metabolic end-products. The pattern of the mature plant model can vary in terms of ear number, ear size, spikelet fertility and grain size, and selection pressure can change the contribution of the individual components. In the final analysis the amount of carbohydrate that is stored in the grain is determined almost entirely by the photosynthetic activity of the ears and flag leaf after anthesis, and measurement of this activity has been shown to agree well with the grain yield. However, varietal differences do exist in this respect in so far as there is evidence that some varieties do not photosynthesize so actively at lower light intensities as do others.

The significance of the developmental characteristics, in terms of phenological phases and length of life cycle, is manifest with regard to varietal adaptation and yield potential. The major controlling influence in cereals is the response of the variety to length of day, or photoperiod, and this response is expressed in the length of the life cycle, the rate of production and number of growing points, and the initiation of ears and the size of the leaves. As also the dry weight and tillering behaviour are affected by photoperiod, physiological response to photoperiod, and its quantitative measurement in controlled environments are useful for predicting the performance of individual genotypes and for exploiting such genotypic variation in breeding. This knowledge is of particular relevance in devising breeding programmes directed towards the shortening of life cycles and the production of varieties of the alternative type which are sufficiently winter-hardy for autumn sowing, but which are suitable also for sowing in the spring.

These considerations of genotype-environment interaction are essential features of the understanding of the physiological components of varietal adaptation and therefore of crop performance in terms of yield level and consistency under different ecological conditions. Such understanding is essential for a rational approach to the continued improvement of yield levels and also for regional adaptation which may probably become a critical factor of increasing importance as varietal productivity rises. As the grower becomes more ambitious, or as economic circumstances demand, husbandry practices will become more delicately balanced and their effects become more critical in affecting individual varietal performance. Such a situation could well have a significant effect on the breeder's concepts of the most suitable genotypes in terms of physiological attributes and mature plant models, and as there is no reason to feel that the current concepts will continue to meet the future requirements, it is as well to explore the possibilities. It is important, for example, to know how particular genotypic patterns behave under such obviously varying conditions of plant density, soil fertility and availability of water. The available data indicate that maximum grain yield in barley is obtained at the plant density at which the total dry matter first reaches a maximum, and the pattern of behaviour for wheat and barley is very similar. A detailed analysis of the pattern of behaviour indicated that the most important effect of plant density variation on growth is through net assimilation rate, and differences could not be ascribed to leaf area differences, nor could any support be found for the sink hypothesis. The data available are, however, still limited regarding varietal response to the effects of variation in plant density, and a wider range of varieties grown under different conditions might reveal differential reaction. Controlled field experiments have, however, demonstrated varietal differences in response to differing water regimes in both wheat and barley, and whereas wheat varieties showed individual responses to water and nitrogen applications, barley varieties did not respond differentially to varying nitrogen regimes.

The physiological picture is still fragmentary in relation to the interpretation of the manifestation of yield in cereal crops, but the position is sufficiently clear for the breeder to assess with some accuracy the basis for the development of selection criteria. The analysis of yield components morphologically and physiologically and their separate contribution to performance in terms of productivity has indicated definite means by which genotypic improvement can be envisaged. This analysis has also demonstrated the practical limitations of applying the accumulating knowledge on a comprehensive scale to a plant breeding programme. But the integration of new morphological and physiological concepts can have profound effects on the genotypic and phenotypic constitution of future cereal varieties and yield potential may be considerably raised, particularly if improved resistance to leaf diseases can be incorporated. Already new winter wheat hybrids of dwarf stature, and with a considerably increased ratio of grain to straw, have been produced, and the grain yields of the best of these are comparable with an advance on the present standards of the most widely grown varieties (table 5). In the meantime, however, the most recently introduced varieties of winter and spring wheat, and of spring wheat are providing the farmer with the means of improving the performance of the varieties that have been in current use for a number of years. The extent of this advance may be judged by the position in winter wheat and spring barley, where in the former the newest varieties are 14% higher than the most widely grown variety Cappelle Desprez, while in the latter there is an advance of 11% over Proctor. The situation in spring wheat is even more impressive with an improvement of 11% over the standard control which was recommended to growers as recently as 1965, while the variety Maris Ensign, like Maris Widgeon has 'field' resistance to yellow rust in addition to resistance to races of this disease.

TABLE 5. 1966. WINTER WHEAT HARVEST ANALYSIS

	grain yield (g/m²)	grain wt. / straw wt.	height (cm)
Cappelle-Desprez	481	0·61	87
Professeur Marchal	507	0·69	88
TJB 32/219	587	0·75	85
Marne × VG 9144	543	0·85	58

Grain quality

The problem of grain quality in cereals looms very large in the breeder's compendium of economic characters, and can be regarded as a complex of components as is yield, and similarly susceptible to analysis. Grain quality is of particular significance in wheat and in barley by reason of the processing requirements; milling and baking for wheat, and malting and brewing for barley. Both cereals are, of course, also used for animal feeding, but this form of utilization has not received anything like the attention of the breeder compared with the other expressions of quality, although greater emphasis on feeding quality in barley is now necessary.

Undoubtedly one of the most significant advances in barley breeding techniques has centred round the development of adequate selection criteria for malting quality. The matter was successfully resolved by identifying the significant quality components, by devising an accurate micro-test with a quantitative basis, and by showing that the components and the final quality expression were hereditary characters. The demonstration of these facts provided for the first time the means of handling malting quality during the routine handling of hybrid progenies instead of having to wait until adequate amounts of grain were available for experimental malting on a semi-commercial scale. The development of a micro-malting test also demonstrated another and vitally important fact: the criteria used gave an accurate measurement of the brewing value of the material without having to produce beer.

A very effective quality model was evolved based on 1000 grain weight, nitrogen percentage, insoluble carbohydrate percentage, predicted extract, actual extract, index of modification, permanently soluble nitrogen, cold water extract and malting loss which could be used for comparative purposes on which to base hybrid selection. Subsequent developments have made it possible to use more critical criteria based on the genotypic performance over a range of nitrogen levels in the grain when tested under different conditions of steeping and flooring times. By the use of this test, and by applying an established formula by which the relative extract (the amount of extract obtained in relation to the predicted extract) improved malting potential was achieved in the variety Maris Badger.

Although malting quality is a heritable character it has a large environmental component in its final expression, and in the handling of this character in breeding it is valuable to know the level of environmental tolerance on which adaptability or stability of the genotype must depend when grown under different conditions. Attempts have been made to measure some of the components of malting quality in respect to its heritability and to study any interaction of these components which could influence the effect of selection pressure on one or more of them. This study not only provided a means of establishing an ideal model in relation to three quality components, but also showed that they are inherited additively without any dominance effects. There were, however, close genetic associations between the inheritance of some of the components which would complicate selection based on individual components. To ease the task of dealing with the large effect of environment on quality, a technique has been developed of comparing genotypes when grown under controlled nutritional conditions in a glasshouse so that comparable grain samples can be produced under different regimes.

Similar investigations on grain quality in wheat have resulted in suitable tests for selecting in hybrid populations, although the basis of quality in terms of grain characters and utilization are entirely different from those in barley. As in barley, though, the expression of quality in wheat is a complex of a number of component characters, which determine the suitability of the grain for processing, but in this case for milling and for producing flour for specific purposes. Milling quality can be measured quantitatively, but the essential differentiation is on hard milling for bread wheat and soft milling for biscuits, a character that is determined by actual milling of small samples in specially designed equipment. This quality expression is determined by the nature of the endosperm based on cellular organization and structure which results in hard milling wheats forming distinct particles largely composed of whole cells, but in soft wheats the cells break down and release the starch grains. The characters which determine the flour type, however, are derived from physico-chemical features which affect the behaviour of the dough in terms of water absorption capacity, and the extensibility and elasticity of the dough when stretched. Good bread-making doughs have high elasticity and intermediate extensibility, and good biscuit doughs have low elasticity and great extensibility. Such characteristics are conferred by the nature of the gluten, while a high protein content allows of the best expression of good bread-making characteristics and low protein for biscuit making.

By the use of suitable equipment, in this case an extensometer, the gluten characters can be accurately determined and used as an effective selection criterion in spite of genotypic interaction with environment. Hardness has a high heritability factor and the stress/strain characters of the dough are somewhat more susceptible to growing conditions. This gives a good basis for breeding, but unfortunately protein percentage, which is inversely related to grain yield, is but weakly inherited. It is this relationship that has made it so difficult to produce the highest expression of yield, with a stable bread-making quality that is acceptable to the miller and baker. Maris Widgeon was the best compromise obtainable in an extensive breeding programme, but the problem of a good bread-making wheat will only be solved if a more stable expression of quality can be obtained combined with a higher yield.

Disease

It is undeniable that cereal diseases have in recent years assumed a far greater importance as a factor limiting grain yield and restricting the life of new varieties. There are a number of reasons for this situation, some due to the changes in the structure of cereal growing and others due to the approach to breeding for resistance on the part of the plant breeder. Intensive cereal growing with high levels of nitrogenous fertilizers, the continuous cultivation of one cereal with no breaks, and the expansion of the barley crop into areas conducive to certain diseases are all predisposing to the development of disease epidemics. But these in themselves would not be responsible for the greater incidence of the leaf diseases which are the real menace, and there would be no disease problem if the varieties being grown were sufficiently resistant.

Breeding for resistance to disease offers special problems because there are two genetic systems involved—that of the host and that of the pathogen, and it is the host/pathogen relationships that determine the levels of resistance that can be achieved and the stability of that resistance. The position in effect is decided by the effectiveness of the host resistance genes in combating the virulence genes of the pathogen. A particular host species may carry a very effective genetic system, or even more than one kind of genetic system, to meet the attacks of a pathogen. On the other hand there may be no protection or very poorly expressed protection which may give no opportunity for exploitation by breeding. Finally, the genotypic constitution of the pathogen may be very restricted and incapable of any virulence changes in the case of certain diseases, while alternatively, and more usually, the pathogen is capable of producing new genetic variation in the form of biologic races which carry different virulence genes.

It is the latter situation which is causing the present serious situation regarding the leaf diseases of wheat and barley particularly, and of oats to a lesser extent.

The diseases causing most concern are caused by the fungi *Erysiphe graminis* (powdery mildew), *Puccinia striiformis* (yellow rust), *P. recondita* (brown rust) and *Rhynchosporium secalis* (leaf blotch). Evidence is available to show that powdery mildew in barley, which is the most damaging of the diseases in this crop can cause losses of yield in farm crops of 15%, and the reduction in the national crop is considerable. Yellow rust is the worst of the leaf diseases in wheat and almost certainly causes similar losses at equivalent levels of infection, but the whole complex of leaf diseases must be considered as a major problem affecting cereal productivity.

The position can only be improved by a most intensive and extensive campaign of research and investigation which will cover all aspects of the problem, and for the most part the work is under way. Race surveys on a co-operative basis are now being undertaken and will be extended to deal with race incidence and typing with the rapid identification of new races. All varieties must be typed also for the resistance genes that they carry, and every possible source of alternative and new genes must be investigated. This type of resistance, called major gene, or specific, resistance is the quickest to exploit by hybridization and backcrossing, but of course has grave limitations when the pathogen is active in new race differentiation, but it must be an intrinsic part of any programme.

Recent research has considerably clarified the position with regard to yellow rust in wheat and powdery mildew in barley so that the occurrence of resistance genes in the host and of virulence genes in the pathogen are very much better appreciated. Identification of races and their virulence genes is conducted by the use of differential host testing varieties, while the identification of the resistance genes in host varieties is being carried out on an international basis. Breeding work must be based on this kind of information in order to avoid the marketing of new varieties which are relying on the same gene or genes to confer resistance. The danger of such a situation is obvious in that new varieties tend to select their own races as such arise and to which they show susceptibility. In this respect it should be noted that out of eleven spring barley varieties recommended for cultivation in this country, and claimed to be resistant to powdery mildew, six carry the same gene, two carry another, one carries a combination of both, and two are of unknown constitution. This is dangerous.

If diseases of this kind are to be dealt with more effectively, there are two other forms of genetic exploitation that may be used, both of which are much more difficult than intervarietal hybridization involving major genes. First, there is the use of non-specific or field resistance which confers an adequate level of protection to a pathogen irrespective of race as far as is known. The genetic basis of this resistance is not completely understood but it is generally thought to depend on a number of genes and exists in some cultivated varieties. It is difficult to work with but the techniques must be found. Secondly, every effort must be made to introduce sources of resistance outside the cultivated varieties thereby utilizing new forms of genetic control. This means in effect using 'alien' genetic variation where available as has already been accomplished in wheat by introducing a resistance gene from the grass *Aegilops comosa* into a wheat background. The process has involved chromosome substitution and a breeder's variety, named Compair, has been produced which can be used as the basis of a breeding programme as it carries resistance to all known isolates of *Puccinia striiformis*, which is the fungus responsible for yellow rust (figure 6).

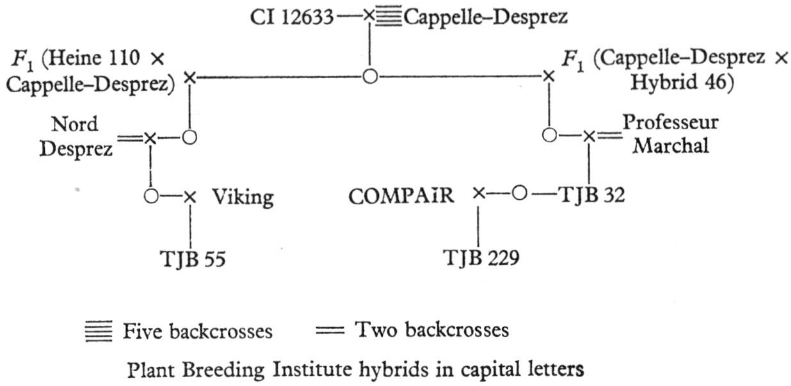

FIGURE 6. Pedigree of Cambridge hybrids: winter wheat

Although such techniques are of the utmost value in resistance breeding, less extreme forms of the use of alien genetic variation have considerable potential and have been used for some years. A detailed and continuous programme of synthesizing disease-resistant lines for wheat breeding is in operation and the genetic resistance of *Triticum spelta* to yellow rust and of *T. timopheevi* to powdery mildew has been incorporated in breeding lines. It is only by such methods that the widest coverage can be given to meet future requirements and provide a new basis for resistance breeding. Breeding for disease resistance, with special reference to pathogens of genetic variation and capable of producing new expressions of virulence, requires this kind of approach. In the meantime, such wheats as Maris Widgeon, which carries only two major genes for resistance to yellow rust as well as a degree of field resistance to yellow rust, appears to be achieving some stability because of non-specific resistance, while Maris Ranger has so far shown resistance to all races of yellow rust known to occur in this country. The value of this resistance to yellow rust in enhancing grain yields in response to high levels of nitrogen fertilizer may be seen from the performance of Maris Widgeon, Maris Ranger and the new dwarf hybrid TJB 32/219 (table 6).

TABLE 6. EFFECT OF NITROGEN ON YELLOW RUST AND YIELD, 1967

N_1 30, N_2 90 lb N per acre

	yellow rust % flag leaf area		yield % Cappelle N_1	
	N_1	N_2	N_1	N_2
Cappelle-Desprez	0·7	16·1	100	106
Maris Widgeon	0·0	0·2	99	121
Maris Ranger	0·0	0·0	103	127
TJB 32/219	0·0	0·0	113	145

Herbage plants

The breeder is presented with particularly difficult problems of selection criteria and plot techniques in the handling of herbage plants because isolation and testing of individual genotypes has to be conducted in the early stages under conditions that are further removed from agricultural practice than is the case with a crop such as a cereal. In addition, the breeding structure and genetic conditions of grass species introduce peculiar difficulties in the initial stages of selection, and consideration of the traditional methods of grass breeding, in spite of the outstanding success of the achievements, indicates certain limitations. Special point is also given to such a reappraisal in view of the fact that it is becoming increasingly apparent that greater precision is required to meet the more specialized objectives in grass breeding which must match the farmer's needs in the exploitation of intensive grassland management.

Experiments have been conducted under varying environmental and management conditions in the field, and also under controlled environmental conditions in growth cabinets in order to analyse the morphological and chemical components of the plant characters which are economically important. Small-scale plots, or even micro-plots, are used and the behaviour studied in logarithmically related series of densities under different regimes of water and nitrogen fertilization. The yield, obtained from regular systems of cutting, is then analysed in terms of unit area and per plant, and the relationship of the components, in this case tiller number and tiller weight, to yield and composition of the green matter calculated. It is clear from the results of this work that a better definition of criteria for separate objectives is necessary and both must be considered in close relation to management and growing conditions. Quality assessment, in terms of feeding value and acceptability are an essential feature of the selection criteria and a modified form of measuring *in vitro* digestibility combined with protein content, soluble carbohydrate and pepsin-soluble components has been decided upon.

It is by such means that future improvements in grass as a herbage and forage crop may be made practicable. Increased productivity from grass is essential and this can be achieved at the moment by management, as there is no doubt that the varieties available are not the limiting factor. But in order to make the most of improved efficiency in management, grasses must be bred for specified purposes and in particular conservation must play an increasingly important role. In this regard whole-plant silage is assuming greater significance and in conjunction with grass there are increasing possibilities for the use of cereal crops including maize. Investigations are in progress on the possibilities of breeding cereals specifically for this purpose, and such work must involve physiological and growth analysis studies on similar lines to those being used for grain production.

There is a considerable range of possibilities for improving the maximum production of dry matter and digestible nutrients per acre by utilizing grass and cereal species, and the most effective exploitation will derive from types bred for special purposes. Attention is being given to this aspect of breeding by examining the possibilities of utilizing exotic forms of Tall Fescue, a range of Italian ryegrass types and forms of forage rye. Breeding work in these species has resulted in experimental varieties in Tall Fescue and forage rye adapted particularly to winter production. The extension of such a combined attack on maximum productivity potentials, with special reference to out-of-season production, could have important repercussions on increasing production of grass and cereal forage for intensive livestock management. However, the exploitation of the most intensive systems of herbage crop management must depend on producing more efficient and potentially higher yielding genotypes, and there is well-supported evidence that clones of perennial ryegrass can be selected which are capable of greater response to high levels of nitrogen fertilizers.

The use of artificially induced tetraploids for obtaining higher yielding herbage plants is not a new procedure. However, a distinct advance in red clover performance has been achieved by combining this technique with resistance to stem eelworm, and to *Sclerotinia* stem rot. A new broad red clover variety has been produced which has given significantly higher yields than the diploid material from which it was derived (table 7).

TABLE 7. RESISTANT TETRAPLOID RED CLOVER

	yield of dry matter (% of B.R. diploid)		stem rot	eelworm
	1st cut	all cuts	(% of row destroyed)	(% swollen seedlings)
Essex B.R. diploid	100*	100†	15·7	70
resistant tetraploid	114	108	11·4	2
sig. diff. ($P = 0{\cdot}05$)	8·7	7·5	3·1	—

* 51·2 cwt/acre, † 95·1 cwt/acre.

POTATOES

Potato growing in this country has been in a unique position up to very recent times in that, in spite of the large number of new varieties that have been bred during the last 20 years, the farmer has been very largely dependent on three varieties—Majestic, King Edward and Arran Pilot—which have been in cultivation for 50, 60 and 40 years respectively. The situation has changed radically during the last 10 years in that improved replacements are now available and the yield potential of the crop has risen significantly. A great deal of this success has been due to the breeder coming to terms with the problem of blight disease by a better understanding of what can and should be achieved in breeding for resistance, and also to the incorporation in breeding programmes of related botanical forms such as the subspecies *andigena* and to a lesser extent by the use of distinct species. It is, of course, not a new concept in potato breeding to use *Solanum* species outside the *tuberosum* species as for many years the work was dominated by *tuberosum* hybrids with *S. demissum* in order to incorporate major gene resistance to blight disease, but other species are of more recent involvement.

The breeding of potatoes has always been, and continues to be, dominated by disease problems, as the crop is subject to a number of diseases, fungous and virus, which affect yield and tuber marketability. In addition, the crop is one for direct human consumption and questions of culinary quality, keeping powers and more recently of suitability for various kinds of processing are of vital significance to the breeder. The long popularity of King Edward, Majestic and Arran Pilot is due partly to their market suitability as culinary potatoes but also to the schemes for maintaining and producing virus-free and certified seed.

The most significant contributions by potato breeders have not been achieved by any radical changes in breeding methods, but rather by a more catholic approach made possible by a release from the blight disease straight-jacket and the handling of more material with a wider range of genetic variation. The development of hybrid seedlings derived from intervarietal crosses gradually widened the genetic background, as sexual sterility, which has been such a limiting factor, ceased to have such a restrictive effect. The choice of parents is also complicated by the fact that any variety or clone is highly heterozygous and its breeding potential cannot be determined, or even assessed with reasonable accuracy, by its phenotypic appearance or performance. It is for this reason that intervarietal hybridization based on new and improved varieties, as is used in self-pollinating cereals, cannot be used in potatoes with any assurance.

The considerable recent advances have thus been made in the face of great technical difficulties and the range of new varieties is a significant contribution to agriculture. Of the seventeen varieties recommended to growers in this country, nine have been introduced since 1955 and the national average yield has been increasing at a most satisfactory rate. Although the increased yield potential of such new varieties as Maris Peer and Maris Page, which are both second earlies with yields of 16 and 26% respectively over the older variety Craigs Royal, has still to be fully exploited agriculturally, they can be the basis of a more ambitious national performance in this class of potato. Of even greater significance, however, is the position in the maincrop class where Pentland Dell and Pentland Crown have undoubtedly made the yield level of Majestic appear quite inadequate.

The breakthrough on yield in potatoes is a matter of great agricultural significance in that it has provided a new basis on which to measure national productivity. For the breeder, however, although there is the satisfaction of removing what appeared to be a ceiling in yield potential, this advance in yield now means turning to make good other deficiencies. In particular, the blight situation is far from satisfactory while the several tuber diseases—dry rot, common scab, and skin spot—which are assuming increasing importance in the modern methods of handling and marketing the crop need urgent attention. An important recent advance in dealing with a pathogen of considerable agricultural significance has been the breeding of the variety Maris Piper which is resistant to pathotype A of the potato root eelworm. The contribution made by this variety is that in the presence of high populations of this pathotype yield losses are greatly reduced, while even in its absence economic yields of good culinary quality are obtained. This resistance was obtained from a clone of *andigena* potatoes, but unfortunately if resistance to pathotype B is to be obtained it will be necessary to transfer it from the wild species *S. multidissectum* or *S. sanctae-rosea*. Looking further to the future if pathotype C is to be dealt with, an even more difficult task has to be faced because resistance will need to be transferred from the thoroughly unacceptable diploid species *S. vernei*. As cultivated potatoes of the subspecies *tuberosum* and *andigena* are tetraploids, attempts are being made to breed at diploid level by the use of dihaploid *tuberosum* forms for hybridizing with *S. vernei*.

The development of various kinds of processing—crisping, dehydration, frozen chips and canning—is putting a premium on new tuber characters and no longer can the breeder restrict quality criteria to cooking and culinary characters. Quite apart from the necessity to produce tubers free from disease and mechanical damage blemishes, it is essential to pay attention to biochemical characters as well as laying greater emphasis on tuber size and conformation. The biochemical basis of tuber blackening on cooking has been known for some years, and the characters associated with good crisping—high dry matter and low reducing sugars—have recently been defined, so that these aspects of quality can be dealt with adequately. It is, however, a much more difficult problem to meet the requirements of resistance to mechanical damage due to harvesting and bulk handling, and it seems more profitable to solve the problem by modifying the machines and showing greater care in handling.

Sugar beet

Although the national average yield of sugar beet had been increasing steadily for the last 20 years there is no good evidence to ascribe this to improvement in the varieties available. Certain standard varieties have been grown over this period—and even longer—but as it is customary in this crop to select within the varieties, thereby changing the genetic constitution, while retaining the same name, it is difficult to apportion the basis of the improvement in yield. It is certain that husbandry methods have improved and the crop is more efficiently handled; thus, for lack of any evidence to the contrary the yield increases must be ascribed to the farmer and not to the breeder. Analysis of the data during the last 20 years indicates that the 50% increase in beet sugar production can be attributed to such factors as an increase in the number of roots per acre reaching the factory, to improved control of the aphid vectors of the yellowing viruses and to increased use of nitrogenous fertilizers. But it seems probable that the breeder has made a contribution by improving the performance of the varieties, while it is quite certain that he has produced varieties that respond to the more intensive husbandry. Thus, it is most unlikely that the varieties have been, and are, limiting the yield of beet sugar in this country.

There have undoubtedly been improvements in certain characters which contribute indirectly to yield. First, intensive selection against bolting—the tendency of plants to produce inflorescences in the first year—has improved field performance in this respect. A study of the physiology of this character some years ago showed how this could be achieved, and an extremely bolting-resistant variety named Cambro, later replaced by an improved similar type named Camkilt, was marketed. This set new standards for commercial varieties and there has been a general improvement in this respect. Secondly, intensive work on the yellowing viruses has resolved the component viruses responsible, interpreted the biology of these viruses and their vectors, and explained their action in reducing yield. It has proved possible as a result of these investigations and a close study of the host/pathogen relation to produce the variety Maris Vanguard which is sufficiently tolerant of the virus to reduce significantly the losses under conditions of severe infection. But important developments have emanated from the virus researches in regard to the recognition of different forms of resistance to the aphid vectors and the complicating effects of simultaneous infection of the plants with the fungal diseases caused by *Alternaria*, *Erysiphe* and *Peronospora*. Effective genetic control of this syndrome of virus, insect and fungus will achieve a great deal to improve the efficiency of sugar beet production, and must be regarded as an important breeding objective.

TABLE. 8. VIRUS YELLOWS TRIAL, CAMBRIDGE 1967

	Sharpe's Klein E		Maris Vanguard	
	yield of sugar (t./acre)	cash return (£/acre)	yield of sugar (t./acre)	cash return (£/acre)
not inoculated	3·14	£129. 10*s*.	3·15	£130. 4*s*.
inoculated with virus yellows in June	1·72	£70. 5*s*.	2·28	£93. 7*s*.
loss	1·42	£59. 5*s*.	0·87	£37. 17*s*.

L.S.D. ($P = 0{\cdot}05$) between yields of sugar = 0·38 t.; between cash returns = £15. 8*s*.

But the most significant breeding activity is in relation to the exploitation of monogermness—the single seeded fruit which makes possible the effective mechanization of singling the crop in the field by drilling to a stand. Monogerm varieties are now available commercially, but they are only beginning to reach the standard of the best multigerm varieties so that they have not achieved the position of contributing to higher yields. The monogerm character is, however, an outstanding example of a significant contribution to the economics of husbandry which can have vitally important repercussions on the future of beet sugar production. But an equally important aspect of sugar beet improvement, economically speaking, is the quality of the root in terms of processing requirements. These requirements are reflected in two components—the percentage of sugar and the purity of the extracted juice from which the sugar is crystallized. Total sugar production per acre is an expression of weight of roots and sugar percentage and the two characters are inversely related. The factory obviously desires to process the minimum amount of roots and the farmer is paid on root weight and sugar percentage. The problem is to increase the sugar percentage while maintaining the root weight and the best compromise has to be exploited. Probably the best way of achieving an improvement is through polyploidy and the use of hybrid varieties in place of multi-line varieties. Improvement of juice quality can be effected by selection against the most important components of impurity which have been identified by the industry as α-amino nitrogen, betaine, sodium, potassium and raffinose.

Each of these has been demonstrated as having an important genetic component, but is greatly susceptible in expression to growing conditions.

By using a suitable computer program it has been possible from field trials on breeding material, and involving the five purity components, to calculate the weighted impurity values, and by regression of purity on impurities to estimate juice purity. The relationship between the components is understood insofar as the effect of selecting on single components on the other components is concerned, and there is no doubt that juice purity can be improved by component selection. Apart from the effect of growing conditions on juice purity components, it has also been shown that in virus-tolerant lines, infection with the virus can affect the expression of quality components, as well as of the total impurity value. Fortunately, selection for root quality in terms of high sugar content automatically involves selection towards high juice purity, but there is evidence that juice purity can be improved within the compass of genetically controlled sugar percentage at all levels by selecting the individual component.

The manipulation of breeding systems

Every plant species is characterized by a particular kind of breeding system with reference to the degree of self-fertilization and cross-fertilization for seed production. Some species are essentially self-pollinating and others essentially cross-pollinating, whereas others show intermediate expressions of selfing and crossing, and these differences in breeding systems characterize the genetic populations with which the breeder works and the methods open to him for handling breeding programmes. In other words, the breeding systems of plants determine for all practical purposes the objectives in plant breeding and have a decisive effect on the kind of variety in terms of genetic constitution. Thus, in a self-pollinating species, the breeder's basic material is homozygous, he creates heterozygosity by hybridization, and then returns to homozygosity in the final product. But for a cross-pollinating species, the basic material is heterozygous, some degree of homozygosity is produced by artificial selection, and then heterozygosity is re-established in the final product by hybridization.

In out-pollinating species which show varying degrees of self-pollinating, or in completely out-pollinating species, genetic improvement can be achieved by controlling the parental combinations in the final genotypic synthesis. The classical examples of this procedure are in the exploitation of hybrid vigour, as in hybrid maize, and the selection of parents for the expression of other desirable characters which by hybridization offer improvements in the ultimate variety produced. The important consideration in all such breeding procedures is to ensure as near 100% cross-pollination as is possible, and there are various means of achieving this, the simplest being when the sexes are separated—unisexual flowers—as in maize, although even in this case artificial control of pollination on a large scale is necessary.

There has been a considerable increase in recent years in the possibility of exploiting hybridity in other species of crop plants by various means, the most widely applicable of which is by using genetic or cytoplasmically inherited male sterility. This character, which is simply the failure of development of the anthers, and consequently of pollen production, has already been used successfully in sugar beet and field beans and is receiving much attention in wheat and barley. Sugar beet is predominantly an out-pollinating species, the field bean is peculiar in that it is approximately two-thirds cross-pollinating and one-third self-pollinating, while wheat and barley are predominantly self-pollinating.

Hybrid varieties of sugar beet, based on cytoplasmic male sterility, are not cultivated in this country, nor indeed in Western Europe; they are, however, the basis of sugar beet growing in the U.S.A. where the breeding is dominated by the necessity to control certain diseases and the concentration on monogerm varieties. In this country the circumstances are different as far as the major breeding objectives are concerned, although monogermness is being achieved by different means. There are also difficulties of seed production in the climatic environment of a maritime country. Nevertheless, there is much to commend the concept of the hybrid sugar beet variety and research is being conducted on the use of male-sterility for producing monogerm varieties and also for multigerm varieties of greater uniformity than current synthetic multi-line varieties. The procedure also has advantages for the production of triploid varieties, for which there is evidence of superiority over tetraploid varieties in field performance, because by using male sterility 100% cross-pollination can be achieved if conditions are suitable.

Cytoplasmically controlled male sterility is, however, being used successfully in the breeding of field beans, and hybrid varieties are now being tested on a field scale. All the experimental data so far available indicates that the exploitation of hybrid vigour by this means can raise the yield levels of beans in a most significant manner. Comparison of the best of these hybrids with the highest yielding commercial varieties has confirmed in many trials a superiority of 20 to 30% in yield at control yield levels of 35 cwt per acre. Even with increased costs of seed, which are inevitable because of the expense of producing hybrid seed, such increases in yield are well worth while economically, and could make an important contribution to the growing of field beans in this country.

TABLE 9. YIELD OF HYBRID BEANS IN THE PRESENCE AND ABSENCE OF BEES (cwt/acre)

	open field		cage	
	1965	1966	1966	1967
hybrid varieties	44·0	28·6	41·3	50·7
open-pollinated varieties	35·2	22·8	22·4	28·6
difference	8·8	5·8	18·9	22·1
L.S.D. (P = 5%)	5·2	3·6	13·4	16·2

Investigations on the economic possibilities of hybrid wheat and hybrid barley are very much more in the experimental stage than are those with the other crops referred to above. The scientific implications and the technical problems are considerably greater in converting a self-pollinating cereal to a cross-pollinating condition on a commercially viable scale than they are with a species that is already naturally a cross-pollinator. On the other hand, genetic male sterility is now available in both wheat and barley, and cytoplasmic male sterility has been extensively studied in wheat and is in the early stages of investigation in barley. It is anticipated that there will be considerable technical advantage in using genetic male sterility compared with the cytoplasmically controlled, but further work is necessary.

If the technical difficulties of handling male sterility in wheat and barley, and the grave handicap of achieving adequate out-pollination under the climatic conditions of this country, are overcome it has still to be demonstrated that hybrid wheats and barleys are a worthwhile objective. Although hybrid vigour has been shown to occur in wheat, and increases in yield due to heterosis are well documented it is at least questionable whether it is necessary to concentrate on hybrid varieties for the maximum exploitation of yield under all circumstances. Standard breeding procedures devoted to the production of so-called pure line varieties are being very effective in raising yield potential even at high levels of productivity and it is questionable whether it can be claimed that there will be an advantage in hybrid varieties. The answer to the question whether comparable increases in yield can be obtained more effectively and economically by devoting the research effort to hybrid varieties can only be given by explaining the genetic basis of heterosis for yield. It can, of course, be claimed that other advantages may accrue from hybrid variety breeding in cereals, such as a greater manoeuvreability in breeding for disease resistance, and a simplification in breeding for grain quality. Whatever the outcome, male sterility is a very useful tool for handling large-scale breeding programmes on traditional lines as well as for genetic analysis of parental breeding material.

An alternative method of manipulating breeding systems has been exploited in marrow-stem kale which has a refined genetic system of controlling parental mating. This crop is almost completely cross-pollinating, and out-breeding is ensured by a high degree of self-incompatibility combined with a cross-compatibility which operates with most other plants in any variety. Detailed genetic analysis of the breeding system of marrow-stem kale demonstrated conclusively that its incompatibility system is of the sporophytic type controlled by a series of genes, or S-alleles, which exhibit dominance. To exploit 100% cross-pollination to produce a hybrid variety it is necessary to inbreed separate lines of each which carries a different S-allele in the homozygous condition. Thus, if a single cross is the objective, the two lines must carry S_1 allele and S_2 allele, and if a double cross is to be produced the F_1 of constitution S_1S_2 must be hybridized with another F_1 of constitution S_3S_4. This has been the basis of the breeding of the first hybrid commercial marrow-stem variety named Maris Kestrel, which has proved to be a considerable improvement in many characters on other available varieties (table 10).

But the improvement is not simply a matter of hybridity as such. The breeding objective was originally to produce kales for special purposes and of much more uniform character in terms of growth type than the standard commercial varieties.

To achieve uniformity it was necessary to inbreed and select for the desired characters, and the final hybridization restored the vigour which had been lost in inbreeding. But the inbreeding uncovered a considerable amount of genetic variation that was hidden in the heterozygous state, and it was possible to improve such characters as winter hardiness, leafiness, edibility, digestibility, standing

TABLE 10. PERFORMANCE OF THE DOUBLE-CROSS HYBRID, MARIS KESTREL, IN THE N.I.A.B. TRIALS OF MARROW-STEM KALE

varieties	relative yield of digestible dry matter: total	leaf	stem	percentage digestibility of stem dry matter
mean of 12 marrow-stem kale stocks	100	34	66	72·4
Maris Kestrel	110	46	63	83·6
normal thousand-head kale	83	45	37	66·4
dwarf thousand-head kale	78	53	24	73·9

Based on results of N.I.A.B. trials, 1962–65. Yields relate to the period October to December.

capacity and resistance to certain fungal and virus diseases as well as to some aphid species. Thus the development of this method of breeding, resulting directly from the unravelling of the genetic basis of the reproductive system, has opened up entirely new prospects for the continued improvement of not only marrow-stem kale, but of thousand-head kale also. It has also indicated how related botanical varieties such as Brussels sprouts and curly kale can be incorporated into the breeding programme for improving such characters as winter hardiness and improved growth characters in terms of rapid dry matter accumulation. Finally, it has led to a more effective and economic method of seed production, based on six inbred lines instead of four, with the ultimate synthesis of a triple-cross for commercial exploitation.

THE MANIPULATION OF CHROMOSOMES

The cultivated bread wheats, which botanically are referred to as the species *aestivum* of the genus *Triticum*, show a unique combination of features which has made it possible to develop extremely delicate techniques in breeding and in genetic analysis. *Triticum* contains a series of species which are divided into three groups according to their chromosome number—diploids (14 chromosomes), tetraploids (28 chromosomes) and hexaploids (42 chromosomes). *T. aestivum* is a hexaploid which behaves cytologically as a diploid due to the genetic control of chromosome pairing known to reside in chromosome 5B. This genetic control by chromosome 5B of chromosome pairing restricts certain forms of chromosome pairing and in particular prevents recombination of homoeologous chromosomes. The effect of this suppression is to put a major obstacle in the way when attempts are made to incorporate into wheat desirable genetic variation from sources outside the genus *Triticum*. Such alien genetic variation occurs in related genera such as *Aegilops* (goat grasses) and *Secale* (rye) which can be hybridized with wheat. The earlier attempts to exploit such alien genetic variation were based on the production of hybrids which, however, were sterile unless the chromosome number was doubled by treating the F_1 plants with colchicine. Such doubled plants—amphiploids—possess the complete genotypic complements, or genomes, in terms of the chromosome complements of the two parents and many such combinations have been studied. Certain of these amphiploids have been used in hybridization programmes for wheat improvement, but nothing of economic value has resulted, although wheat-rye amphiploids (Triticales) have been successfully exploited in Hungary.

But the polyploid structure of *T. aestivum*, which is derived from its phylogenetic origin, and the detailed cytogenetic study to which it has been subjected, have made possible more delicate and precise techniques of chromosome manipulation. These are based on the practicability of manipulating the chromosome complements of individual parents in controlled hybridization programmes so that individual chromosomes can be transferred from one parent to another. This makes it possible to incorporate into wheat desirable variation from alien sources in a more effective way than the production of amphiploids, as it allows the major features of the genotype to be maintained while introducing the desired variation known to reside on any individual chromosome. The technique has two obvious applications: first, the production of valuable breeding material, and, secondly, the production of material for genetic analysis of quantitatively inherited characters.

The breeder's variety, Compair, has already been referred to, and is a unique example of the successful transfer of alien genetic variation by chromosome substitution. This achievement is particularly significant because it was necessary to overcome the genetically controlled restricted chromosome pairing that is characteristic of wheat before the chromosome carrying the yellow rust resistance of *Aegilops comosa* could be incorporated. The hybridization programme had, therefore, to be conducted in two steps. The first involved the *T. aestivum* (variety Chinese Spring) and *Aeg. comosa* from which an addition line with 44 chromosomes —42 from the wheat and two from the *Aegilops*—was derived. Then this line was hybridized with *Aeg. speltoides*, which has the capacity to suppress the inhibiting of chromosome pairing which is controlled by chromosome 5B of wheat. From a series of backcrosses a normally pairing 42-chromosome line was selected which carried the rust resistance in a heterozygous condition. Compair was selected from among derived lines homozygous for rust resistance and it carries a chromosome which gives protection against all known biologic races of the fungus.

The use of chromosome substitution for genetic analysis has been successful without the involvement of alien genetic variation complications. The procedures are based on the ability of a polyploid organism to tolerate the loss of single chromosomes and exist in the so-called aneuploid state. Hexaploid wheat with its 42 chromosomes can thus be perpetuated with 41 chromosomes, and any genotype can provide 21 aneuploid forms deficient in single different chromosomes. By substituting single pairs of chromosomes from a donor variety for their homologous counterparts in a recipient aneuploid, the genetics of individual chromosomes and genetic basis of individual characters can be studied. Thus, it has been demonstrated that resistance to brown rust disease is controlled by at least four genes situated on the three chromosomes of group 7 and a firm basis is given for selecting resistant lines based on the possible cumulative effect of the main gene on chromosome 7B and its modifiers on the same chromosome and the other two in the group.

But the use of such sets of aneuploid lines within one genotype is particularly valuable for the study of quantitatively inherited characters such as yield and the components of yield. The possibilities for plant breeding are provided by the practicability of studying single-chromosome effects in the same genetic background, thereby being able to concentrate on one chromosome at a time and localizing the genetic effect. Thus chromosome assay becomes an effective tool in the biometrical analysis of quantitative inheritance which is especially applicable to an economic character such as yield. If the identification of the chromosomes involved is then associated with a determination of the genetic factors, as is practicable by the use of the substitution technique, the ultimate refinement in genetic analysis is achieved.

Two examples of such genetic analysis in hexaploid wheat have already provided valuable information which can be used to good effect in breeding work. In the first example the character studied was the number of days to ear emergence, a physiological character already referred to, which is controlled by genetic factors which respond to changes in daylength and to temperature. This character has a high heritability and, of the many different factors involved, may show a considerable difference in their contribution to controlling this aspect of plant behaviour. Apart from confirming evidence that chromosome 5D plays a major part, although a number of other chromosomes are also involved, the new evidence shows that 5D increases the time to ear emergence in some genetic backgrounds while in others it decreases the time due to between-chromosome interactions. Further analysis has indicated that this major effect of chromosome 5D can be explained in terms of allelic action at one locus on the long arm of the chromosome. The second example is concerned with yield, ear number, 500-grain weight and plant height and these were considered in relation to the three genetic parameters—additive chromosome effects, within-chromosome interactions and between-chromosone interactions. Yield, height and ear number showed a positive between-chromosome interaction and a negative within-chromosome interaction, while grain weight differed from the other characters in having a predominantly additive inheritance. In this analysis, owing to the characteristic genetic structure of wheat, with its well-defined component A, B and D genomes, it was possible to ascribe a dominating contribution by the B genome to chromosome interaction, whereas the D genome participated in an excess of between-chromosome interactions. No other crop species is as susceptible to such fine cytogenetic analysis as is wheat and few crops offer such possibilities for the exploitation of available genetic variation.

Conclusion

It is commonly asserted that British agriculture has been passing through a technological revolution for the last few decades and this is true insofar as there has been an intensive application of the results of scientific and technical knowledge and the tangible results of research. The industry is in a very receptive state and its place in the national economy is acquiring greater significance and importance and more will be expected in terms of home-grown food supplies. As the natural resources in terms of available land for agriculture contract due to absorption for other purposes, so will productivity need to rise. Further, the natural evolution of a diverse national economy normally involves a reduction in the available manpower for agriculture and greater all round efficiency becomes imperative.

Crop production is basic to the structure and economy of British agriculture, and in spite of the increasing specialization of individual farming enterprises there is a continuing considerable diversity in the range of crops grown. The principal features of the crop range will not change, but there may be adjustments in cropping balance with the possible addition of crops of minor importance agriculturally but of significance for rotational purposes. There will also be a greater emphasis on crop utilization for specialized consumer purposes, whether for direct human and animal consumption or for processing. This will result in growing individual crops more specifically for definite markets with a consequent development of a qualitatively specified product.

Agriculture is completely dependent for its development economically on the more efficient use of plant and animal resources within the context of the natural environment as determined by climate and soil. Amelioration of the adverse effects of environment has been one of the outstanding achievements of agriculture and the acquiring of some degree of independence of the environment by mechanization, fertilizers, protective chemicals and buildings has been a feature of agricultural progress. A most important additional factor has been, however, as far as crops are concerned, the availability of new and improved varieties. The successful accomplishment of the required advance in agricultural efficiency and productivity must depend inevitably on the plant breeder being able to meet the demands of the industry.

There is no reason why this should not be accomplished. Varieties with improved yield potential, of better quality and of greater suitability for changing and more efficient methods of crop husbandry are being made available. Limitations to the fullest exploitation of yield are often due to deficiencies in growing conditions, but farmers commonly raise their standards commensurate with varietal potential and improved yield expectation depends on husbandry as well as on variety. It is the breeder's responsibility to ensure that it is not the variety that is the limiting factor as far as this is practicable. There are, however, undoubtedly circumstances where this is so because changed circumstances agriculturally may occur which have not been anticipated by the breeder in time to meet the occasion. It is certainly true that changes in husbandry practice or in specialized market requirements may forestall the breeder who obviously needs time to meet a new demand if he has not anticipated it.

The more demanding agriculture becomes the greater is the scope for crop improvement and there is no major crop which cannot still be exploited genetically for advancing the level of varietal performance. The greater concentration of scientific effort, and the closer integration of this effort to provide a wider spectrum of problems to be dealt with, is allowing more ambitious objectives and greater precision in methods and techniques. Similarly, the wider range of genetic variation that is being used and the introduction of improved techniques for measuring genetic potential and for isolating valuable genotypes for breeding and improved genotypes in hybrid population are providing a new level of effectiveness in genetic control. Selection criteria are being improved by the resolution of complex plant attributes into manageable components which can be studied genetically and physiologically, while the interaction of such components and environmental interaction with them is being resolved.

Such knowledge as it accrues provides an effective means of creating new genetic models which in turn define morphological and physiological models on which improved field performance must depend. While the agricultural and other economic requirements must define the broad context of varietal requirements, the plant breeder must pioneer the means of achieving the desired ends and also provide new models which open up fresh possibilities in crop exploitation. These are not new concepts, of course, but the opportunities are greater and the means for coping with them are more potent.

The Bakerian Lecture, 1967

Reversals of the Earth's magnetic field

By Sir Edward Bullard, F.R.S.

Department of Geodesy and Geophysics, University of Cambridge

(*Delivered* 15 *June* 1967—*MS. received* 28 *March* 1968)

Philosophical Transactions of the Royal Society, A, *volume* 263, *pp.* 481–524, 1968

CONTENTS

This paper is an account of the Bakerian Lecture given to the Royal Society on 15 June 1967. Reversals of the Earth's magnetic field can be studied in the magnetization of lavas and sediments on land, in the magnetization of deep sea cores and in the magnetic pattern on the ocean floor. The lavas give radiometric dates but not a continuous sequence; the cores give continuity, great detail and a resolution as fine as 1000 years; the magnetic pattern gives information all through the Tertiary and connects the spreading of the ocean floor with the radiometric time scale.

The dynamo theory of the Earth's magnetic field may be able to account for reversals as an instability in the dynamo, but only models with a finite number of degrees of freedom have been investigated. Spreading of the ocean floor is believed to be associated with convective motions in the upper mantle, although there are difficulties connected with the equality of the oceanic and continental heat flows.

There is some evidence for the extinction of radiolaria at times of reversal of the magnetic field; it has been suggested that this is due to the effect of the field on cosmic rays but this appears impossible. If the extinctions are due to the reversals, the mechanism is unknown.

Reversely magnetized rocks are more highly oxidized than normally magnetized ones. The cause of this is unknown and is one of the outstanding problems of Earth science.

1. Introduction

Sixty years ago, Brunhes (1906) showed that some rocks are magnetized in a direction rather accurately opposite to that of the present magnetic field of the Earth. This striking phenomenon was almost totally neglected for 40 years (but see the references given by Koenigsberger (1938), especially Mercanton (1910), Matuyama (1929) and Gelletich (1937)) and it is only very recently that it has been realized that these reversals of magnetization are connected with many of the fundamental problems of earth science. The purpose of this lecture is to review these developments in both their geological and their physical aspects.

2. Reversed and normally magnetized rocks

Objects that have been baked by man, such as bricks, pottery and kilns, are invariably magnetized in the direction of the field at the time of the baking; deviations from the direction of the present field are due mainly to the secular variation of the field and enable that variation to be investigated both in magnitude and direction for periods before the first instrumental measurements, which go back only to the sixteenth century (Bucha 1967; Chevalier 1925; Thellier & Thellier 1959). Similar results are obtained from Recent and Pleistocene lavas (with rare exceptions, which will be discussed in §3). The explanation of these facts is very simple: when any ferromagnetic material with randomly oriented crystals is heated above its Curie point and cooled in a magnetic field, it becomes magnetized in the direction of the field. If the susceptibility is not too great, the field within the body is in almost the same direction as the applied field and is independent of the shape of the body; this condition is satisfied for igneous rocks whose susceptibility is typically of the order of 10^{-3}. Similar phenomena occur with sediments. Here there may be more than one process of magnetization. As ferromagnetic grains sink in water they may be in some degree oriented by the magnetic field, giving a nett magnetization in the direction of the field. Another process which is less well understood is the magnetization of chemical compounds formed in a magnetic field; in particular it seems that if sand grains are coated with iron oxides the coating becomes magnetized as it is formed, or perhaps as it is oxidized.

3. Reversals in lavas

On examining older rocks a new phenomenon is found: for Tertiary rocks, that is rocks formed between 1 and 70 million years before the present, about half are found to be reversely magnetized. If ten or twenty samples are taken covering a time span of 1000 years or more and the directions of magnetization are averaged, the effects of secular variation are largely eliminated and the mean approximates to the geographical north or south direction with the accuracy to be expected from the scatter of the individual measurements. Older, pre-Tertiary, rocks show directions of magnetization that depart from the present north and south directions; such departures are the result of polar wandering and continental drift and will not be further considered here. They do not obscure the existence of two opposite directions of magnetization and from the continuity of the motions of the continents it is usually possible to say which is the normal and which the reversed direction.

Irving (1964) has collected all measurements of the direction of magnetization up to 1963. Table 1 gives the number of his rock groups in each geological period that give normal (N) and reversed (R) polarity and also the number (M) giving sometimes one and sometimes the other polarity. A rock group usually includes many specimens taken from several sites; its formation often covered a period of several million years as, for example, did the Old Red Sandstone of England. In the light of the discussion later in this paper the occurrence of 'mixed' polarities is certainly due to this coarseness of the sampling and also, perhaps, to a proportion of unreliable results. In table 2 the results are combined in larger groups to allow a more reliable assessment of the proportion of normal and reversed rocks. The Quaternary results have been omitted from this table to avoid combining a large number of results from so short a period with the other more or less comparable groups.

The last column of table 2 gives the probability of finding a deviation from equal numbers as great as that observed if the results are a random sample drawn from a population in which the two polarities are equally probable.

Table 1. Numbers of rock units in each geological period having normal (N), reversed (R) and mixed (M) polarities

period	time interval (My)	numbers N	R	M
Precambrian	> 600	49	23	16
Cambrian	600–500	7	15	5
Ordovician	500–425	5	2	3
Silurian	425–405	7	4	3
Devonian	405–345	12	16	22
Carboniferous	345–280	24	34	15
Permian	280–230	2	50	8
Trias	180–230	21	9	17
Jurassic	135–180	37	5	12
Cretaceous	63–135	13	3	3
Tertiary	63– 2	37	42	43
Quaternary	2– 0	51	5	11

Table 2. Results from table 1 combined

	N	R	N/R	probability
Precambrian	49	23	2·1	0·0015
Cambrian to Devonian	31	37	0·84	0·27
Carboniferous	24	34	0·70	0·12
Permian	2	50	0·02	3×10^{-13}
Trias to Cretaceous	71	17	4·2	3×10^{-9}
Tertiary	37	42	0·88	0·33
total	214	203	1·1	0·31

It is clear that the total numbers of reversed and normal rocks are about equal and that this is also true of the Palaeozoic (without the Permian) and of the Tertiary. All but two of the Permian rocks are reversed and the bulk of Mesozoic ones are normal. The pre-Cambrian shows an excess of normal magnetization but there may well be doubt about many of these ancient specimens. The uniformity of the reversal in the Permian is very striking; one of the two exceptions is from the Pyrenees and is a sediment which may be Triassic and may have acquired its magnetization in the Tertiary; contemporary igneous rocks from the same area are reversely magnetized. The other normally magnetized Permian rock is from Szechwan, the results for it have a large scatter but show no other sign of unreliability.

The central problem is to discover whether the reversals of magnetization are due to reversals of the field in the past or to some chemical or mineralogical peculiarity of the rocks.

In 1951 it was discovered that a lava, a dacite from the volcano of Haruna in Japan, became magnetized in a direction opposite to the ambient field when it was cooled from a temperature above its Curie point (Nagata 1951, 1961; Nagata, Uyeda & Akimoto 1952). A number of mechanisms that would explain this bizarre behaviour were suggested by Néel (1951) and Uyeda (1958). For a while it seemed quite uncertain whether all, or most, of the reversely magnetized rocks found in nature had behaved in this way or were indicators of field reversals in the past (Blackett 1962). The ensuing controversy brought out a number of lines of evidence suggesting that field reversals were almost always the cause of reversed magnetization. First, self-reversing rocks have proved extremely rare; in fact the only example, other than the Haruna dacite, is a basalt dredged from the Pacific (Ozima & Ozima 1967), this acquires a reversed polarity only when heated for less than 150 mins to a temperature between 300 and 330 °C; other temperatures or longer heating show no reversal. Self-reversing mineral grains can be separated from some rocks (Uyeda 1958) and other rocks can be rendered self-reversing by suitable heat treatment (Carmichael 1961; Everitt 1962).

The universal normal magnetization of recent and Pleistocene rocks, except for the Haruna dacite, also speaks strongly against spontaneous self-reversal as a common process. Perhaps the most convincing of the early arguments was the observation that, when a lava or an intrusive rock has baked another lava or a sediment, the baked rock invariably has the same direction of magnetization as the baking rock; over one hundred examples are known (Wilson 1962*a*, *b*, 1966) and it would be very difficult to explain this except by supposing that the magnetization represents the direction of the Earth's field at the time of the baking. Other arguments are given by Wilson & Haggerty (1966).

These arguments were strong but did not seem conclusive to all. What was required was a study of the distribution of normal and reversed magnetization in time and space. If all rocks formed during certain periods have a normal polarity and all rocks formed during other periods have a reversed polarity, then the cause can hardly be other than a reversal of the Earth's field; it is inconceivable that there can be world-wide periods when the chemistry of all rocks, igneous and sedimentary, changes in such a way that they become spontaneously reversing. The great preponderance of reversely magnetized rocks during the Permian suggests that there are, in fact, world-wide periods of one polarity. It is, however, important to look more closely and to examine periods when there are both reversely and normally magnetized rocks.

In the 1950s many workers (for example, Hospers 1951, 1953*a*, *b*, 1954; Khramov 1957; Creer 1958; Girdler 1959) found evidence of reversals at intervals of a few hundred thousand years in Tertiary and Mesozoic rocks, but the dating of the rocks was not sufficiently accurate to show that the changes of polarity were simultaneous in widely separated places. The ages of lavas are nearly always determined by the potassium–argon method. Since the ages have an uncertainty of, at best, 3%, it is desirable to investigate lavas not more than 4 My old; for older rocks the uncertainty in the ages may make it impossible to correlate reversals between different regions (Cox & Dalrymple 1967*a*; Dalrymple, Cox, Doell & Grommé 1967).

The first studies of lavas in which both the polarities and the ages were determined with adequate precision on carefully chosen material, were those of Cox, Doell & Dalrymple (1963*a*, *b*) and McDougall & Tarling (1963). The results strongly suggest a consistent pattern of ages which has been confirmed and extended by later work (Grommé & Hay 1963; Cox, Doell & Dalrymple 1964*a*, *b*, 1965; McDougall & Tarling 1964; Dalrymple, Cox & Doell 1965; Chamalaun & McDougall 1966; Cox & Dalrymple 1966, 1967*a*, *b*; Cox, Hopkins & Dalrymple 1966; Doell & Dalrymple 1966; Doell, Dalrymple & Cox 1966; McDougall

& Wensink 1966; Cox, Dalrymple & Doell 1967; Ozima *et al.* 1967). For both the magnetic and the age determinations it is desirable to set strict requirements for the specimens; many lavas have suffered chemical changes after solidifying and may have lost argon, some contain overwhelming amounts of atmospheric argon, some may not have been free from argon when formed, some are magnetically unstable, some are magnetized by lightning and a few

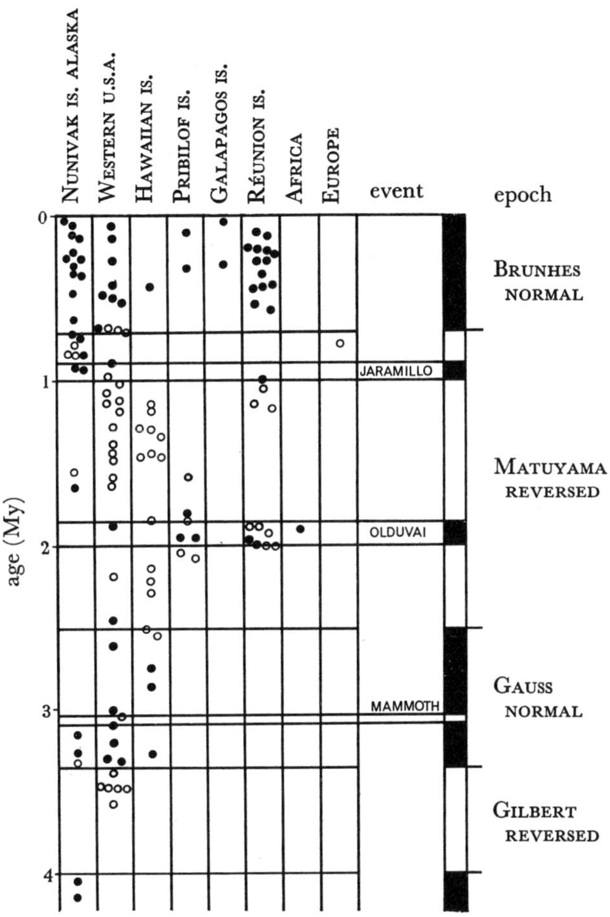

Figure 1. Ages and direction of magnetization of lavas (adapted from Opdyke & Foster 1967). A specimen from Nunivac Island which may represent the doubtful Gilsa event is shown at 1·65 My.

are spontaneously reversing. Criteria have been set out by Cox *et al.* (1963*b*), Doell *et al.* (1966) and Cox *et al.* (1966). One hundred and twenty-one of the most reliable results are given in figure 1; each point represents a number of concordant samples from one or more lava flows. During the last 3·6 My there have been four main periods of alternating normal and reversed magnetization which have been named the Brunhes, Matuyama, Gauss and Gilbert epochs. There are also three shorter periods called the Jaramillo, Olduvai and Mammoth events after the places where they were first found. It will be seen from figure 1 that the reality of the first two of these brief events is quite well established by samples from several parts of the world. Naturally the uncertainty in the ages causes rocks of both polarities to be intermixed for 100 ky or so on either side of a boundary between a normally and a reversely magnetized period; however the epochs, which last for about a million years, are clearly demarcated. The events are only 100 or 200 ky in duration and therefore are not clearly separated; during them the uncertainties of the ages cause an apparent mixing of polarities even if they are in reality periods of consistent reversal. The extent of the mixing allows the uncertainty of the potassium–argon dates to be estimated; the result, $3\frac{1}{2}\%$ or 100 ky, whichever is the greater, is consistent with estimates based on the errors of measurement and on the agreement of several specimens from the same lava flow (Cox & Dalrymple 1967*a*). It is, of course, possible that there are other as yet undiscovered events, in fact there is a flow in Iceland that has a normal polarity and an age of 1·6 My (McDougall & Wensing 1966). This flow may represent an event which has been named the Gilsa event, however, Icelandic lavas have an undesirably low content of potassium and this flow contains some glass; it is possible that the age is in error, though another example of a normally magnetized flow of the same age has been found recently on Nunivak Island by Cox & Dalrymple (1967*b*). There is also a reversely magnetized Haiwaiian lava with an age of 2·8 My which may represent another event, the Katna event (McDougall & Chamlaun 1966).

The data from lavas leave something to be desired. The outflow of lava from a volcano is an intermittent process and a pile of flows does not give a continuous sequence; the interval between two successive flows may be 6 months or a 100 ky and it is not easy to estimate how large it is. Moreover, it is not possible to say which of two flows is the older unless one lies on top of the other or unless the age difference is sufficient to be detectable by potassium–argon dating. Fortunately the examination of sediments from the deep sea gives an independent method of studying reversals which has confirmed the results from the lavas.

4. Reversals in deep-sea sediments

Sediments accumulate in the deep ocean at rates of 1 to 10 mm in 1000 years (Goldberg & Koide 1962; Dymond 1966). The most recent reversal, which occurred about 700 ky ago, will therefore be found at a depth of 0·7 to 7 m and the transition between the Gauss and Gilbert epochs will be at a depth between 3 and 35 m. If a core of ocean sediment is to contain a record of the Pleistocene and Pliocene reversals found in lavas it should be as long as possible and should be taken in sediment that is deposited slowly and contains a fair proportion of magnetic material. These criteria suggest the use of cores of red clay. Since the inclination of the magnetization is easier to measure than the declination and is less likely to be disturbed in coring, it is desirable to avoid cores taken near the equator.

Early work showed no reversals, probably because the cores were too short (McNish & Johnson 1938; Johnson, Murphy & Torreson 1948; Keen 1960, 1963). The first observation of a reversal in a core was by Harrison & Funnell (1964) who observed five reversals in cores of clayey radiolarian ooze from the equatorial Pacific. More detailed observations on longer cores have been made by Dickson & Foster (1966); Harrison (1966); Harrison & Somayajulu (1966); Ninkovich, Opdyke, Heezen & Foster (1966); Opdyke, Glass, Hays

& Foster (1966); Hays & Opdyke (1967); Watkins & Goodell (1967*a*, *b*, *c*) and Berggren, Phillips, Bertels & Wall (1967); Goodell & Watkins (1968). Results from a core 13 m in length taken in the North Pacific about 1000 km south of the Aleutians are shown in figure 2; the figure shows the magnetic dip measured on samples taken at intervals of 10 to 20 cm in depth (2 cm in the neighbouhood of reversals). For the upper 5 m of the core the

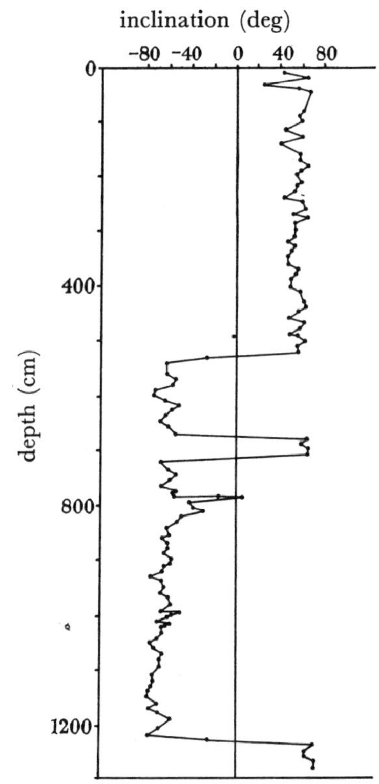

FIGURE 2. Dip of the magnetization of a core from the North Pacific (from Ninkovich *et al.* 1966).

dip has the normal, downward, direction and fluctuates between about 40 and 60° with a mean of about 52°, which is not far from the present dip of 57°. At a depth of 525 cm the dip changes sign and remains negative till 678 cm where it again reverses to the normal direction where it remains for the next 34 cm. It then reverses again and remains so till near the bottom of the core where a further transition to the normal direction occurs. All these changes are shown by other cores from the same area and also by cores from other parts of the world; they clearly represent a general phenomenon and not a peculiarity of the individual core. The spike at about 800 cm is peculiar to this core and presumably represents a disturbance produced in coring or, conceivably, a disturbance of the sediment while on the sea floor.

It is natural to correlate the upper transition with the Brunhes–Matuyama transition found from the lavas at 700 ky before the present, the brief return to normal with the

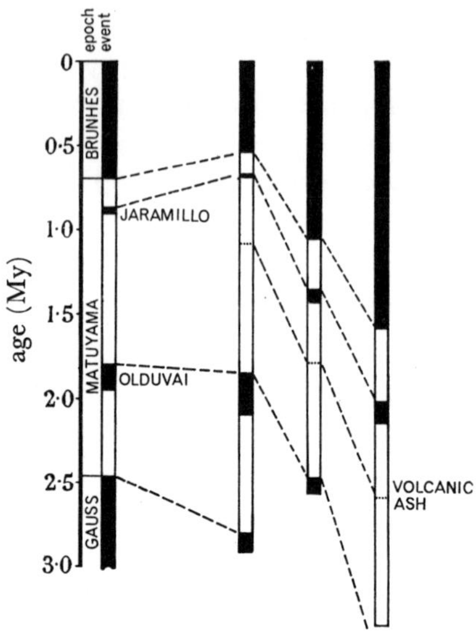

FIGURE 3. Comparison of three cores from the North Pacific with the time scale of reversals deduced from lavas (from Ninkovich *et al.* 1966). Black sections are normally magnetized, white ones are reversed.

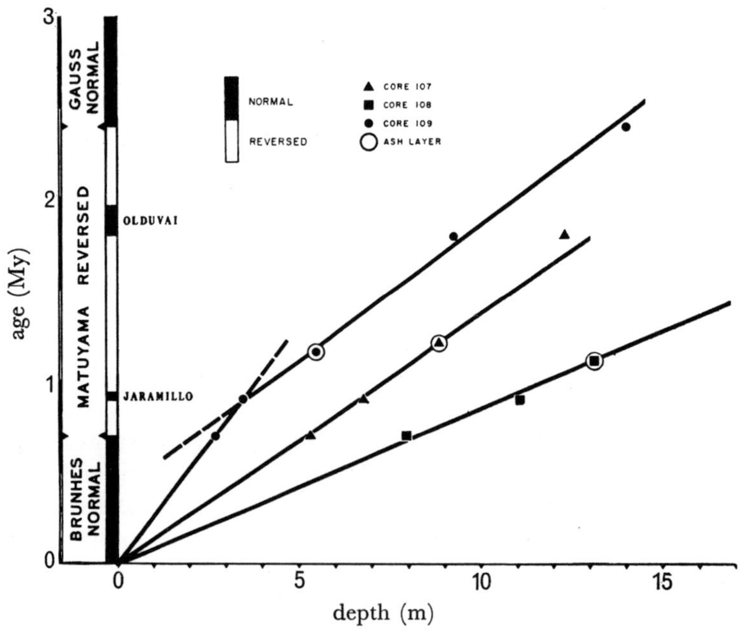

FIGURE 4. Rate of sedimentation from the cores shown in figure 3 (from Ninkovich *et al.* 1966).

Jaramillo event and the final short piece of normally magnetized sediment at the bottom with the Olduvai event at 1·8 My. These correlations for this and two other cores from the same region are shown in figure 3. The progress of sedimentation in three of the cores is shown in figure 4; two of the cores give constant rates of 0·75 and 1·13 cm/ky, the third gives 0·36 in the upper part and 0·80 in the lower. The correctness of the correlations is confirmed by the occurrence of a well-marked ash band at depths corresponding to about 1·2 My in the three cores. Measurements going back about 5 My, that is to a time before the start of the Gilbert epoch, have been obtained by Hays & Opdyke (1967), they find three well marked events

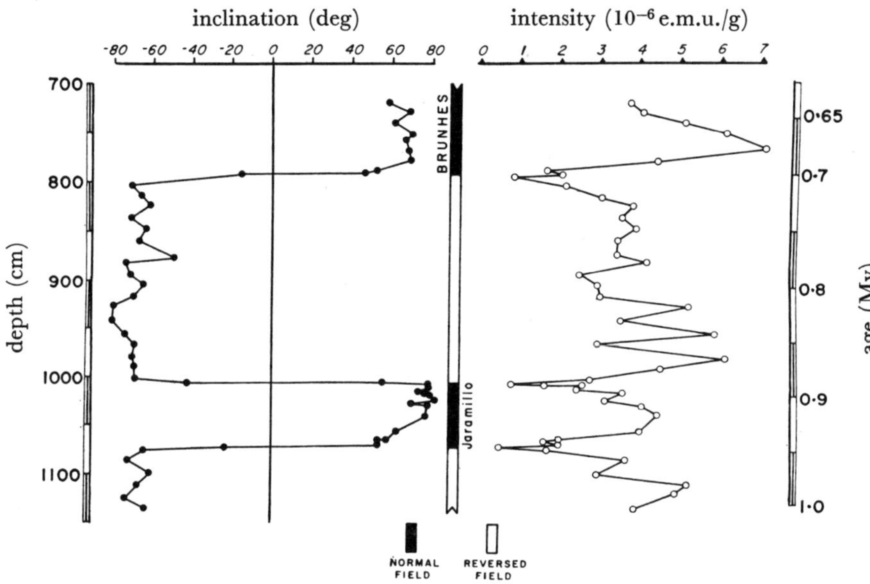

FIGURE 5. Dip and intensity of magnetization of a portion of a core from the North Pacific (from Ninkovich *et al.* 1966).

of normal magnetization in the Gilbert epoch. The results constitute a striking confirmation of the sequence of reversals deduced from the magnetization of lavas. No two substances could be more different or have more different histories than the lavas of California and the pelagic sediments of the Pacific. The lavas were poured out, hot and molten, by volcanos and magnetized by cooling in the Earth's field; the ocean sediments accumulated grain by grain by slow sedimentation and by chemical deposition in the cold depths of the ocean. If these two materials tell the same story then it must be the story of an external influence working on both and not a story of recurrent synchronous change in the two materials. The evidence compels belief in reversals of the Earth's field as the cause of these reversals of magnetization.

The accuracy, continuity and detail available in data such as that in figure 3 can be used in many ways. Clearly the reversals are nearer to sudden transitions than to simple harmonic oscillations. This is shown still more clearly in figure 5, which shows the dip and intensity of magnetization for a portion of a core taken at a point about 250 km to the north of that shown in figure 3; for this core samples were taken at 1 cm intervals in the neighbourhood of the reversals. It seems that a large part of the change in dip takes place in a time interval of 1 ky. The change in intensity of magnetization of the sediment is shown on the right of figure 5; in a uniform sediment this provides a measure of the magnetizing field. The magnetization is very variable but shows its lowest values at the times of reversals; this had previously been demonstrated by van Zijl, Graham & Hales (1962*a*, *b*) for the Stormberg lavas of the Transvaal. At these minima the magnetization is 10 to 20% of its normal value. The intensity of the field begins to fall before there is any large change in dip and continues to increase after the dip has completed its transition; the total time taken for the magnetization to decrease and recover is about 40 ky and is several times larger than the time required for the reversal of the dip.

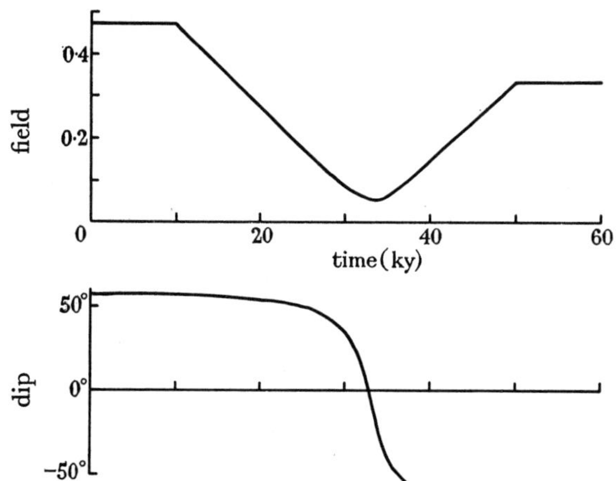

FIGURE 6. Change of dip and field intensity if a field of 0·4 G reverses in 40 ky in the presence of a constant field whose north, east and vertical components are equal and are all 0·05 G.

There are two extreme ways in which a vector field can reverse: it can swing round without change in magnitude or it can decrease without change in direction, pass through zero and then grow in a direction opposite to its intial direction. The observed behaviour is nearer the second of these, which would give a discontinuous change in dip as the field passed through zero however slowly the field decayed. The observed behaviour is compatible with a reversal of the dipole field accompanied by the usual fluctuating non-dipole field. Figure 6 shows the course of events if a field of 0·4 G decays linearly and grows again in 40 ky in the presence of a small constant field. More realistic calculations with a fluctuating non-dipole field are easily made.

The reversal of dip in about 1000 years is surprising from another point of view. Any mixing of sediment by marine animals, such as mud-eating worms, will make the transition less sharp than it would be in undisturbed sediment. It is usually supposed that the mottled appearance of the material in most cores is due to this mixing process and indicates that it extends over about 3 cm (see, for example, the reproduction by Ninkovich *et al.* (1966) of photographs of parts of the cores from which the data of figures 2 and 5 were taken). It is also surprising that the dip is so close to that measured today; it would be expected that compaction would reduce it. It is conceivable that the magnetization occurs some time after burial either by the magnetization of crystals forming slowly in the sediment or by the grains remaining mobile till they are locked by compaction.

Comparison of the depths of magnetic transitions with the dates from lavas on land allows rates of sedimentation to be determined. They are of the same order (1 to 10 mm/ky) as those found by other methods and vary from one core to another; they do not, however, show the variation that might be expected from the large fluctuations in the climate during the Pleistocene. It is likely that the climate varies on a shorter time scale than the magnetic reversals and that the fluctuations in rate of sedimentation are averaged out. The durations of the 'events' are too short to be found with any accuracy from the dating of lavas, better estimates can be obtained from the cores, assuming a constant rate of sedimentation. The Jaramillo event seems to have lasted about 50 ky and the Olduvai event about 140 ky.

In the detail of the results there are a number of perplexing discrepancies which have been discussed by Evans (1969); Dickson & Foster (1966) get a poor correlation between measurements of declination and dip, Ninkovich *et al.* (1966) find the Olduvai event to be double, whereas Dickson & Foster do not, Opdyke *et al.* (1966) have a core that shows neither the Jaramillo nor the Olduvai event, Watkins & Goodell (1967*b*) have found evidence for the Gilsa event in their cores which others failed to observe. These discrepancies are presumably due to defects in coring technique and, perhaps some of them, to irregularities in sedimentation. In principle, coring is simple: a tube is driven into the sea floor and brings up an undisturbed sample of sediment. In practice, it is a skilled and somewhat chancey operation. Cores taken with a free fall corer are usually too short to reach the first reversal and almost always give a core that is shorter than the depth of penetration, the shortening is not uniform and may be achieved by the squeezing out of soft bands (Emery & Dietz 1941; Ross & Riedel 1967). Piston corers (Kullenberg 1947) can give a much greater penetration but frequently 'suck in' sediment which is grossly disturbed and may be out of place if the suck occurs as the corer is being withdrawn. There is some risk of the polarity-time scale being reduced to chaos by attempts to correlate spurious events produced by uncertainties in potassium argon dating of lavas with apparent reversals of magnetization in sediments due to disturbances in coring. There are also possibilities of disturbances of the sediment on the sea floor due to the removal or addition of material by slumping or turbidity currents. In view of the importance of getting reliable results a substantial effort to improve the techniques of taking cores and dating lavas would be justified. Even if no improvement were made in the technique of coring, a recorder that gave the motion of the coring tube and of the piston relative to the sediment outside the tube as a function of time would make it easier to select cores meeting the very exacting requirements for the study of the history of reversals. A reliable method of dating cores independently of correlation with the lavas would remove many doubts. It seems unlikely that the ionium method can give reliable results for material more than 500 ky old (Goldberg & Koide 1962), perhaps the best prospect is the dating of bands of volcanic ash by the potassium–argon method (Dymond 1966).

5. The magnetic pattern on the ocean floor

A large part of the floor of all the oceans is covered by a magnetic pattern in the form of parallel stripes. One of the most significant discoveries of recent years is that this pattern repeats again the same story of reversals in the Earth's magnetic field as is told by the lavas on land and the sediments beneath the ocean floor. This zebra-like pattern was first seen in the magnetic surveys off the Pacific coast of North America conducted by the Scripps Institution and the U.S. Coast and Geodetic Survey (Mason 1958; Mason & Raff 1961; Raff & Mason 1961; Vacquier, Raff & Warren 1961; Raff 1962, 1966; Peter & Steward 1965; Talwani, Le Pichon & Heirtzler 1965; Peter 1966; Elvers, Mathewson, Kohler & Moses 1967). The pattern shows a marked north-south lineation extending 3700 km from Mexico to the Aleutians. It is broken into blocks by 'fracture zones', but, in spite of this, individual stripes can be traced over the whole distance and probably extend further into unsurveyed areas. The fracture zones are clearly faults which appear as scarps on the sea floor. At these scarps the magnetic pattern is offset, sometimes by as much as 1200 km. Unlike the fracture zones, the magnetic pattern itself is not related to the topography of the ocean floor (figure 7) or of a surface buried beneath the sediments. For reasons that have only recently become clear, some critical features of the pattern do not appear off the Californian coast. It was the measurements of Heirtzler, Le Pichon & Baron (1966) over the Rykjanes Ridge to the south of Iceland which first showed the phenomenon in its clearest and simplest form. Here, as is shown in figures 8 and 9, the magnetic stripes run parallel to the axis of the mid-Atlantic Ridge and are almost exactly symmetrical about it. Similar patterns have been found in all the oceans (Heirtzler & Le Pichon 1965; Matthews, Vine & Cann 1965; Bowin & Vogt 1966; Cann & Vine 1966; Pitman & Heirtzler 1966; Demenitskaya 1967; Elvers *et al.* 1967; Grimm & Erickson 1967; Heirtzler & Hayes 1967; Schlich & Patriot 1967; Rassokho *et al.* 1967; Uyeda *et al.* 1967). There is some difficulty in adequately representing these patterns. Individual profiles, such as those of figure 8 contain noise from the magnetic effects of the topography and magnetic variations during the survey. On the other hand, maps, such as that of figure 9, showing merely areas of excess and defect of field often break up the stripes into blobs because a peak just fails to reach the zero line; this fault is accentuated by a slight general depression of the field over the crest of the ridge, perhaps due to the Curie point being nearer the surface there than elsewhere (Heirtzler & Le Pichon 1965). A detailed contoured map requires more information and better navigation than is usually available.

The straightness, symmetry, extent and ubiquity of these patterns is without parallel in geology and their discovery was completely unexpected. Nothing like them is known on the continents and they must be an expression of some world-wide oceanic process. The nature of this process was first suggested by Vine & Matthews (1963) largely on the basis of measurements made from H.M.S. *Owen* in 1962 during the International Indian Ocean Expedition. An area about 50×40 miles on the crest of the Carlsberg Ridge in the north-west Indian Ocean was surveyed in detail (Matthews *et al.* 1965); it was found that the magnetic effects of individual topographic features could be picked out and that some were normally magnetized and some reversely (Cann & Vine 1966). With this as a clue, Vine & Matthews (1963) suggested that the pattern of linear anomalies is due to strips of the sea

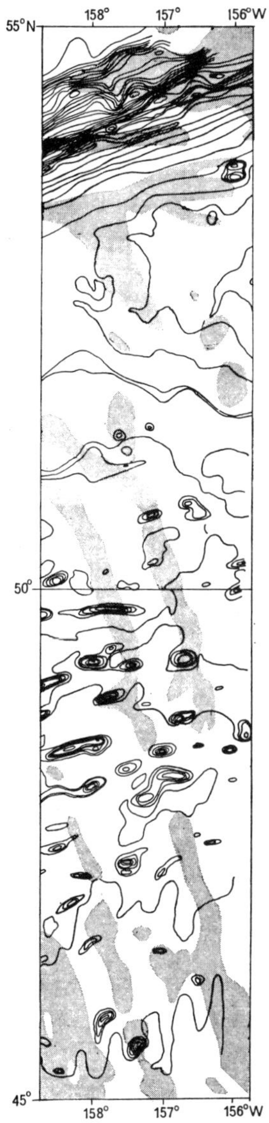

FIGURE 7

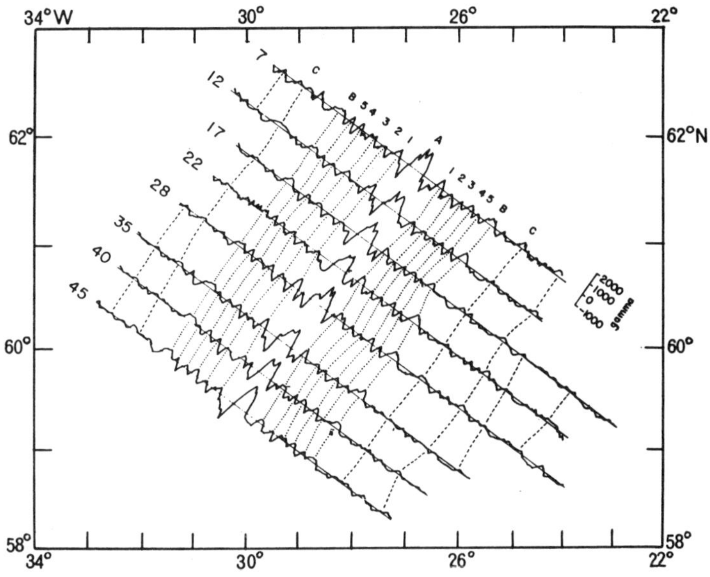

FIGURE 8

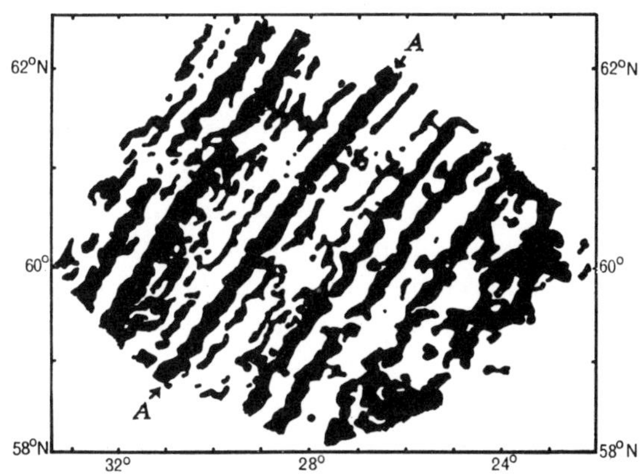

FIGURE 9

FIGURE 7. A portion of the magnetic pattern off the west coast of North America showing the lack of relation to topography. The contours represent the topography and the shading the regions of high magnetic field (from Peter & Stewart 1965).

FIGURE 8. Eight of the 58 profiles of total magnetic force across the Reykjanes Ridge to the south of Iceland projected on lines at right angles to the ridge. Note the continuity of the anomalies along lines parallel to the axis of the ridge (marked *A*) and the symmetry about this axis (from Heirtzler *et al.* 1966).

FIGURE 9. The magnetic pattern over the Reykjanes Ridge. Areas of positive anomaly in the total force are black. *A–A* is the axis of the ridge (from Heirtzler *et al.* 1966).

floor being magnetized in opposite directions. They explained the existence of such strips by combining the reversals of the magnetic field with the ideas of sea floor spreading.

Sea floor spreading was first suggested by Dietz (1961) and Hess (1962, 1965), the claim that the idea can be found in much earlier papers (e.g. Holmes 1931) seems to me not to be well founded. Dietz and Hess suggested that mantle material rises under the mid-ocean ridges, spreads out horizontally near the surface and produces a drag on the surface material which causes cracks along the axis of the ridge as is shown in figure 10. The crack is filled with igneous rocks from below, which are again cracked by the same process. Thus

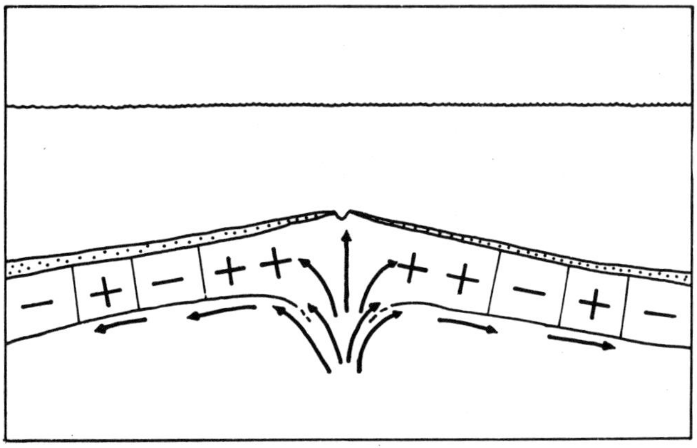

FIGURE 10. Cracking at the axis of the ridge and the production of oppositely magnetized stripes.

at any time the youngest material is on the axis of the ridge and older material is carried away on both sides. Such a picture unifies many aspects of marine geology. The mid-ocean ridges do very commonly have a central valley floored by young igneous rocks and young sediments and the sediments on either side do increase in age as one recedes from the ridge (Burckle, Ewing, Saito & Leyden 1967; Funnell 1968). The flow of heat from the interior of the Earth is exceptionally high over the axis of the ridge. The majority of mid-ocean earthquakes are rather accurately located in the central valleys of the mid-ocean ridges and give 'first motions' which show that they are associated with the opening of a crack (Sykes 1967); the remainder of the earthquakes occur on the fracture zones which traverse and offset the ridge axis, the first motions show them to be transform faults (Sykes 1967) as predicted by Wilson (1965*a–c*). In recent years the idea has been elaborated in connexion with continental drift; spreading of the ocean floor does not imply continental drift, but is compatible with it. A convection current rising under a ridge might dive down again at the edge of a continent, without moving it, but producing stresses, an ocean deep and deep earthquakes; that is, producing a 'Pacific' type of coast line in the sense of Suess (1888). On the other hand the horizontal limb of the convection current might continue under the continent and only sink when it encountered an oppositely moving limb under the ocean on the far side of the continent. The continent would then be transported, as on a conveyor belt, with no large stresses, except perhaps at its leading edge.

There are difficulties in these views. To discuss them in detail would take us too far from our main theme. Perhaps the most serious is the dredging of Miocene globigerina ooze incorporated in a piece of volcanic glass from a fracture zone near the axis of the ridge in the South Atlantic (Ewing, Le Pichon & Ewing 1966; Saito, Ewing & Burckle 1966). If the fracture zone is a transform fault as in figure 11, rocks of all Tertiary ages will have moved past it and it is not impossible that pockets of older rocks will survive along the fault. Alternatively, it is a little suspicious that a specimen of an identical, very rare rock should have been dredged on the same expedition 130 km from the axis of the ridge; it is perhaps conceivable that both bits came from the latter haul. A basalt with an age of 27 My was dredged from Cobb seamount. According to Budinger & Enbysk (1967) this seamount is associated with the axis of the Juan da Fuca Ridge, however their diagram shows it to be 100 km to the west of the axis. In any case, the rock may well be an erratic since, according to Heirtzler *et al.* (1968), drilled samples from the seamount give an age of 2 My. Such explanations may appear a little forced, but the few discrepancies must be viewed against the background of hundreds of cores and dredge hauls that give the expected results; odd things must be expected to occur occasionally.

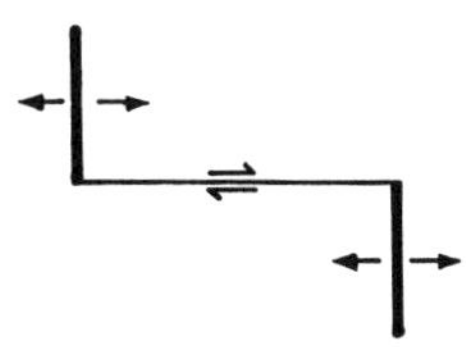

FIGURE 11. A transform fault joining two sections of a ridge.

Another unexpected result is the lack of disturbance of the sediment in the ocean trenches (Ewing & Ewing 1967*a*; Huene *et al.* 1967; Scholl *et al.* 1967), if the material spreading from the East Pacific Rise is really diving beneath South America and if South America is advancing into the Pacific it is difficult to see how the sediments can remain undistorted for any long period. It would be interesting to know the age of the undistorted material.

The essential suggestion of Matthews and Vine was that the new material on the axis of the ridge would always be magnetized in the direction of the magnetic field at the time of injection and that the history of the reversals would be spelt out by the horizontal transport of the magnetized rock, in a way similar to the recording of a message by a tape recorder (figure 10). This hypothesis has been brilliantly verified by Vine & Wilson (1965), Vine (1966) and by their followers (Pitman & Heirtzler 1966; Heirtzler 1967; Krause 1967; Phillips 1967; Dickson *et al.* 1968; Heirtzler *et al.* 1968; Le Pichon & Heirtzler 1968; Pitman *et al.* 1968). The last four papers collect and summarize all the material; I am grateful to Dr Heirtzler for letting me see them before publication.

If the hypothesis is correct it should be possible to predict the results of magnetic surveys from the sequence of reversals. Four examples from widely separated parts of the ridge are given in figure 12, in which the second line shows the observed variation of the magnetic field across the ridge. The top line is identical to the second one turned end to end, the close similarity of the two shows the high degree of symmetry about the axis of the ridge. At the bottom of the figure is the pattern of reversals deduced from the magnetization of lavas on land, above this is the magnetic profile calculated on the assumption that there are blocks of normally and reversely magnetized lava under the ocean floor corresponding

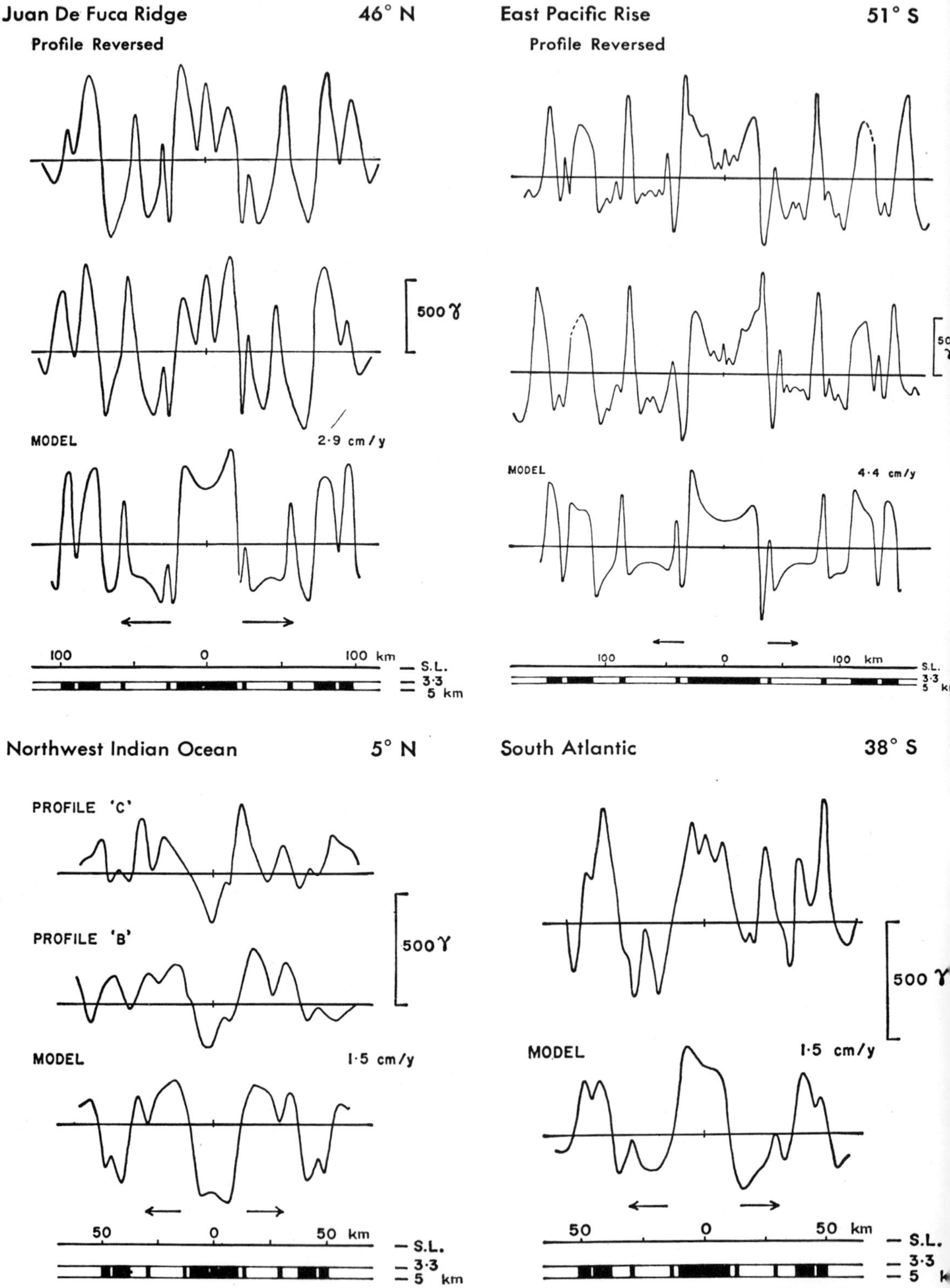

FIGURE 12. Magnetic profiles (total force) along tracks crossing the mid-ocean ridges. The top line in each section is the observed profile reversed (note the symmetry). The third line is the profile calculated from the pattern of reversals deduced from the study of lavas on land, which is shown at the bottom of each section on an appropriate scale (from Vine 1966).

to the pattern of field reversals. The agreement of these calculated profiles with the observed ones is very striking and about as good as that between the direct and reversed profiles. Exact symmetry would only be expected where the ridge runs in a north–south direction, but in fact the departures from symmetry are quite small in both the calculated and the observed profiles; they are most easily seen in the peak over the central valley, the form and amplitude of which have been investigated in detail (Heirtzler & Le Pichon 1965) and are fully accounted for by a block under the valley magnetized in the direction of the present field. As would be expected a north–south ridge gives no coherent pattern in the neighbourhood of the equator (Vacquier 1965).

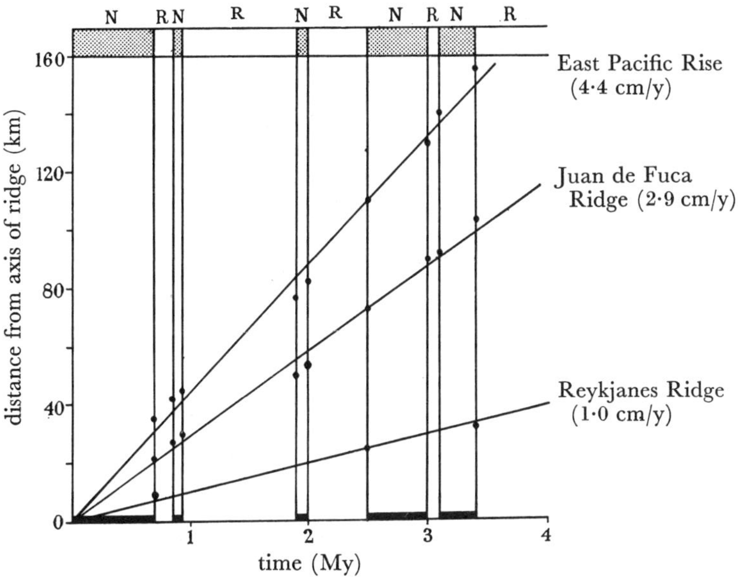

FIGURE 13. Plots of the distance of magnetic stripes from the ridge axis against times for the corresponding field reversals determined from the ages of lavas on land (from Vine 1966).

To predict the results of magnetic surveys from measurements on rocks thousands of miles away is a claim so exorbitant in a subject where things are rarely clear-cut or predictable, that it is necessary to be very sure that we are not deceiving ourselves. The agreement of the calculated and observed curves is so good that it seems impossible to ascribe it to chance. One could wish in this and other geophysical contexts that there were objective criteria for describing the resemblance of curves; the matter is one of some difficulty, the obvious criteria, using cross-correlation, are not very useful because they are dominated by the central peak and give little weight to the agreement of small peaks. Also cross-correlation is grossly disturbed by a little irregular stretching of the curves such as might be produced by slight variations in the velocity of spreading of the ocean floor. Similarity is not a simple idea, and is only partly metrical; it is concerned with the number and sequence of maxima, minima and zero crossings as well as with their positions and amplitudes. It may be quite difficult to formalize intuitive impressions about it, perhaps ideas from the detection of bomb tests by seismic arrays and from numerical taxonomy may be fruitful.

It is important also to consider how many adjustable parameters are involved in the calculations; anything can be fitted convincingly to almost anything if enough arbitrary constants are available. In comparisons such as those of figure 12 there are two essential quantities to be determined in making a fit, the horizontal scale, that is the velocity of spreading, and the intensity scale, that is the intensity of magnetization of the rock. It might be imagined that the velocity of spreading would vary from time to time and that it would be necessary to assume a more or less complicated and arbitrary velocity–time curve. In fact, this is not so over the period of $3\frac{1}{2}$ My shown in the comparisons of figure 12. In that figure a constant velocity has been assumed for each profile; if the distances from the axis of the ridge are plotted against the times of reversal the points lie very well on straight lines (figure 13).

In the comparisons of figure 12 the intensities of magnetization of all blocks, except the central one on each profile, have been taken to be 0·005 e.m.u. The central block has been assumed to have twice the intensity of the other blocks, this assumption is necessary since the magnetic anomaly over the central valley is nearly always much greater than the peaks in the rest of the pattern. An increase for the central valley is plausible since there will not, in reality, be a succession of completely separate blocks but a series of more or less overlapping lava flows fed by many feeder dykes; the central valley of the ridge may be floored entirely by recent lavas but on each side there will be overlapping with older lavas. Matthews & Bath (1967) and Vine & Morgan (1967) have made calculations of the effect of a statistical distribution of intrusions and conclude that the feeder dykes may be spread over a few kilometres. Mudie & Harrison (1967) have made measurements of the magnetic field near the ocean floor using a towed, deep-diving magnetometer; they find a fine structure in the pattern which probably represents the effects of individual dykes and flows. It is perhaps fortunate that this fine structure is not apparent in observations made near the surface of the sea; had it been, the elegant regularities of the large-scale pattern would have been much less apparent.

The depths of the upper and lower surfaces of the magnetized blocks are also, in principle, variable, but in practice these parameters are of little significance. The thickness of the sediment on the deep ocean floor is usually only a few hundred metres and is almost always much less than the depth of water; the upper surface of the blocks can therefore be taken as at the sea floor or a few hundred metres beneath it. The position of the lower surface of the blocks has little effect on the form of the curves. In figure 12 the upper and lower surface of the magnetized blocks have been assumed to be 3·3 and 5 km below the surface of the sea. It is possible that allowance for the actual form of the upper surface would improve the agreement, for example Vine & Wilson (1965), by allowing for the topography of the Juan da Fuca Ridge, get a triple central peak, as in the observed profile of figure 12. It might be argued that the lower surface of the blocks should be deeper, the 5 km depth was chosen under the influence of Hess's (1962, 1965) opinion that on the ocean floor a few kilometres of basalt overlie almost non-magnetic serpentine, if the seismologists' 'layer 3' is basalt a greater depth should be taken.

Other adjustable parameters are involved in removing the 'trend' from magnetic surveys to get anomalies; the procedures commonly used are not very satisfactory but the arbitrariness can be completely avoided by using a world-wide reference field to remove trend or by numerical smoothing over distances large compared to the features being studied (Bullard 1967*a*). In summary the comparisons in figure 12 involve one adjustable constant per profile which fixes the horizontal scale and two more for the whole set, one of which fixes the vertical scale and the other the factor of two in the magnetization of the central blocks. Such an exiguous number of adjustable quantities, which are just those essential to the theory, can scarcely have produced the numerous coincidences of peaks seen in the figures. We therefore conclude that the hypothesis of Vine & Matthews, that the magnetic pattern found at sea is due to field reversals and spreading of the ocean floor, is verified.

The hypothesis was, very naturally, at first received with some scepticism (see, for example, Talwani, Le Pichon & Heirtzler 1965). Most of the original doubters have now been converted, but there have been some recent, though rather half-hearted, attempts at alternative theories by Watkins (1968) and Ozima *et al.* (1968). They suggest that the magnetic pattern may be due to the intrusion of dykes along a system of fractures in a 'crustal plate'. I find it difficult to imagine a system of tensile fractures so regular that a series of thirty or more of them will run parallel at a separation of a few tens of kilometres for a distance of thousands of kilometres without running into each other. The main difficulty is not, however, these mechanical improbabilities, but the impossibility of explaining the bilateral symmetry of the patterns, their similarity all over the world and the detailed agreement with Vine's models. A real reason for doubt is the difficulty of explaining by ocean floor spreading the sharp bend in the anomalies in the northeast Pacific, where they swing from roughly north-south to east-west (figure 14) with no magnetic or topographic signs of disruption or crumpling. On the Matthews–Vine hypothesis these anomalies are probably over 60 My old and long antedate the formation of the Aleutian arc and the arrival of America in its present position. It is far from clear what has happened to the associated ridge, which now extends only to the latitude of the north end of Vancouver Island, or where it was when the pattern was formed. The detailed publication of the results, particularly of the individual magnetic and topographic profiles, is clearly of high importance. The geometry of the bend does pose a serious problem, but it seems rash to throw over the very clear conclusions from so large a body of data on this ground alone.

If the Vine–Matthews hypothesis is accepted the whole history of the ocean floors is open to investigation. The rates of spreading to each side of the ridge during the last $3\frac{1}{2}$ My for the sections of ridge so far investigated are given in table 3. The rate of spreading from the East Pacific Rise averages 3·8 cm/y and is two to three times greater than that from the ridges in the Atlantic and Indian Oceans. There are a number of other ways in which the East Pacific Rise differs from the mid-Atlantic and Carlsberg Ridges (Phillips 1967), but the significance of these is not clear. From a comparison of the magnetic pattern off the coast of California with that farther north it seems that only the west part of the pattern is present and that this part of the East Pacific Rise has been overrun by the westward drift of North America and now lies under Utah and Arizona (Vine 1966). A similar conclusion had been reached earlier by Menard (1964) from a study of the topography, which is characteristic of the western flank of a ridge.

Very recently Heirtzler *et al.* (1968) have attempted to correlate all known magnetic profiles across the ridges. Here the emphasis has shifted, the primary object is no longer to confirm or refute the Vine–Matthews hypothesis; the hypothesis and the spreading of the ocean floor are assumed to be of general application. The object is to see further into the past and to get an idea of the rates of spreading and of the ages of the ocean floor for the last 50 to 80 My. Often the correlations are very clear even in the oldest part of the pattern, for example, figure 15 shows that the curves from the North Pacific are similar in almost every detail to those taken south-east of New Zealand, these parts of the pattern are believed to have been formed 60 to 80 My ago. In many places the correlations are not so clear, here what seems the most plausible choice has been made. If these correlations are accepted it seems that the whole of the east and south Pacific, the southern and western parts of the Indian Ocean and a central strip of the Atlantic are covered by a correlatable pattern.

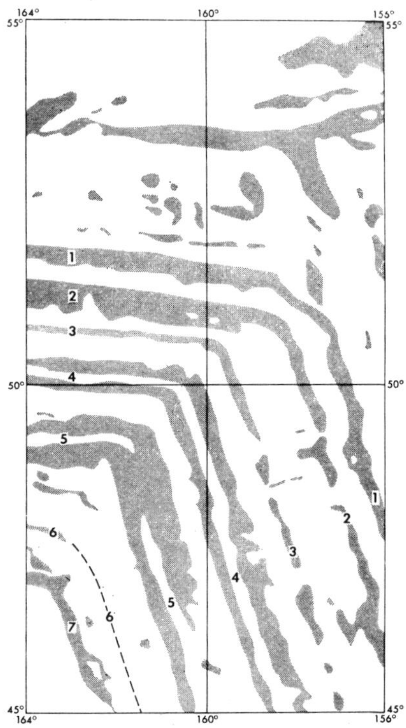

FIGURE 14. Bend in the pattern of anomalies to the south of the Aleutians (from Elvers *et al.* 1967).

The eastern sector of the Arctic Ocean has a remarkable pattern (Demenitskaya 1967; Rassokho *et al.* 1967) which no one has attempted to correlate with that in the other oceans. There is also a pattern in the north-west Pacific (Uyeda *et al.* 1967). The north-eastern Indian Ocean appears to have no comparable pattern; it is not clear whether there is a pattern in the Indian Ocean south of Africa. Unpublished profiles from operation NAVADO (a joint project of the British and Dutch navies) show an extensive pattern in the North Atlantic but, on a cursory examination, some of them show no clear symmetry. Ewing & Ewing (1967 *b*) have suggested, from a study of the sediments, that spreading in the Atlantic may have been intermittent. Clearly the exact delineation and interpretation of the pattern over all the oceans is an enormous undertaking that will take decades to complete, most of the present information is in the form of isolated profiles separated by hundreds of kilometres, the surveys of the north-west Pacific and of the Rykjanes Ridge show how much more instructive are surveys in which the lines are 10 km or less apart.

TABLE 3. RATE OF SPREADING FROM THE RIDGE AXIS DURING THE LAST 3·5 MY

ocean	lat.	long.	rate (cm/y)	ref.	notes
Atlantic	60° N	28° W	1·0	1	Reykjanes Ridge
	27° N	44° W	1·25	2, 5	—
	22° N	45° W	1·40	2, 5	—
	25° S	13° W	2·25	5	—
	28° S	13° W	1·95	5	—
	30° S	14° W	2·0	5	—
	38° S	17° W	1·5	1	—
	38° S	17° W	2·0	5	—
	41° S	18° W	1·65	5	—
	47° S	14° W	1·60	5	—
	50° S	8° W	1·53	5	—
Indian	19° N	40° E	1·0	1, 5	Red Sea
	16° N	41° E	1·0	1	Red Sea
	13° N	50° E	1·0	5	Gulf of Aden
	7° N	60° E	1·5	5	Carlsberg Ridge
	5° N	62° E	1·5	1, 5	Carlsberg Ridge
	22° S	69° E	2·2	5	Chagos-St Paul Ridge
	30° S	76° E	2·4	5	Chagos-St Paul Ridge
Pacific	46° N	130° W	2·9	1	Juan de Fuca Ridge
	42° N	127° W	1·0 to 3	4	Gorda Ridge
	17° S	113° W	6·0	5	E. Pacific Rise
	40° S	112° W	5·1	5	—
	45° S	112° W	5·1	5	—
	48° S	113° W	4·7	5	—
	51° S	117° W	4·9	5	—
	51° S	117° W	4·4	1, 3	—
	58° S	149° W	3·8	5	—
	60° S	150° W	4·0	5	—
	63° S	167° W	2·3	5	—
	65° S	170° W	2·0	5	—
	65° S	174° W	2·8	5	—

1, Vine (1966); 2, Phillips (1967); 3, Pitman & Heirtzler (1966); 4, Menard (1967); 5, Heirtzler *et al.* (1968)

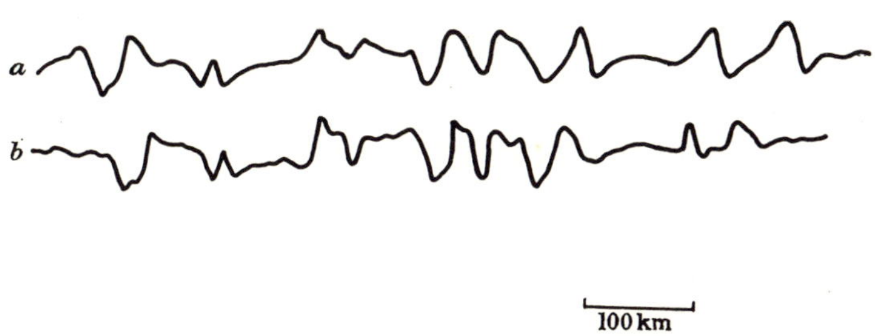

FIGURE 15. Comparison of a magnetic profile (*a*) in the North Pacific in latitude 41° N with one (*b*) in the South Pacific in latitude 50° S to the east of New Zealand. The section lies about 1500 km from the axis of the ridge and is believed to cover ages from about 60 to 80 My (from Pitman *et al.* 1968).

The rates of spreading are only determined for the last 3·6 My, which is the period during which the radiometric dates on the continental lavas can be correlated confidently with the magnetic reversals. If the ages of lavas beneath the magnetic stripes on the sea floor could be determined this time scale could be pushed back to perhaps 80 My. Unfortunately the outer parts of the ridge are largely covered by sediment and where igneous rocks are visible they are so decayed as to be unsuitable for dating (see, for example, Matthews 1961); even if they were fresh it is doubtful whether seamounts projecting through the sediments are contemporaneous with the underlying rocks. It may be possible to

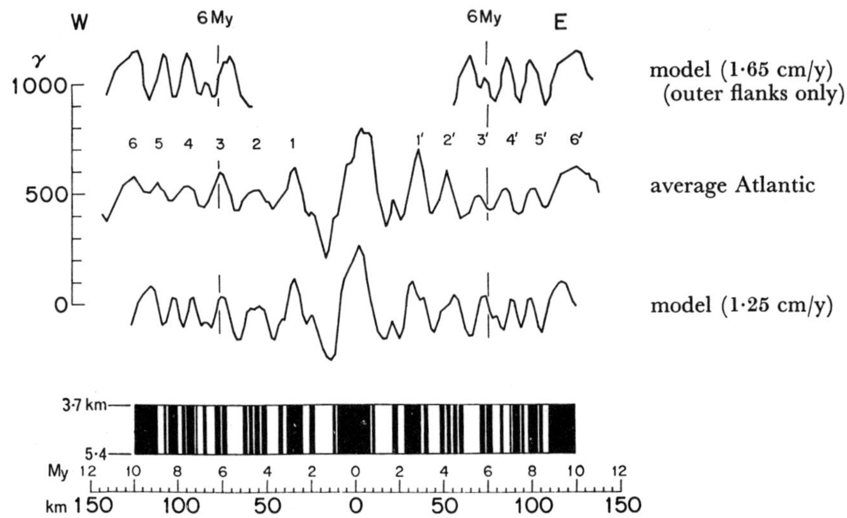

FIGURE 16. Comparison of magnetic profiles in the South Pacific and the Atlantic during the last 10 My (from Phillips 1967).

obtain suitable samples by drilling and to do so is one of the main objects of the JOIDES project (a cooperative drilling project by five universities in the United States). In the meanwhile it is possible to obtain some evidence on the constancy of the spreading rate in the past by comparing sections of the magnetic patterns in different areas. Owing to the effects of differences in latitude and in the strike of the axis of the ridge it is not desirable to compare two magnetic profiles directly. A history of reversals is derived from one profile and the variation of field to be expected along the other is calculated from this and compared with observation; if the rate of spreading has been uniform in both areas, or if the two rates have always been proportional, then the observed curve will agree with the calculated one when the scales are suitably adjusted. Figure 16 shows an example of a comparison of this kind (Phillips 1967), the history of reversals is taken from Pitman & Heirtzler's (1966) analysis of profiles in the South Pacific and the expected magnetic profile 11000 km away at 27° N in the Atlantic is calculated from it. The agreement with observation is almost perfect and removes any lingering doubt that might exist as to the general correctness of the principles on which the calculations are made. A close examination shows that if the scale of the model is adjusted to fit the first three main peaks on each side of the central maximum then the peaks further out fall a little nearer the ridge than do the observed ones. This is most naturally interpreted either as a decrease in the rate of spreading in the Atlantic 5 My ago from 1·65 to 1·25 cm/y or as an increase of the rate in the South Pacific from 3·4 to 4·5 cm/y. An example of a much more extended comparison is shown in figure 17; here the distances of corresponding magnetic stripes in the north and south Pacific are compared. Clearly one, or both, rates of spreading have varied in the past; if the spreading rate in the South Pacific has had its present average value (table 3) of 4 cm/y then the figure covers a period of 50 My. Doubtless such comparisons can be made in many areas, a few age determinations will then remove the ambiguities and give a fairly complete history of reversals and of the spreading of the ocean floor all through the Tertiary.

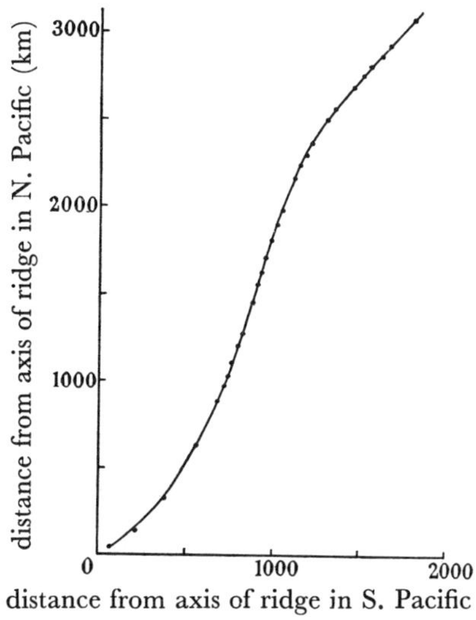

FIGURE 17. Comparison of spreading in the North and South Pacific during the last 50 My (from Pitman *et al.* 1968).

All sufficiently long profiles show a change in character at a distance from the ridge axis corresponding to about 25 My; at this point the anomalies rather suddenly become smoother, of longer wavelength and of greater amplitude than they are nearer the axis (Heirtzler & Le Pichon 1965). The natural explanation is to suppose that at this time there was an increase in the frequency of reversals which has been maintained ever since. In view of the long period during the Permian when the field had a single direction such a change of rate is not improbable and seems more likely than the alternative explanation of a world-wide simultaneous change in the velocity of spreading. A few dates from the outer flanks of the ridge would decide between the two hypotheses.

Around the Atlantic there is a zone extending 100 to 400 km beyond the continental edge in which there is no magnetic pattern parallel to the ridge (Heirtzler & Hayes 1967); it is conceivable that this represents the absence of reversals in the Permian, alternatively it may indicate that this outer rim of the deep Atlantic has a different history from the rest or that the magnetized material is here depressed beneath the thick sediments at the foot of the continental slope and has been heated sufficiently to destroy the magnetization.

The rates of spreading in table 3 are compatible with the usual ideas about continental drift, for example, at 38° S the corresponding points on the edges of the continental shelves of South America and Africa are separated by 6400 km; spreading at 1·5 cm/y on both sides of the ridge would produce this separation in 210 My, that is, in the interval since the middle of the Triassic.

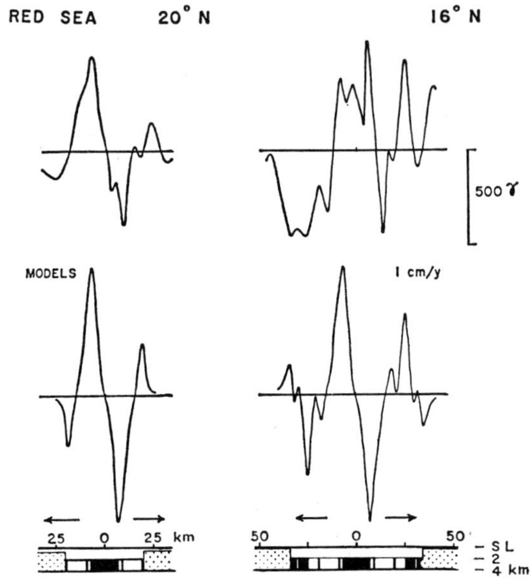

FIGURE 18. Two magnetic profiles across the Red Sea, the upper curves are the observed profiles, the lower those calculated from the models shown at the bottom (from Vine 1966).

There are several instances where magnetic surveys have revealed symmetrical linear patterns which confirm ideas derived from continental fits and palaeomagnetism. For example, Matthews has found a magnetic pattern parallel to a line bisecting the angle made by the continental edges off the north coast of Spain and the west coast of France. This confirms the suspicion (Carey 1958; Bullard, Everett & Smith 1965), that the Bay of Biscay has been formed by a rotation of Spain.

The Red Sea (figure 18) and the Gulf of Aden have well-marked magnetic patterns which confirm the view, based on other evidence (Girdler 1962, 1967, 1968; Drake & Girdler 1964), that they are embryo oceans which have only started to crack open in the last 3 My, though there may have been a period of stretching and marine transgression prior to this (Vine 1966). There is also a magnetic pattern in the Labrador Sea (Vine 1966) which probably indicates a buried ridge and which should give information about the separation of Greenland and Canada. It would be of great interest if magnetic patterns were found in the landlocked seas, particularly in the Mediterranean, the Black Sea and the Caribbean, the origin of these seas is not clear and may well be different from that of the Atlantic and Indian Oceans.

The motion of the floor of the whole North Pacific Ocean as a rigid plate has recently been confirmed by a study of the 'first motions' of earthquakes around its edge (McKenzie & Parker 1967).

It is interesting to consider which of the three methods is likely to produce the most detailed history of the reversals of the field, making the most optimistic assumptions. The dating of lavas is unlikely to have an uncertainty of less than 50 ky for very young rocks and 3 % of the age for older ones. Moreover, it is not usually possible to establish the correct sequence for a series of lava flows unless they lie one on top of the other. By itself, therefore, the method is of little use beyond 10 My where the uncertainty is unlikely to be less than 0·3 My. It is, however, the only method that can give absolute dates and it is to be hoped that occasional conspicuous sequences of reversals established by other methods will be recognizable in the lavas all through the Tertiary.

The examination of ocean sediments gives an unambiguous temporal sequence and, at a rate of deposition of 2 cm/ky, might give a resolution as fine as 1000 years over the whole of the Tertiary. The resolution greatly exceeds that of the other methods; it is of the greatest importance to push the technique to its limits (not all results are as clear as those shown in figures 2 and 5) and, if possible, to obtain absolute dates from the cores.

The magnetic pattern on the sea floor can give a resolution limited by the width of the region covered by the dykes on the axis of the ridge. If this is 3 km and the rate of spreading is 5 cm/y, a resolution of 6 ky might be attainable, though this may be over-optimistic due to the overlap of lavas fed by the dykes.

6. Theory of reversals

In addition to their connexion with some of the major problems of geology, the reversals of the Earth's magnetic field pose a difficult physical problem of wide interest in magneto-hydrodynamics, geophysics and astrophysics. There are strong reasons for supposing that the Earth's magnetic field is produced by motions in the liquid core which act as a self-exciting dynamo (Rikitake 1966). Such a dynamo, consisting of a simply connected body of fluid, is called a homogeneous dynamo; it is governed by Maxwell's equations and by the equations of hydrodynamics. Certain terms in the full equations are negligible and will be omitted here though their inclusion would cause no difficulty in the present argument. Among these are viscous forces, the displacement current and the field due to the motion of charges. With the omission of these the equations, for an observer on the rotating Earth, are:

$$\operatorname{curl}\mathbf{B} = \mu_0\mathbf{I} = \mu_0 k(\mathbf{E}+\mathbf{v}\times\mathbf{H}), \qquad \operatorname{curl}\mathbf{E} = -\dot{\mathbf{B}},$$
$$\operatorname{div}\mathbf{B} = 0, \qquad \operatorname{div}\mathbf{E} = c^2\mu_0\sigma,$$
$$\rho\dot{\mathbf{v}}+\rho((\mathbf{v}+\mathbf{\Omega}\times\mathbf{r}).\nabla)(\mathbf{v}+\mathbf{\Omega}\times r) = -\nabla p+\mathbf{I}\times\mathbf{B}+\mathbf{F}, \qquad \operatorname{div}(\rho\mathbf{v}) = 0, \tag{1}$$

where $\mathbf{B}$ and $\mathbf{E}$ are the magnetic flux density and electric field, $\mathbf{I}$ is the current, σ the charge density and k the electrical conductivity, all in SI units, $\mathbf{v}$ is the velocity of the fluid relative to axes rotating with the earth, ρ its density, p the pressure, $\mathbf{F}$ the driving force per unit volume (e.g. buoyancy in the Boussinesq approximation), $\mathbf{\Omega}$ the angular velocity of the Earth, $\mathbf{r}$ the vector from the centre of the Earth, c the velocity of light and μ_0 is the permeability of free space ($4\pi\times10^{-7}$). If the electrical conductivity of the mantle is neglected, the boundary conditions at the surface of the core are the continuity of **B** and of the tangential components of **E** and the vanishing of the normal components of **I** and **v**. At infinity **B** and **E** vanish at least like $1/r^3$.

Maxwell's equations and the boundary conditions are linear and homogeneous in the fields, currents and charges **E**, **B**, **I** and σ and are therefore unaffected by a reversal of the signs of all these quantities. The hydrodynamic equations contain the field only through the product $\mathbf{I}\times\mathbf{B}$ and are therefore also unaffected by reversal. Thus if a dynamo can maintain a magnetic field it can equally well maintain one that is everywhere equal to the first but in the opposite direction; this applies both to a steady field and to a field that varies with time. The velocity and the applied forces can not in general be reversed, though there are cases in which they can, for example the Herzenberg dynamo.

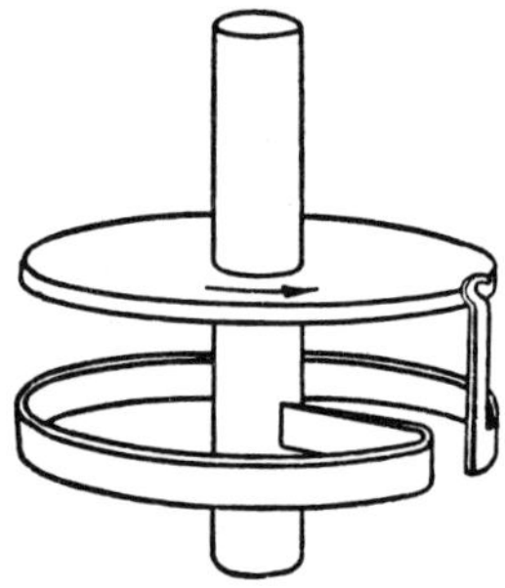

FIGURE 19. A disk dynamo.

This theorem merely demonstrates the existence of reversed solutions; it does not establish that a transition from one state to the other will occur. A man can stand on his head, but it does not follow that he will. The question of whether a homogeneous dynamo will reverse is a very difficult one with which almost no progress has been made. In view of the difficulty of the real problem it has seemed worth while to investigate certain simple systems with a finite number of degrees of freedom that bear some remote resemblance to a homogeneous dynamo. The simplest of these is shown in figure 19, it consists of a conducting disk driven by a constant couple and surrounded by a coil which is connected by brushes to the axle and to the periphery of the disk. If a magnetic field is present a radial e.m.f. is produced in the disk, a current flows in the coil and, if the rotation is in the right direction, the field is maintained. The system has three equilibrium states: one with no current and no field and two with currents and fields in opposite directions and of such magnitudes that the ohmic dissipation of energy equals the work done by the couple. The state with no field and no current is unstable; any disturbance will lead to a permanent current and field. If a shunt is placed across the coil the solutions giving a steady field and current perform the most curious oscillations when disturbed; peaks of current occur, separated by periods with almost no current and, as time passes, all solutions tend to a Dirac comb with ever higher peaks and greater spacing (Bullard 1955*a*). If two of these dynamos are connected together so that the disk of one feeds current to the coil of the other, reversals occur (Rikitake 1958; Allan 1958, 1962; Mathews & Gardner 1963); examples from a digital solution of the differential equations are given in figure 20. Surprisingly, the solutions are quite similar to the observed reversals of the Earth's field such as those in figure 2. The field fluctuates many times about its equilibrium value and then suddenly flips over to fluctuate about the reversed equilibrium state. The reversals are of two kinds: sometimes after a reversal the field returns to its original direction after a time of the order of the fluctuation period and sometimes it remains reversed for a much longer time. For most sets of initial conditions the behaviour is quite irregular, a long period near one equilibrium state may be followed by numerous irregularly spaced reversals. The calculations of Mathews & Gardner (1963) were made with an analogue computer and show the solutions to be unstable when both the currents are near zero. It is not impossible that some of the variety of behaviour is not, strictly speaking, a property of the solutions of the equations,

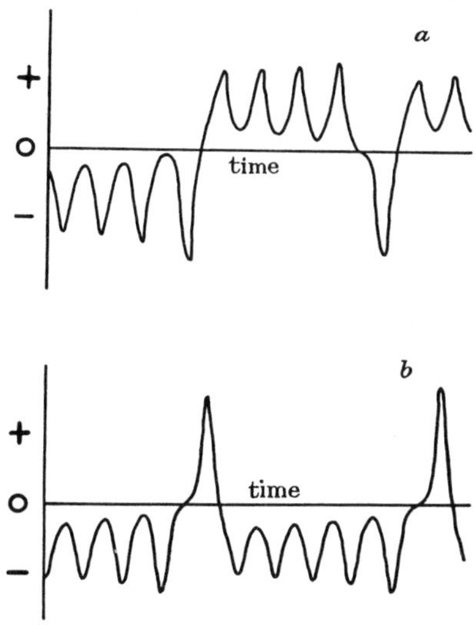

FIGURE 20. One of the currents in a pair of coupled disk dynamos. In (*a*) the current oscillates about one equilibrium value and then reverses and oscillates about the other. In (*b*), which is an earlier part of the same solution, the reversal only lasts for a short time (from Allan 1962).

but is due to the effects of imperfections in the analogue computer. The single disk dynamo shows a similar instability and will reverse in the presence of small disturbances; it is conceivable that the apparent randomness of the Earth's reversals may be due to a similar instability.

It is possible to suggest reasons for the success of the double dynamo model. We know that at the Earth's surface the field is not far from that which would be given by a suitable dipole at the centre of the Earth. If the field is extrapolated downward to the core the dipole field is still the predominant component. We have no direct knowledge of the field within the core but believe that the dipole field is distorted by the motion in such a way that it is reinforced sufficiently to compensate for its natural decay by ohmic dissipation. The possible ways in which this can occur have been catagorized by Bullard & Gellman (1954) by expansion in spherical harmonics and representation in Feyman diagrams (they are not called this in the paper since the similarity was only recently noticed when reading

Hasselmann (1966)). The interactions of the various possible distortions turn out to be quite complicated and all the plausible systems seem to involve two stages: first the dipole field is distorted by being wound round the axis to give a toroidal field, then this is pushed out into a loop and twisted into a meridian plane to give a dipole field. The velocity field proposed by Bullard & Gellman for doing this is inefficient and there is some suggestion from the work of Braginski (1964) and Tough (1967) that it may not, in fact, work as a dynamo. This defect can probably be remedied. A similar mechanism has been proposed by Babcock (1961) for the Sun; here its plausibility is much increased by the possibility of explaining the laws of sunspot polarity (Bullard 1955*a*). In the Bullard–Gellman dynamo the dipole field produces the toroidal field and the toroidal field produces the dipole field; Babcock suggested that the dipole field might first destroy the toroidal field and then build it up in the opposite direction, the toroidal field would then destroy and reverse the dipole field in the same way. No detailed theory exists in which such a behavior is shown to be a consequence of assumed forces or motions in the Sun. It seems, however, quite likely that some such mechanism is at work and might also occur in the Earth's core (the fields of the Earth and the Sun have been compared by Bullard (1966)). It is interesting that these theories divide the Earth's field into two parts, each of which destroys and creates the other just as the fields of the two halves of the double dynamo do. It is possible that the essential electromagnetic features of the mechanism producing the Earth's field are simulated or realistically caricatured in the double dynamo. On the dynamical side the double dynamo is certainly deficient; in it the applied couples work against electrodynamic forces and any unbalance accelerates the disk; in the Earth the inertia term ($\rho\dot{\mathbf{v}}$ in (1)) is negligible but the Coriolis forces (terms containing the product of $\mathbf{v}$ and $\mathbf{\Omega}$ in (1)) are comparable with the applied forces and the electrodynamic forces. The double dynamo contains no analogue of the Coriolis forces.

In stationary electrical conductors a time of the order $\mu_0 kl^2$ is required for any large change in the field, here k is the electrical conductivity and l is the length in which the field changes by a large fraction of itself. In a fluid system whose parts move with a velocity v there is another time constant l/v which is less than $\mu_0 kl^2$ if

$$v > 1/\mu_0 kl,$$

if k is 3×10^5 ohm^{-1} m^{-1} which is commonly taken as the conductivity of the Earth's core (though Stacey (1967) argues for a value ten times greater) and l is the radius of the core (3400 km), $1/\mu_0 kl$ is 10^{-4} cm/s which is less than the velocities in the core deduced from the westward drift of the field (Bullard, Freedman, Gellman & Nixon 1950). It is therefore quite possible for the Earth's field to change by a large amount in a time much shorter than the time needed for the currents to decay. The physical meaning of this result is that even in a perfect conductor the field at a point can change arbitrarily rapidly since the field is anchored to the fluid and may be carried from place to place or increased by stretching of the lines of force. As a trivial example: suppose currents to be circulating around the equator of a solid metallic sphere. If the sphere is suddenly turned over, the field seen by a fixed observer is reversed. In the sun the changes of the field are very rapid compared to the decay time. A sunspot grows and dies in a few weeks, whereas its electromagnetic decay time would be some years. The main field, both poloidal and toroidal, reverses approximately every eleven years, whereas the decay time of a field spread through the whole Sun exceeds 10^9 y. It is therefore quite possible that the Earth's field reverses in less than the decay time of the core, which is about 15 ky, though it is not clear that it actually does so.

7. Causes of spreading of the ocean floor

Belief in spreading of the ocean floor comes fairly directly from the observational evidence and is not dependent on hypotheses about processes occurring deep in the Earth. The occurrence of spreading is, however, a new piece of evidence which must be accommodated in any account of the evolution of the earth. The only plausible cause that has been suggested for it is the occurrence of convection currents in the Earth's mantle; it is not possible here to do more than sketch the leading ideas; many papers on the subject will be found in books edited by Tozer (1967*a*) and Gaskell (1967).

Convection in the mantle has been postulated for a variety of reasons by many people (e.g. Holmes 1931; Meinesz 1952; Hess 1962; Runcorn 1962) and has been declared unnecessary and impossible by others (e.g. Jeffreys 1964; Beloussov 1967; Lyustikh 1967). The main objection has been that the material of the mantle is solid and will therefore have a finite strength and will not distort under small forces. Even if some form of imperfection in elasticity is assumed it is difficult to find a mechanism that will allow thermal convection in the mantle and also account for the numerous other anelastic phenomena, such as the absorption of seismic waves, the slowing of the Earth's rotation and the damping of the Eulerian nutation. The difficulties probably arise from too naïve a view of the solid state. There are many ways in which a solid can depart from Hook's law; which one is important depends on the temperature, pressure and time-scale of the phenomena considered. It is not necessary that the absorption of seismic waves with a period of a few seconds should involve the same mechanism at the atomic level as does convection in the mantle continuing for tens of millions of years. The evidence has recently been reviewed by McKenzie (1968) who concludes that, at the pressures and temperatures occurring in the mantle, diffusion creep will be the predominant mechanism of yield under small, long continued stresses. The rate of strain is then proportional to the shear stress and the material behaves like a viscous liquid. In the lower mantle the resulting viscosity is believed to be too high to allow convection at a significant rate (this is necessary if the non-hydrostatic part of the earth's ellipticity is to be regarded as a fossil relic of a time when the rotation was faster (McKenzie 1966, 1967*a*)). In the upper mantle the viscosity may be of the order of 10^{20} kg $m^{-1}s^{-1}$ which would allow thermal convection and is also compatible with the rise of Scandinavia after the unloading at the end of the ice age. Tozer (1967*b*) has pointed out that, although the viscosity varies rapidly with temperature, there is a certain degree of self regulation; a region of high viscosity moves slowly and gets heated by its own radioactivity until its viscosity is reduced, it is possible that the viscosity may be approximately constant through a layer hundreds of kilometres in thickness, the convective motion may then extend through the whole layer.

There seems to me to have been some misapprehension about the way in which convection is likely to be driven in the mantle. A layer of fluid uniformly heated from below is in equilibrium and, if the temperature gradient does not exceed the adiabatic, the equilibrium is stable. It was shown by Benard, Rayleigh and many later authors (e.g. Chandrasekhar 1961) that instability does not set in immediately the adiabatic gradient is exceeded; the gradient must exceed the adiabatic by a finite amount proportional to $1/\kappa\eta$, where κ is the thermal conductivity and η the viscosity, before motion will occur. If this excess gradient is not present viscosity slows down the motion to such an extent that thermal conductivity has time to remove the temperature excess from the incipient rising current. This excess over the adiabatic gradient is very small in large systems and is quite negligible in the Earth. Many authors (e.g. Runcorn 1962) have, however, assumed that this type of calculation will give the size and arrangement of the convection cells. This seems to me unlikely: Rayleigh–Benard convection is the last resort of a system that has nothing else to determine its motion; it is in equilibrium but the equilibrium is unstable and, of all the possible modes of motion, the one that predominates is the one that grows most quickly. In practice Rayleigh–Benard convection is quite hard to observe in the laboratory and is not common in nature. When a layer of fluid is heated from below there are usually horizontal temperature gradients; the system is then not in equilibrium and it is these gradients that determine the motion, the hot parts rise and the colder parts sink, the growth of disturbances from an unstable equilibrium is irrelevant (Allan, Thompson & Weiss 1967). It would be interesting to inquire how large the horizontal gradient must be before it overwhelms the Rayleigh–Benard cells but, so far as I know, this has not been done.

In the Earth there will be deep-seated temperature differences between continents and oceans. The heat flow from the interior of the Earth is very variable from place to place, but on the average is the same under continents and oceans. Since the rocks of the continental crust are several times more radioactive than those of the ocean floor it is to be supposed that the heat reaching the floor of the ocean has come from greater depths than that under the continents. If there were no motion and the thermal conductivity is the same beneath the continents and the oceans then the temperature would be about 200 °C higher at depths of a few hundred kilometres beneath the oceans than it is at the same depth beneath the continents; there would therefore be a tendency for convection currents to rise under the oceans. If motion in fact occurs and the vertical gradient in the mantle is super-adiabatic, the temperature difference will be accentuated by the transport of hot material from the interior to the surface.

Thus it is natural to suppose that the pattern of convection is controlled by the distribution of continents and oceans. If the convection currents are able to move the continents and the distribution of continents determines the convection currents, we have a coupled system whose properties have not been explored. There will, presumably, be a tendency for the continents to congregate over regions of convergence of flow, but the thermal time constants are so long (of the order of 100 My for conduction through 40 km of rock) that more complicated things may happen, for example, a continent might over-run a rising current which continues to rise beneath the continent and causes rifting. It is conceivable that this is at present happening beneath Arizona and Utah and perhaps beneath the African Rift Valleys.

The igneous activity and high heat flow along the axis of the mid-ocean ridges gives some confirmation of the existence of the rising limb of a convection current under the ridges and some success has been achieved by McKenzie (1967*b*) in calculating the distribution of heat flow as a function of distance from the axis of the ridge. Many things are left unexplained: why does the convection occur in long lines and not in discrete cells, or is it perhaps in discrete cells distributed along the ridges? Is the existence of discrete cells indicated by the fracture zones which split up the ridge into sections of a few hundred kilometres? Above all, why are the oceanic and continental heat flows the same? If a piece of ocean has a heat flow of 0·05 J m^{-2} s^{-1} and is over-run by a moving continent, then the heat flow through the continent should be the sum of the heat generated in the continental crust, say, 0·05 J m^{-2} s^{-1} and that from the erstwhile ocean floor giving a total of 0·1 J m^{-2} s^{-1} which is twice as great as that assumed for the ocean. Alternatively, if the continent travels on the horizontal limb of a convection current it may retain the same heat flow as it moves, but why should the heat flow from the ocean forming behind it be the same as for the continent? Such arguments are too simple because the time taken for the heat flow through a continent to adjust itself to changes below the Moho is of the order of 100 My. It is not, however, obvious that this provides an escape from the difficulty. Elsasser (1967) has suggested that the convection will automatically even out differences of temperature gradient and heat flow between oceans and continents but the precise way in which this happens is not clear to me from his brief note.

8. The extinction of species

Three independent investigations have shown that the extinction of certain species of marine organisms occurred at about the time of the most recent reversal of the Earth's magnetic field (Harrison & Funnell 1964; Opdyke *et al.* 1966; Hays & Opdyke 1967; Watkins & Goodell 1967*a*). Typical results from Opdyke *et al.* for an Antarctic core are given in figure 21; six species of radiolaria are found to become extinct and three to make their appearance within 30 cm of the polarity change. The column on the right gives the ranges of four faunal zones named Ω, Ψ, X and Φ by Hays (1965). These are based primarily on the sequence of radiolaria in eighty Antarctic deep-sea cores; the boundary between Ψ and X lies near the magnetic transition but the Ω to Ψ boundary does not, the older transition, X to Φ, is rather uncertainly correlated with the Olduvai event.

It is difficult for one who is not a palaeontologist to assess the significance of such results. Species usually do not vanish suddenly, for example, *Discoaster brouweri* which disappeared about 700 ky ago was the last surviving species of a varied assemblage which flourished in the Miocene and became less numerous and less varied during the Pliocene. The species whose disappearance is noted because it coincides with the magnetic transition is one of many about whose relation to transitions we know nothing. The question is, in principle, statistical but data are almost wholly lacking. Even for a single species it is not clear whether the observed discrepancy of 30 cm or so in depth is significant or a result of uncertainty in the determinations; in the Antarctic cores it corresponds to between 50 and 150 ky, which is not small in relation to the time scale of magnetic changes. It is not even certain that the extinction of a single species is simultaneous all over the world.

Whatever doubts there may be about the reality of the correlation between extinction of species and magnetic transitions it is desirable to consider possible explanations. The occasional sudden extinction of large parts of the whole fauna of the Earth is one of the great mysteries of natural history and has defied explanation, or rather has attracted a multiplicity of unproved explanations, for over a hundred years (Newell 1963); no opportunity of throwing light on it should be neglected. In fact a correlation between biological change and magnetic reversals was predicted by Uffen (1963, 1965), who suggested that at times when the magnetic field was zero or small, the solar wind and a larger proportion of

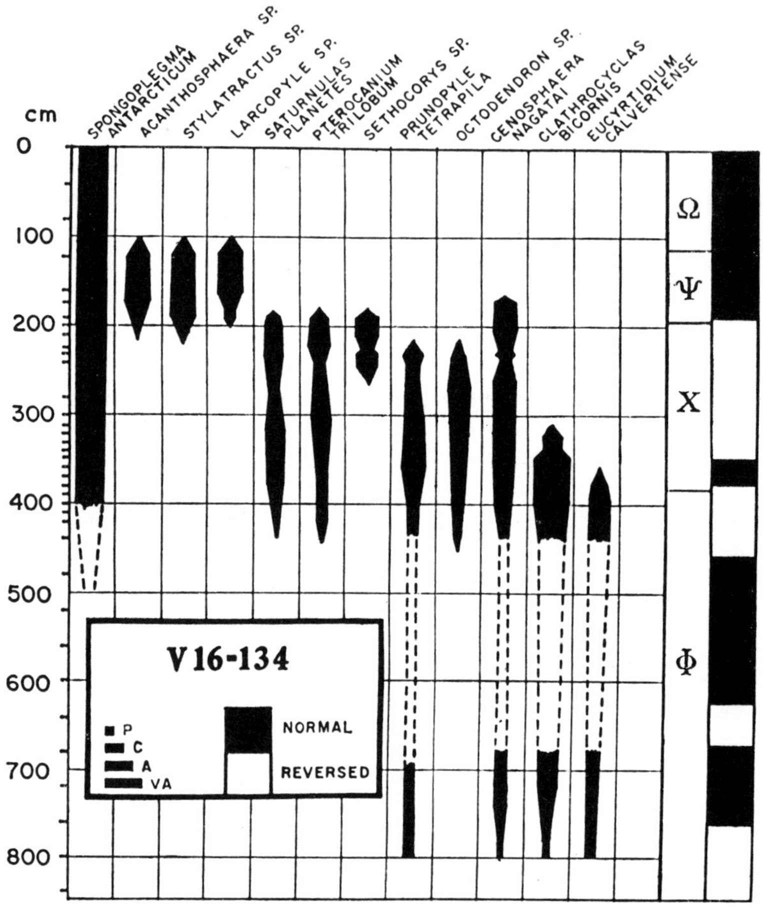

FIGURE 21. Disappearance of radiolaria near the time of the last reversal of the Earth's magnetic field. The width of the vertical bars indicates the abundance of the species on a scale: P, present; C, common; A, abundant; VA, very abundant. The magnetic stratigraphy is shown in the right-hand column (from Opduke *et al.* 1966).

the cosmic rays would be able to reach the surface of the Earth and also that the particles trapped in the radiation belts would be dumped on the Earth. Uffen supposed that the biological effects of the increase in ionization might affect the course of evolution and bring about the extinction and transformation of species at the time of a magnetic reversal. The verification of the consequences of this hypothesis is striking, but Sagan (1965), Black (1967), Waddington (1967) and Harrison (1968) have pointed out its implausibility. The Earth is well protected from ionizing particles by its atmosphere. Even in the absence of a magnetic field the solar wind and particles dumped from the radiation belts would be stopped high in the atmosphere; only high energy cosmic rays can reach the surface, most of the flux being secondary particles generated in the atmosphere. The magnetic field has no effect at the poles but excludes all particles with energy less than 15 GeV at the equator; if the field were turned off the ionization at the Earth's surface near the poles would be unchanged and that at the equator would be increased by about 10 % (the larger effect quoted by Simpson (1966) seems to be incorrect). At the equator the increase would be nullified by descending 1 m deeper into the ocean and is therefore much less than the variations to which the animals are subjected over the 100 m range of depth in which they normally live; over most of this range the whole dose from cosmic rays is unimportant relative to that from the potassium in sea water. In any event, the change in the cosmic rays at the high latitudes of the cores of Opdyke *et al.* is very small.

A number of other effects have been considered by Black (1967). Among these is the ionization due to radioactive atoms produced by cosmic rays. At the top of the atmosphere the softer cosmic rays are present and will be more affected by the field than are those that are able to reach sea level. Thus the production of radioactive isotopes, such as ^{14}C, which can be formed at high altitudes will increase by a much greater factor than does the ionization due to cosmic rays at sea level. For some isotopes this factor is as high as 5 but for ^{14}C, which is the most important, it is 2. The dose from cosmic-ray produced isotopes incorporated in the organism is more important than that from those dissolved in sea water but even at zero field this is less than half that from the potassium in sea water and in the organism, which itself is only 20 % of that from cosmic rays at sea level. It seems impossible that the total ionization within an organism at sea level can be increased by more than 12 % when the Earth's magnetic field is reduced to zero.

Another possibility is that the solar wind might produce ozone in the high atmosphere which would absorb radiation and produce large changes in climate. Black shows that the ozone produced in this way is a very small fraction of that produced by ultraviolet light.

The only possibility of a causal connexion between magnetic reversals and the extinction of species is that ionization in the high atmosphere has an unexpected effect on climate. In spite of our ignorance of many aspects of the general circulation of the atmosphere this seems unlikely, particularly as the time scale of climatic variations seems to be a good deal shorter than the interval between reversals.

Two suggestions have been made of catastrophes that might cause both extinctions and reversals. Glass & Heezen (1967*a*, *b*) have noticed that the great field of tektites covering Australia, Indonesia and a large part of the Indian Ocean, fell 700 ky ago, at about the time of the last magnetic reversal. They suggest that the fall of the body from which the tektites were formed killed the now extinct radiolaria and gave a jolt to the Earth which disturbed the motion in its core and caused the dynamo to reverse. I find it difficult to believe that a fall of a large meteorite could selectively kill certain species of radiolaria all over the world and yet spare the kangeroos near its point of fall.

Another rather similar suggestion is that the Earth is occasionally visited by showers of magnetic monopoles which kill animals and whose field penetrates to the core and causes the dynamo to reverse (Kolm 1967). The monopoles are supposed to be capable of surviving indefinitely after landing; this is necessary to the theory since a magnetic field would take something of the order of 30 y to penetrate the mantle (Bullard 1967*b*). It would also be necessary for the monopoles to be predominantly of one sign. At present the contribution of monopoles to the Earth's field is less than 0·1 % of the whole. If a shower of monopoles occurred there would presumably be a sudden large change in the field at the surface; this has not occurred during the last 400 y. In more remote times the field seems either to have fluctuated about a constant value or to have reversed as is shown in figure 2, one does not get disturbances of all sizes. It is difficult to assess so vague a theory based on the hypothetical properties of particles whose existence is doubtful, but on the whole the hypothesis does not seem either necessary or particularly plausible in relation to the events occurring 700 ky ago. It is, perhaps, more attractive as an explanation of the great extinctions such as those at the end of the Permian and of the Cretaceous.

9. Oxidation and reversed magnetization

The evidence that some rocks are magnetized in a direction opposite to the present field because the field was reversed when they were formed, seems overwhelming; yet there are facts that appear incompatible with this view and for which no convincing explanation can be offered. In order not to interrupt the main argument with inconclusive discussion, this matter has been left till the end.

Table 4. Iron-titanium oxides in order of increasing oxidation

formula	name	O/(Fe+Ti) ratio by atoms
$(Fe, Ti)_3O_4$	titano-magnetite	1·33
$(Fe, Ti)_2O_3$	ilmenite-haematite	1·50
$(Fe, Ti)_3O_5$	pseudo-brookite	1·67

The ferromagnetic minerals in reversely magnetized lavas are more highly oxidized than in normally magnetized ones. This was first shown by Balsey & Buddington (1954, 1958) in the pre-Cambrian rocks of the Adirondacks; it has since been confirmed by several workers in rocks from Mull, Siberia, Iceland, Japan and elsewhere (references to the, in part, rather inaccessible literature are given by Wilson & Watkins (1967); I have also had the advantage of seeing an unpublished manuscript by Professor P. M. S. Blackett). The minerals concerned are mixed oxides of iron and titanium which are shown in table 4 in order of increasing oxidation. The ilmenite first appears as submicroscopic lamellae in the magnetite which, as oxidation proceeds, become larger and more numerous until they replace a large part of the magnetite grains. The degree of oxidation may be estimated by chemical analysis of separated minerals or by a qualitative judgement based on the appearance of polished sections in reflected light (coloured photographs are reproduced in Wilson & Haggerty (1966)). The latter method is more convenient, it enables the rock to be placed in one of five (or sometimes six) classes representing increasing oxidation.

The Columbia Plateau basalts, which are of Miocene age, have been studied in more detail than any other sequence (Wilson & Watkins 1967). Here there are five series each of up to twenty lavas with reversed magnetization which are predominantly, but not exclusively, highly oxidized alternating with five sequences with normal magnetization and a predominantly low degree of oxidation. The correlation of polarity with state of oxidation is shown in figure 22; it is striking and beyond doubt statistically significant. The relation, is however, only statistical; there are highly oxidized rocks that are normally magnetized and reversely magnetized ones that are lightly oxidized. Not all lava sequences show as clear a correlation as is shown in figure 22; in fact, Watkins & Haggerty (1967) have shown that the state of oxidation often varies greatly in a single flow being higher in the centre than near the upper or lower margin. When this happens the whole flow has a single polarity. The variations in oxidation appear to be an original feature of the flows and to have been produced during the original solidification and cooling, probably by oxygen derived from the decomposition of water.

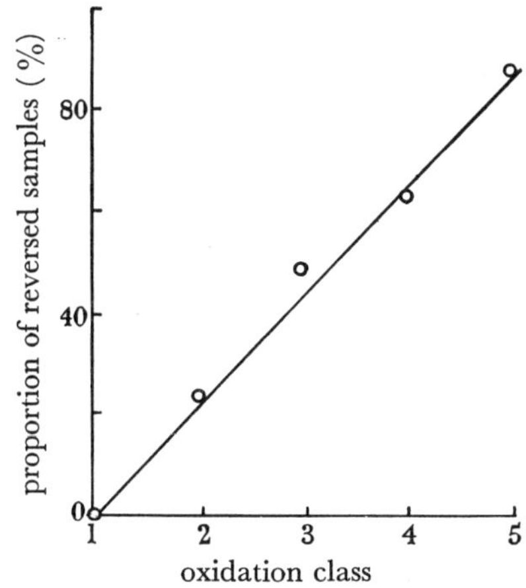

FIGURE 22. Relation between oxidation class and proportion of reversely magnetized samples for the Columbia Plateau Basalts (from Wilson & Watkins 1967).

Several types of explanation of the relation between polarity and oxidation are, in principle, possible; all appear to fall in one of the following classes:

(*a*) The results are due to chance variation and are not statistically significant. This seems demonstrably untrue.

(*b*) The state of oxidation of the rock at the time of cooling determined the direction of magnetization, i.e. the reversely magnetized rocks are self-reversing. This would be the 'natural' explanation if the correlation between polarity and oxidation were all that had to be explained. Against it is the correlation of the polarity of baked and baking rocks (there are several examples of this in the Columbia Plateau basalts), the world-wide consistency of the directions of magnetization during the last 3 My and the agreement between lavas and ocean cores. These latter points do not apply directly to the Columbia Plateau basalts, which are 14 My old. It is very desirable that a detailed study should be made of the state of oxidation of the rocks used to demonstrate the occurrence of the last reversal 0·7 My ago.

(*c*) The field at the time of cooling determined both the polarity and the state of oxidation. The state of oxidation of a lava is determined primarily by its chemical composition, the amount of water present and the rate of cooling. The presence of a magnetic field could have some effect through terms in the free energy proportional to MH where M is the magnetic moment of an oxygen molecule (3×10^{-27} J/G) and H is the Earth's field (0·4 G). The corresponding temperature, MH/k, where k is the Boltzmann constant, is 10^{-4} °K. This is negligible compared to the temperature of the lava, but even if it had been larger it would not have produced the required effect. What is wanted is not merely an effect of field on oxidation, but a differential effect depending on the relation of the field to the rotation of the Earth. Such terms no doubt exist but they will be many orders of magnitude smaller than the direct effect of the field and certainly cannot play any appreciable part in the oxidation of the lavas. Indirect effects, such as an effect of field direction on the weather and the weather on the oxidation, seem most implausible, though it is impossible to demonstrate that none can exist.

(*d*) It might be that the field direction and the oxidation had a common cause. For example, reversal of the earth's dynamo might require a redistribution of motions in the core and this might affect motions in the mantle and the chemistry of lavas. The slowness of the mantle motions makes this implausible, but it would be interesting to know if the brief Jaramillo magnetic event is associated with oxidation changes.

If the reversals were precipitated by catastrophes, such as those mentioned in §8, it might be imagined that these also affected the oxidation of rocks. It is, however, difficult to see how the effect could continue through a period of hundreds of thousands of years after the catastrophe.

(*e*) There might be some systematic effect in the selection of samples. The rocks used for polarity determination are a stringently selected class, altered rocks and magnetically unstable rocks are avoided. It is easy to imagine that the more oxidized rocks are preferentially collected, since they are more stable magnetically, but it is difficult to see why highly oxidized rocks that are reversely magnetized should be collected in preference to similar rocks that are normally magnetized.

(*f*) Oxidized rocks differ in their magnetic properties from less oxidized ones. They have a higher coercive force and are more difficult to demagnetize. If a proportion of the lightly oxidized rocks were magnetically unstable they would become magnetized in the direction of the present, 'normal', field and a correlation between polarity and oxidation would be produced. This, at first sight attractive theory, was proposed by Larson & Strangway (1966) who found no correlation between polarity and oxidation when unstable samples were excluded. Wilson & Watkins (1967), on the other hand, find the correlation shown in figure 22 after removing unstable components of magnetization by a.c. fields of up to 800 G. It might be suggested that some lightly oxidized rocks that pass laboratory tests for stability may yet be unstable when subjected to numerous reversals at long intervals of time as the Columbia Plateau basalts have been. Even this hypothesis will not explain all the facts: in particular, it cannot explain the rarity of highly oxidized, normally magnetized rocks among the Columbia Plateau basalts. Worse still, there exist basalts of Carboniferous age at Kinghorn in Scotland, which show a correlation of polarity with oxidation, but for which both the reversed and normal directions make a large angle with the present field (Wilson 1966). Instability of these rocks in the present field would not turn reversed into normal polarity. The case is not as clear as could be wished since the correlation appears only after large unstable components of magnetization have been removed; in their natural state the rocks are magnetized almost at random (Wilson & Everitt 1963).

None of the proposed explanations is able to explain all the facts, and the correlation of oxidation and magnetic polarity may be seen as one of the major unsolved problems of Earth science. It is probable that things are more complicated than we think and that no one hypothesis can explain all the facts. Perhaps the phenomenon does not occur for rocks of all ages. The most striking examples are for rocks of Eocene and Miocene age. Eocene and Oligocene rocks are predominantly reversely magnetized, Storetvedt (1968) has suggested that the majority of the normally magnetized ones are relatively unstable lightly oxidized ones which have been reversed by subsequent baking in a normal field; baking in a reversed field would produce no change. There is no reason to suppose that this has occurred to the Columbia Plateau Basalts or that the field was predominantly reversed when they were formed in the Miocene. Perhaps the most urgent need is for a study of the state of oxidation of lavas over the period for which good dates and detailed magnetic evidence exist, in particular, for the last million years. Have any highly oxidized, reversed lavas been formed during the last 0·7 My? Are the rocks from the Jaramillo event lightly oxidized compared to those above and below? Do intrusive rocks show a correlation between polarity and oxidation? It is also important to examine the magnetic minerals from ocean sediments; do they show a correlation between polarity and oxidation?

10. Conclusion

The lecture on which this paper is based was given in June 1967; it was a well chosen time, the threefold story of the reversals of the field had just become clear and could be easily and elegantly set out. In the few months needed to write the paper there has been an avalanche of new results which has revealed many discrepancies and many matters needing elucidation; the usual chaos of the Earth sciences is clearly about to be re-established at a higher level of understanding. We can now see many particular facts in the light of a global theory and realize that they are anomalous, whereas previously they appeared merely as isolated facts. The world-wide nature of the reversals can tie together the spreading of the ocean floor and the volcanic and sedimentary history of the last 50 My with a detail never before approached. It is not surprising that ambiguities, difficulties and contradictions are emerging.

References

Allan, D. W. 1958 Reversals of the earth's magnetic field. *Nature, Lond.* **182**, 469–471.

Allan, D. W. 1962 On the behaviour of systems of coupled dynamos. *Proc. Camb. Phil. Soc.* **58**, 671–693.

Allan, D. W., Thompson, W. B. & Weiss, N. O. 1967 Convection in the earth's mantle. *Mantles of the earth and terrestrial planets* (ed. S. K. Runcorn), pp. 507–512. London: Interscience.

Babcock, H. W. 1961 The topology of the sun's magnetic field and the 22-year cycle. *Astrophys. J.* **133**, 572–587.

Balsley, J. R. & Buddington, A. F. 1954 Correlation of reverse remanent magnetism and negative anomalies with certain minerals. *J. Geomagn. Geoelectr., Kyoto* **6**, 176–181.

Balsley, J. R. & Buddington, A. F. 1958 Iron-titanium oxide minerals, rocks and aeromagnetic anomalies of the Adirondack area, New York. *Econ. Geol.* **53**, 777–805.

Beloussov, V. V. 1967 Against continental drift. *Sci. J.* (Jan.), pp. 56–61.

Berggren, W. A., Phillips, J. D., Bertels, A. & Wall, D. 1967 Late Pliocene–Pleistocene stratigraphy in deep sea cores from the south-central North Atlantic. *Nature, Lond.* **216**, 253–255.

Black, D. I. 1967 Cosmic ray effects and faunal extinctions at geomagnetic field reversals. *Earth Planet. Sci. Lett.* **3**, 225–236.

Blackett, P. M. S. 1962 On distinguishing self-reversed from field-reversed rocks. *J. Phys. Soc. Japan* **17** (suppl. B 1), 699–705.

Bowin, C. O. & Vogt, P. R. 1966 Magnetic lineations between Carlsberg Ridge and Seychelles Bank, Indian Ocean. *J. Geophys. Res.* **71**, 2625–2630.

Braginskiĭ, S. I. 1964 O samovozvuzhdenii magnitnogo polya pri dvizhenii khorosho provodiashcheĭ zhidkosti. *Zh. éksp. teor. Fiz.* **47**, 1084–1098.

Bruhnes, B. 1906 Recherches sur le direction d'aimantation des roches volcaniques. *J. Phys. Radium, Paris* (4) **5**, 705–724.

Bucha, V. 1967 Archaeomagnetic and palaeomagnetic study of the magnetic field of the earth in the past 600,000 years. *Nature, Lond.* **213**, 1005–1007.

Budinger, T. F. & Enbysk, B. J. 1967 Late Tertiary date from the East Pacific Rise. *J. geophys. Res.* **72**, 2271–2273.

Bullard, E. C. 1955*a* The stability of a homopolar dynamo. *Proc. Camb. Phil. Soc.* **51**, 744–760.

Bullard, E. C. 1955*b* The magnetic fields of sunspots. *Vistas in astronomy* (ed. A. Beer), pp. 685–691. New York: Pergamon Press.

Bullard, E. C. 1966 Solar and terrestrial dynamos. *Pubblicazione del comitato nazionale per le manifestazione celebrative del IV centario della nascita di Galileo Galilei*, Vol. 2, Tomo 4, 278–283. Florence: G. Barbèra.

Bullard, E. C. 1967*a* The removal of trend from magnetic surveys. *Earth Planet. Sci. Lett.* 2, 293–300.

Bullard, E. C. 1967*b* Electromagnetic induction in the earth. *Q. Jl R. Astr. Soc.* **8**, 143–160.

Bullard, E. C. & Gellman, H. 1954 Homogeneous dynamos and terrestrial magnetism. *Phil. Trans.* A, **247**, 213–278.

Bullard, E., Everett, J. E. & Smith, A. G. 1965 The fit of the continents around the Atlantic. *Phil. Trans.* A, **258**, 41–51.

Bullard, E. C., Freedman, C., Gellman, H. & Nixon, J. 1950 The westward drift of the earth's magnetic field. *Phil. Trans.* A, 243, 67–92.

Burckle, L. H., Ewing, J., Saito, T. & Leyden, R. 1967 Tertiary sediment from the East Pacific Rise. *Science, N.Y.* **157**, 537–540.

Cann, J. R. & Vine, F. J. 1966 An area of the crest of the Carlsberg Ridge: petrology and magnetic survey. *Phil. Trans.* A, **259**, 198–217.

Carey, S. W. 1958 *Continental drift, a symposium.* Hobart.

Carmichael, C. M. 1961 The magnetic properties of ilmenite-haematite crystals. *Proc. R. Soc.* A, **263**, 508–530.

Chamalaun, F. H. & McDougall, I. 1966 Dating geomagnetic polarity epochs in Réunion. *Nature, Lond.* **210**, 1212–1214.

Chandrasekhar, S. 1961 *Hydrodynamic and hydromagnetic stability.* Oxford: Clarendon Press.

Chevalier, R. 1925 L'aimantation des laves de l'Etna et l'orientation du champ terrestre en Sicile du XIIe au XVIIe siecle. *Annls. Phys.* (10), **4**, 5–162.

Cox, A. & Dalrymple, G. B. 1966 Palaeomagnetism and potassium-argon ages of some volcanic rocks from the Galapagos Islands. *Nature, Lond.* **209**, 776–777.

Cox, A. & Dalrymple, G. B. 1967*a* Statistical analysis of geomagnetic reversal data and the precision of potassium-argon dating. *J. Geophys. Res.* **72**, 2603–2614.

Cox, A. & Dalrymple, G. B. 1967*b* Geomagnetic polarity epochs: Nunivac Island, Alaska. *Earth Planet. Sci. Lett.* **3**, 173–177.

Cox, A., Dalrymple, G. B. & Doell, R. B. 1967 Reversals of the earth's magnetic field. *Scient. Am.*, **216**, 44–54.

Cox, A., Doell, R. R. & Dalrymple, G. B. 1963*a* Geomagnetic polarity epochs and Pleistocene geochronology. *Nature, Lond.* **198**, 1049–1051.

Cox, A., Doell, R. R. & Dalrymple, G. B. 1963*b* Geomagnetic polarity epochs: Sierra Nevada II. *Science, N.Y.* **142**, 382–385.
Cox, A., Doell, R. R. & Dalrymple, G. B. 1964*a* Geomagnetic polarity epochs: a lesson in correlation. *Science, N.Y.* **143**, 351–352.
Cox, A., Doell, R. R. & Dalrymple, G. B. 1964*b* Reversals of the earth's magnetic field. *Science, N.Y.* **144**, 1537–1543.
Cox, A., Doell, R. R. & Dalrymple, G. B. 1965 Quaternary paleomagnetic stratigraphy. *Quaternary stratigraphy of the United States* (ed. H. E. Wright and D. G. Frey), pp. 817–830. Princeton University Press.
Cox, A., Hopkins, D. M. & Dalrymple, G. B. 1966 Geomagnetic polarity epochs: Pribilof Islands, Alaska. *Bull. Geol. Soc. Am.* **77**, 883–910.
Creer, K. M. 1958 Preliminary palaeomagnetic measurements from South America. *Annls. Géophys.* **14**, 373–390.
Dalrymple, G. B., Cox, A. & Doell, R. R. 1965 Potassium-argon age and paleomagnetism of the Bishop Tuff, California. *Bull. geol. Soc. Am.* **76**, 665–674.
Dalrymple, G. B., Cox, A., Doell, R. R. & Grommé, C. S. 1967 Pliocene geomagnetic polarity epochs. *Earth Planet. Sci. Lett.* **2**, 163–173.
Demenitskaya, R. M. 1967 *Kora i mantiya zemli.* Moscow: Izdatel'stvo Nedra.
Dickson, G. O. & Foster, J. H. 1966 The magnetic stratigraphy of a deep sea core from the North Pacific Ocean. *Earth Planet. Sci. Lett.* **1**, 458–462.
Dickson, G. O., Pitman, W. C. & Heirtzler, J. R. 1968 Magnetic anomalies in the South Atlantic and ocean floor spreading. *J. Geophys. Res.* **73**, 2087–2100.
Dietz, R. S. 1961 Continent and ocean basin evolution by spreading of the sea floor. *Nature, Lond.* **190**, 854–857.
Doell, R. R. & Dalrymple, G. B. 1966 Geomagnetic polarity epochs—a new polarity event and the age of the Brunhes–Matuyama boundary. *Science, N.Y.* **152**, 1060–1061.
Doell, R. R., Dalrymple, G. B. & Cox, A. 1966 Geomagnetic polarity epochs; Sierra Nevada data 3. *J. Geophys. Res.* **71**, 531–541.
Drake, C. L. & Girdler, R. W. 1964 A geophysical study of the Red Sea. *Geophys. J.* **8**, 473–495.
Dymond, J. R. 1966 Potassium-argon geochronology of deep-sea sediments. *Science, N.Y.* **152**, 1239–1241.
Elsasser, W. M. 1967 Interpretation of heat flow equality. *J. Geophys. Res.* **72**, 4768–4770.
Elvers, D. J., Mathewson, C. C., Kohler, R. E. & Moses, R. L. 1967 Systematic ocean surveys by the USC and GSS Pioneer 1961–1963. *Coast and Geodetic Survey Operational Data Report*, C & GSDR–1.
Emery, K. O. & Dietz, R. S. 1941 Gravity coring instrument and mechanics of sediment coring. *Bull. Geol. Soc. Am.* **52**, 1685–1714.
Evans, P. 1969 Towards a Pleistocene time-scale. *Q. Jl Geol. Soc. Lond.* (In the press.)
Everitt, C. W. F. 1962 Self-reversal of magnetisation in a shale containing Pyrrhotite. *Phil. Mag.* (5), **7**, 831–842.
Ewing, M. & Ewing, J. I. 1967*a* Deep sea sediments in relation to island arcs and mid-ocean ridges. *Trans. Am. Geophys. Un.* **48**, 216 (abstract).
Ewing, M. & Ewing, J. I. 1967*b* Sediment distribution on the mid-ocean ridges with respect to spreading of the sea floor. *Science, N.Y.* **156**, 1590–1592.
Ewing, M., Le Pichon, X. & Ewing, J. 1966 Crustal structure of the Atlantic Ridge. *J. Geophys. Res.* **71**, 1611–1636.
Funnell, B. M. 1968 Pre-Quaternary occurrences of micro-fossils in the oceans. *Micropalaeontology of Oceans* (SCOR Symposium Volume). Cambridge University Press (in the Press).
Gaskell, T. F. (editor) 1967 *The earth's mantle.* London: Academic Press.
Gelletich, H. 1937 Über magnetitführende eruptive Gänge und Gangsysteme in mittleren Teil des südlichen Transvaals. *Beitr. angew. Geophys.* **6**, 337–406.

Girdler, R. W. 1959 Possible reversals of the earth's magnetic field in the Jurassic period. *Nature, Lond.* **184**, 540–541.

Girdler, R. W. 1962 Initiation of continental drift. *Nature, Lond.* **194**, 521–524.

Girdler, R. W. 1967 The role of translational and rotational movements in the formation of the Red Sea and Gulf of Aden. In *The world rift system. Geol. Surv. Can. Paper*, 66–14, 65–77.

Girdler, R. W. 1968 Drifting and rifting in Africa. *Nature, Lond.* **217**, 1102–1106.

Glass, B. P. & Heezen, B. C. 1967*a* Tektites and geomagnetic reversals. *Nature, Lond.* **214**, 372–374.

Glass, B. P. & Heezen, B. C. 1967*b* Tektites and geomagnetic reversals. *Scient. Am.* **217**, 32–38.

Goldberg, E. D. & Koide, M. 1962 Geochronological studies of deep sea sediments by the ionium/thorium method. *Geochim. cosmochim. Acta* **26**, 417–450.

Goodell, H. G. & Watkins, N. D. 1968 The paleomagnetic stratigraphy of the Southern Ocean: 20° west to 160° east longitude. *Deep Sea Res.* **15**, 89–112.

Grim, P. J. & Erickson, B. H. 1967 Marine magnetic anomalies and fracture zones south of the Aleutian Trench. *Geol. Soc. Am. Abstracts*, p. 84.

Gromme, C. S. & Hay, R. L. 1963 Magnetisation of basalt of Bed I, Olduvai Gorge, Tanganyika. *Nature, Lond.* **200**, 560–561.

Harrison, C. G. A. 1966 The paleomagnetism of deep sea sediments. *J. Geophys. Res.* **71**, 3033–3043.

Harrison, C. G. A. 1968 Evolutionary processes and reversals of the earth's magnetic field. *Nature, Lond.* **217**, 46–47.

Harrison, C. G. A. & Funnell, B. M. 1964 Relationship of palaeomagnetic reversals and micropalaeontology in two late Caenozoic cores from the Pacific Ocean. *Nature, Lond.* **204**, 566.

Harrison, C. G. A. & Somayajulu, B. L. K. 1966 Behaviour of the earth's magnetic field during a reversal. *Nature, Lond.* **212**, 1193–1195.

Hasselmann, K. 1966 Feynman diagrams and interaction rules of wave-wave scattering processes. *Rev. Geophys.* **4**, 1–32.

Hays, J. D. 1965 Radiolaria and late Tertiary and Quaternary history of Antarctic seas. In *Biology of the Antarctic seas*, II, Vol. 5. *Antarctic Res. Ser. Am. geophys. Un. Publ.* no. 1297.

Hays, J. D. & Opdyke, N. D. 1967 Antarctic radiolaria, magnetic reversals and climatic change. *Science, N.Y.* **158**, 1001–1011.

Heirtzler, J. R. 1967 Marine magnetic anomalies and a moving ocean floor. *Trans. Am. Geophys. Un.* **48**, 216 (abstract).

Heirtzler, J. R. & Hayes, D. E. 1967 Magnetic boundaries in the North Atlantic Ocean. *Science, N.Y.* **157**, 185–187.

Heirtzler, J. R. & Le Pichon, X. 1965 Crustal structure of the mid-ocean ridges 3. Magnetic anomalies over the mid-Atlantic ridge. *J. Geophys. Res.* **70**, 4013–4033.

Heirtzler, J. R., Le Pichon, X. & Baron, J. G. 1966 Magnetic anomalies over the Reykjanes Ridge. *Deep Sea Res.* **13**, 427–443.

Heirtzler, J. R., Dickson, G. O., Herron, E. M., Pitman, W. C. & Le Pichon, X. 1968 Marine magnetic anomalies, geomagnetic field reversals and motions of the ocean floor and continents. *J. Geophys. Res.* **73**, 2119–2136.

Hess, H. H. 1962 History of ocean basins. *Petrological studies: a volume in honor of A. F. Buddington* (ed. A. E. J. Engel, H. L. James and B. F. Leonard), pp. 559–620. New York: Geol. Soc. Am.

Hess, H. H. 1965 Mid-oceanic ridges and tectonics of the sea-floor. *Submarine geology and geophysics* (ed. W. F. Whittard and R. Bradshaw), pp. 317–333. London: Butterworths.

Holmes, A. 1931 Radioactivity and earth movements. *Trans. Geol. Soc. Glasg.* **18**, 559–606.

Hospers, J. 1951 Remanant magnetism of rocks and the history of the geomagnetic field. *Nature, Lond.* **168**, 1111–1112.

Hospers, J. 1953*a* Reversals of the main geomagnetic field, I. *Proc. Ned. Akad. Wet.* B, **56**, 467–476.

Hospers, J. 1953*b* Reversals of the main geomagnetic field, II. *Proc. Ned. Akad. Wet.* B, **56**, 477–491.

Hospers, J. 1954 Reversals of the main geomagnetic field, III. *Proc. Ned. Akad. Wet.* B, **57**, 112–121.

Huene, R. van, Scholl, D. W. & Ridlan, J. B. 1967 Sediment distribution and deformation in the Peru–Chile trench and the adjacent continental shelf. *Geol. Soc. Am. Abstracts*, 229 (abstract).

Irving, E. 1964 *Palaeomagnetism.* New York: Wiley.

Jeffreys, H. 1964 How soft is the earth? *Q. Jl R. Astr. Soc.* **5**, 10–22.

Johnson, E. A., Murphy, T. & Torreson, O. W. 1948 Pre-history of the earth's magnetic field. *Terr. Magn. Atmos. Elect.* **53**, 349–372.

Keen, M. J. 1960 Magnetisation of sediment cores from the eastern North Atlantic Ocean. *Nature, Lond.* **187**, 220–222.

Keen, M. J. 1963 The magnetisation of sediment cores from the eastern basin of the North Atlantic Ocean. *Deep Sea Res.* **10**, 607–622.

Khramov, A. N. 1957 O paleomagnetizme kak osnove novogo metoda korrelyatsii i rascheleneniya osadochnykh tolshch. *Dokl. Akad. Nauk SSSR* **112**, 849–859.

Khramov, A. N. 1958 *Paleomagnitnaya korrelyachiya osadochnykh tolshch.* Leningrad: Gostoptechizdat (translated as *Palaeomagnetism and stratigraphic correlations* by the Australian National University, Canberra).

Koenigsberger, J. K. 1938 Natural residual magnetism of eruptive rocks. *Terr. Magn. Atmos. Elect.* **43**, 119–130, 299–320.

Kolm, H. H. 1967 Search for magnetic monopoles in deep-sea sediment. *Phys. Today*, **20**, 69–70.

Krause, D. C. 1967 Magnetic evidence for western Caribbean Basin origin by late cretaceous spreading. *Trans. Am. Geophys. Un.* **48**, 133 (abstract).

Kullenberg, B. 1947 The piston core sampler. *Svenska hydrogr.-biol. Kommn. Skr.* (ser. 3, Hydrog.), 1, no. 2, 1–46.

Larson, E. E. & Strangway, D. W. 1966 Magnetic polarity and igneous petrology. *Nature, Lond.* **212**, 756–757.

Le Pichon, X. & Heirtzler, J. R. 1968 Magnetic anomalies in the Indian Ocean and sea floor spreading. *J. Geophys. Res.* **73**, 2101–2117.

Lyustikh, E. N. 1967 Criticism of hypothesis of convection and continental drift. *Geophys. J. R. Astr. Soc.* **14**, 347–352.

McDougall, I. & Chamalaun, F. H. 1966 Geomagnetic polarity scale of time. *Nature, Lond.* **212**, 1415–1418.

McDougall, I. & Tarling, D. H. 1963 Dating of polarity zones in the Hawaiian Islands. *Nature, Lond.* **200**, 54–56.

McDougall, I. & Tarling, D. H. 1964 Dating of geomagnetic polarity zones. *Nature, Lond.* **202**, 171–172.

McDougall, I. & Wensink, H. 1966 Paleomagnetism and geochronology of the Pliocene–Pleistocene lavas in Iceland. *Earth Planet. Sci. Lett.* **1**, 232–236.

McKenzie, D. P. 1966 The viscosity of the lower mantle. *J. Geophys. Res.* **71**, 3995–4010.

McKenzie, D. P. 1967*a* The viscosity of the mantle. *Geophys. J. R. Astr. Soc.* **14**, 297–305.

McKenzie, D. P. 1967*b* Some remarks on heat flow and gravity anomalies. *J. Geophys. Res.* **72**, 6261–6273.

McKenzie, D. P. 1968 The geophysical importance of high temperature creep. NASA conference on the history of the earth's crust (in the Press).

McKenzie, D. P. & Parker, R. L. 1967 The North Pacific: an example of tectonics on a sphere. *Nature, Lond.* **216**, 1276–1280.

McNish, A. G. & Johnson, E. A. 1938 Magnetisation of unmetamorphosed varves and marine sediments. *Terr. Magn Atmos. Elect.* **43**, 401–407.

Mason, R. G. 1958 A magnetic survey off the west coast of the United States between latitudes 32° and 36° N, longitudes 121° and 128° W. *Geophys. J. R. Astr. Soc.* **1**, 320–329.

Mason, R. G. & Raff, A. D. 1961 A magnetic survey off the west coast of North America, 32° N to 42° N. *Bull. Geol. Soc. Am.* **72**, 1259–1265.

Mathews, J. H. & Gardner, W. K. 1963 Field reversals of 'paleomagnetic' type in coupled disk dynamos. *NRL Rep.* no. 5886.
Matthews, D. H. 1961 Lavas from an abyssal hill on the floor of the North Atlantic Ocean. *Nature, Lond.* **190**, 158–159.
Matthews, D. H. & Bath, J. 1967 Formation of magnetic anomaly pattern of mid-Atlantic Ridge. *Geophys. J. R. Astr. Soc.* **13**, 349–357.
Matthews, D. H., Vine, F. J. & Cann, J. R. 1965 Geology of an area of the Carlsberg Ridge, Indian Ocean. *Bull. Geol. Soc. Am.* **76**, 675–682.
Matthews, D. H., Williams, C. & Laughton, A. S. 1967 Mid-ocean ridge in the mouth of the Gulf of Aden. *Nature, Lond.* **215**, 1052–1053.
Matuyama, M. 1929 On the direction of magnetisation of basalt in Japan, Tyôsen and Manchuria. *Proc. Imp. Acad. Japan* **5**, 203–205.
Meinesz, F. A. V. 1952 Convection-currents in the earth and the origin of the continents. *Proc. K. Ned. Akad. Wet.* B, **55**, 427–553.
Menard, H. W. 1964 *Marine geology of the Pacific.* New York: McGraw-Hill.
Menard, H. W. 1967 Sea floor spreading, topography, and the second layer. *Science, N.Y.* **157**, 923–924.
Mercanton, M. P. 1910 État magnétique des diabases de l'Isfjord au Spitzberg. *C. r. hebd. Séanc. Acad. Sci., Paris* **151**, 1092–1094.
Mudie, J. D. & Harrison, C. G. A. 1967 Deep tow profiles across Mason-Raff lineations. *Trans. Am. Geophys. Un.* **48**, 133 (abstract).
Nagata, T. 1951 Reverse thermo remanant magnetism. *Nature, Lond.* **169**, 704–705.
Nagata, T. 1961 *Rock magnetism.* Tokyo: Maruzen.
Nagata, T., Uyeda, S. & Akimoto, S. 1952 Self-reversal of thermo-remanant magnetisation of igneous rocks. *J. Geomagn. Geoelect., Kyoto* **4**, 22–38.
Néel, L. 1951 L'inversion de l'aimantation permanente des roches. *Annls Géophys.* **7**, 90–102.
Newell, N. D. 1963 Crises in the history of life. *Scient. Am.* **208**, 76–92.
Ninkovich, D., Opdyke, N., Heezen, B. C. & Foster, J. H. 1966 Paleomagnetic stratigraphy, rates of deposition and tephrochronology in North Pacific deep-sea sediments. *Earth Planet. Sci. Lett.* **1**, 476–492.
Opdyke, N. D. & Foster, J. H. 1967 Reversals of the earth's magnetic field. *Sci. J.* (Sept.), pp. 56–61.
Opdyke, N. D., Glass, B., Hays, J. D. & Foster, J. 1966 Palaeomagnetic study of Antarctic Deep-Sea cores. *Science, N.Y.* **154**, 349–357.
Ozima, M. & Ozima, M. 1967 Self-reversal of remanent magnetisation in some dredged submarine basalts. *Earth Planet. Sci. Lett.* **3**, 213–215.
Ozima, M., Ozima, M. & Kaneoka, I. 1968 Potassium-argon ages and magnetic properties of some dredged submarine basalts and their geophysical implications. *J. Geophys. Res.* **73**, 711–723.
Ozima, M., Kono, M., Kaneoka, I., Kinoshita, H., Kobayashi, K., Nagata, T., Larsen, E. E. & Strangway, D. 1967 Paleomagnetic and potassium-argon ages of some volcanic rocks from the Rio Grande Gorge, New Mexico. *J. Geophys. Res.* **72**, 2615–2621.
Peter, G. 1966 Magnetic anomalies and fracture pattern in the northeast Pacific Ocean. *J. Geophys. Res.* **71**, 5365–5374.
Peter, G. & Stewart, H. B. 1965 Ocean surveys: the systematic approach. *Nature, Lond.* **206**, 1017–1018.
Phillips, J. D. 1967 Magnetic anomalies over the mid-Atlantic ridge near 27° N. *Science, N.Y.* **157**, 920–923.
Pitman, W. C. & Heirtzler, J. R. 1966 Magnetic anomalies over the Pacific-Antarctic ridge. *Science, N.Y.* **154**, 1164–1171.
Pitman, W. C., Herron, E. & Heirtzler, R. 1968 Magnetic anomalies in the Pacific and sea-floor spreading. *J. Geophys. Res.* **73**, 2069–2085.

Raff, A. D. 1962 Further magnetic measurements along the Murray fault. *J. Geophys. Res.* **67**, 417–418.

Raff, A. D. 1966 Boundaries of an area of very long magnetic anomalies in the northeast Pacific. *J. Geophys. Res.* **71**, 2631–2636.

Raff, A. D. & Mason, R. G. 1961 A magnetic survey off the west coast of North America, 40° N to 52½° N. *Bull. Geol. Soc. Am.* **72**, 1267–1270.

Rassokho, A. I., Senchura, L. I., Demenitskaya, R. M., Karosik, A. M., Kiselev, Yu. G. & Timoshchenko, N. K. 1967 Podvodnyĭ sredinnyĭ Arkticheskiy Khrebet i ego mesto v sisteme khrebtov Severnogo Ledovitogo okeana. *Dokl. Akad. Nauk SSSR* **172**, 659–662.

Rikitake, T. 1958 Oscillations of a system of disc dynamos. *Proc. Camb. Phil. Soc.* **54**, 89–105.

Rikitake, T. 1966 *Electromagnetism and the earth's interior.* Amsterdam: Elsevier.

Ross, D. A. & Riedel, W. R. 1967 Comparison of upper parts of some piston cores with simultaneously collected open-barrel cores. *Deep Sea Res.* **14**, 285–294.

Runcorn, S. K. 1962 Towards a theory of continental drift. *Nature, Lond.* **193**, 311–314.

Sagan, C. 1965 Is the early evolution of life related to the development of the earth's core. *Nature, Lond.* **206**, 448.

Saito, T., Ewing, M. & Burkle, L. H. 1966 Tertiary sediment from the mid-Atlantic ridge. *Science, N.Y.* **151**, 1075–1079.

Schlich, R. & Patriat, P. 1967 Profils magnetiques sur la dorsale 'Indo-Pacifique'. *IUGG Abstracts* **8**, 48 (abstract).

Scholl, D. W., Huene, H. van & Ridlon, J. B. 1967 Deformation of sediments in Peru-Chile trench in relation to possible sea-floor spreading. *Geol. Soc. Am. Abstracts*, p. 196 (abstract).

Simpson, J. F. 1966 Evolutionary pulsations and geomagnetic polarity. *Bull. Geol. Soc. Am.* **77**, 197–203.

Stacey, F. D. 1967 Electrical resistivity of the earth's core. *Earth Planet. Sci. Lett.* **3**, 204–206.

Storetvedt, K. M. 1968 On remagnetisation problems in palaeomagnetism. *Earth Planet. Sci. Lett.* **4**, 107–112.

Suess, E. 1888 *Das Antlitz der Erde*, vol. 2, pt. 3, chap. 4. Vienna: F. Tempsky (translated 1906 by H. B. C. Sollas as *The face of the earth.* Oxford: Clarendon Press).

Sykes, L. R. 1967 Mechanism of earthquakes and nature of faulting on the mid-ocean ridges. *J. Geophys. Res.* **72**, 2131–2153.

Talwani, M., Le Pichon, X. & Heirtzler, J. R. 1965 East Pacific rise: the magnetic pattern and the fracture zones. *Science, N.Y.* **150**, 1109–1115.

Thellier, E. & Thellier, O. 1959 Sur l'intensité du champ magnétique terrestre dans le passé historique et géologique. *Annls Géophys.* **15**, 285–376.

Tough, J. G. 1967 Nearly symmetric dynamos. *Geophys. J. R. Astr. Soc.* **13**, 393–396.

Tozer, D. C. (ed.) 1967*a* The proceedings of the International Upper Mantle Committee Symposium on non-elastic processes in the mantle. *Geophys. J. R. Astr. Soc.* **14**, 1–450.

Tozer, D. C. 1967*b* Some aspects of thermal convection theory for the earth's mantle. *Geophys. J. R. Astr. Soc.* **14**, 395–402.

Uffen, R. J. 1963 Influence of the earth's core on the origin and evolution of life. *Nature, Lond.* **198**, 143–144.

Uffen, R. J. 1965 The evolution of the interior of the earth and its effects on biological evolution. *Upper Mantle Symposium, New Delhi* 1964. IUGS: Copenhagen.

Uyeda, S. 1958 Thermo-remanent magnetism as a medium of palaeomagnetism, with special reference to reverse thermo-remanent magnetism. *Jap. J. Geophys.* **2**, 1–123.

Uyeda, S., Vacquier, V., Yasui, M., Sclater, J., Sato, T., Lawson, J., Watanabe, T., Dixon, F., Silver, E., Fukao, Y., Sudo, K., Nishikawa, M. & Tonaka, T. 1967 Results of geomagnetic survey during the cruise of *R/V Argo* in Western Pacific 1966 and the compilation of magnetic charts of the same area. *Bull. Earthq. Res. Inst.* **45**, 799–814.

Vacquier, V. 1965 Transcurrent faulting in the ocean floor. *Phil. Trans.* A, **258**, 77–81.

Vacquier, V. & Von Herzen, R. P. 1964 Evidence for connection between heat flow and the mid-Atlantic Ridge magnetic anomaly. *J. Geophys. Res.* **69**, 1093–1101.

Vacquier, V., Raff, A. D. & Warren, R. E. 1961 Horizontal displacements in the floor of the Pacific Ocean. *Bull. Geol. Soc. Am.* **72**, 1251–1258.

Van Zijl, J. S. V., Graham, K. W. & Hales, A. L. 1962*a* The palaeomagnetism of the Stormberg lavas of South Africa. I. Evidence for a genuine reversal of the earth's field in Triassic-Jurassic times. *Geophys. J. R. Astr. Soc.* **7**, 23–39.

Van Zijl, J. S. V., Graham, K. W. T. & Hales, A. L. 1962*b* The palaeomagnetism of the Stormberg lavas, II. The behaviour of the magnetic field during a reversal. *Geophys. J. R. Astr. Soc.* **7**, 169–182.

Vine, F. J. 1966 Spreading of the ocean floor: new evidence. *Science, N.Y.* **154**, 1405–1415.

Vine, F. J. & Matthews, D. H. 1963 Magnetic anomalies over ocean ridges. *Nature, Lond.* **199**, 947–949.

Vine, F. J. & Morgan, W. J. 1967 Simulation of mid-ocean ridge magnetic anomalies using a random injection model. *Geol. Soc. Am. Abstracts* p. 228 (abstract).

Vine, F. J. & Wilson, J. T. 1965 Magnetic anomalies over a young ocean ridge off Vancouver island. *Science, N.Y.* **150**, 485–489.

Vlasov, A. Ya. & Kovalenko, G. V. 1963 Magnetizm perekhodn sloev mezhdu zonami s pryamoĭ obratnoĭ namagnichennosty. *Izv. Akad. Nauk SSSR* (Ser. Geofiz.), pp. 552-560.

Waddington, C. J. 1967 Paleomagnetic field reversals and cosmic radiation. *Science, N.Y.* **158**, 913–915.

Watkins, N. D. 1968 Comments on the interpretation of linear magnetic anomalies. *Geofiz. pura appl.* **69**, 111–123.

Watkins, N. D. & Goodell, H. G. 1967*a* Geomagnetic polarity change and faunal extinction in the Southern Ocean. *Science, N.Y.* **156**, 1083–1087.

Watkins, N. D. & Goodell, H. G. 1967*b* Confirmation of the reality of the Gilsa geomagnetic polarity event. *Earth Planet. Sci. Lett.* **2**, 123–129.

Watkins, N. D. & Goodell, H. G. 1967*c* Inconsistencies resulting from a limited study of magnetic polarity variation in submarine sedimentary cores. *Trans. Am. Geophys. Un.* **48**, 90. (abstract).

Watkins, N. D. & Haggerty, S. E. 1967 Primary oxidation variation and petrogenesis in a single lava. *Contr. Miner. Petr.* **15**, 251–271.

Wilson, J. T. 1965*a* A new class of faults and their bearing on continental drift. *Nature, Lond.* **207**, 343–347.

Wilson, J. T. 1965*b* Submarine fracture zones, aseismic ridges and the International Council of Scientific Unions Line: proposed western margin of the East Pacific Ridge. *Nature, Lond.* **207**, 907–911.

Wilson, J. T. 1965*c* Transform faults, oceanic ridges and magnetic anomalies southwest of Vancouver Island. *Science, N.Y.* **150**, 482–485.

Wilson, R. L. 1962*a* The palaeomagnetism of baked contact rocks and reversals of the earth's magnetic field. *Geophys. J. R. Astr. Soc.* **7**, 194–202.

Wilson, R. L. 1962*b* The palaeomagnetic history of a doubly-baked rock. *Geophys. J. R. Astr. Soc.* **6**, 397–399.

Wilson, R. L. 1966 Further correlations between the petrology and the natural magnetic polarity of basalts. *Geophys. J. R. Astr. Soc.* **10**, 413–420.

Wilson, R. L. & Everitt, C. W. F. 1963 Thermal demagnetisation of some Carboniferous lavas for palaeomagnetic purposes. *Geophys. J. R. Astr. Soc.* **8**, 149–164.

Wilson, R. L. & Haggerty, S. E. 1966 Reversals of the earth's magnetic field. *Endeavour* **25**, 104–109.

Wilson, R. L. & Watkins, N. D. 1967 Correlation of petrology and natural magnetic polarity in Columbia plateau basalts. *Geophys. J. R. Astr. Soc.* **12**, 405–424.

The Leeuwenhoek Lecture, 1968

The physiology of an algal nuisance

By G. E. Fogg, F.R.S.

Department of Botany, Westfield College, London, N.W. 3

(*Lecture delivered* 7 *November* 1968—*MS received* 27 *November* 1968)

Proceedings of the Royal Society, B, *volume* 173, *pp.* 175–189, 1969

In his letter of 7 September 1674, addressed to Mr Oldenburg, Leeuwenhoek wrote as follows:

> About two Leagues from this Town [Delft] there lyes an Inland-Sea, called Berkelse-Lake, whose bottom in many places is very moorish. This water is in Winter very clear, but about the beginning or in the midst of Summer it grows whitish, and there are then small green clouds permeating it, which the Country-men, dwelling near it, say is caused from the Dews then falling, and call it *Honey-Dew*. This water is abounding in Fish, which is very good and savoury. Passing lately over this Sea at a time, when it blew a fresh gale of wind, and observing the water as above described, I took up some of it in a Glass-vessel which having view'd the next day, I found moving in it several Earthy particles, and some green streaks, spirally ranged, after the manner of the Copper or Tin-worms, used by Distillers to cool their distilled waters; and the whole compass of each of these streaks was about the thickness of a man's hair on his head.

This passage is well known as containing the first description of the alga *Spirogyra*, some species of which do indeed form green 'clouds' in peaty water (Prescott 1951), and testifies to Leeuwenhoek's wide-ranging interest in micro-organisms. But it might well be used as a basis for claiming Leeuwenhoek as the father of limnology as well as of microbiology—he certainly seems to have possessed the ability to retain an interest in science in the face of physical discomfort, which is essential in the study of lakes—and its reassures one who feels diffident in describing himself as a microbiologist that he has some justification in accepting an invitation to deliver the Leeuwenhoek Lecture.

The Berkelse lake was drained by the engineer Krukius and the polder on its site yielded its first harvest in 1777 (Dr H. Golterman, personal communication). Had the lake persisted to the present it is doubtful whether it would still be in the clear and wholesome condition described by Leeuwenhoek. Like fresh-waters in all habitable parts of the world it would in all probability have undergone the change now known as 'eutrophication'. The term eutrophication really means nothing more than enrichment with nutrients and consequent increase in biological productivity, but as generally used it implies over-enrichment with development of objectionable growths of plants (Lund 1967). It is a particular type of algal growth characteristic of eutrophic waters, namely dense blooms of planktonic blue-green algae (Myxophyceae), that I shall discuss in this lecture. Such blooms have always occurred—for example in 1188 Giraldus Cambrensis gave an unmistakable description of one in Llangors Lake, Brecknockshire (Griffiths 1939)—but they show a special correlation with human activities and in recent years have become more prevalent and persistent. Sometimes they are welcome; they are, after all, the result of high biological productivity, and in Lake George, Uganda, for example, evidently support high rates of fish production. More often they are a nuisance because they make water unfit for drinking and bathing, because their decomposition produces unpleasant smells, because they cause mass mortality of fish, or because they occasionally produce toxins which kill birds and cattle and cause gastro-intestinal disorders in man (Schwimmer & Schwimmer 1964). Control of these blooms is now a general problem (Brook 1959; Stewart & Rohlich 1967) but one example—from Leeuwenhoek's homeland—must suffice here to indicate its seriousness. With the completion of the present works closing the mouths of the Rhine–Meeuse delta great expanses of water will become fresh and almost static. Rhine water is known to be highly favourable for the growth of myxophycean blooms and since this is a farming area their development on a large scale could have serious economic consequences. It is imperative that some means of control be found before the works are completed in 1974 (Peelen 1969).

A great deal of miscellaneous information is on record about the chemical and physical characteristics of bodies of water in which myxophycean blooms develop, but no definite relation of blooms to any particular combination of factors is evident. The only generalization which appears valid is the paradoxical one that although characteristic of highly productive waters, these algae develop most rapidly when inorganic nutrients such as nitrate and phosphate are present in minimal concentrations (Hutchinson 1967). In the English Lakes Pearsall (1932) found a close correlation between the percentage frequency of blue–green algae in the plankton and the concentration of organic nitrogen in the water in the previous month. Hutchinson (1967) points out that such a correlation does not always hold, but the general impression is, nevertheless, that there is some causal relationship between organic pollution and myxophycean blooms (Singh 1955; Brook 1959; Vance 1965).

Another approach to defining the conditions which lead to the development of blooms is by study of the physiological characteristics of the organisms themselves. The forms with which we are concerned are species belonging to the genera *Microcystis*, *Coelosphaerium*, *Oscillatoria*, *Anabaena*, *Aphanizomenon* (figures 1 and 2, plate 3) and *Gloeotrichia*. In discussing their physiology it will have to be assumed that there are certain general similarities between them and other kinds of blue–green algae which have been more intensively studied in the laboratory.

Despite the apparent importance of organic substances for planktonic blue–green algae it seems certain that these are not essential for their growth. Various species have been grown for long periods in artificial media in which the only organic ingredients are chelating agents which can be shown not to play any direct nutrient role (Allen 1952; Gerloff, Fitzgerald & Skoog 1950; Tischer 1965;

Eberley 1967). In contrast to algae of other groups, 70 % of which tested *in vitro* have been found to have a requirement for one or more vitamins, Myxophyceae mostly have no such requirements and their growth is not enhanced if these substances are supplied (see, for example, Kiyohara, Fujita, Hattori & Watanabe 1962; Provasoli 1966). One of the few exceptions to this is a marine species, *Phormidium persicinum*, which requires vitamin B_{12} (Pintner & Provasoli 1958).

Before proceeding further it is worth considering whether high concentrations of organic substances in natural waters may sometimes be the effect, rather than the cause, of the growth of blue–green algae. It is a general characteristic of these algae that during healthy growth they liberate extracellular nitrogenous substances, mainly polypeptides, which, particularly in the initial stages of growth in fresh medium or in medium deficient in certain mineral nutrients, may sometimes account for more than half of the nitrogen assimilated (Fogg 1952, 1966). Nevertheless, in the polluted waters with which we are chiefly concerned extracellular products of bloom algae must constitute only a small proportion of the total dissolved organic matter, which even in unpolluted waters may exceed that in the plankton itself by 5 to 20 times, and the only quantitative data available, those of Pearsall (1932), do suggest that it is the concentration of organic matter present *before* the bloom develops which is of importance.

It is probable that to some extent organic matter affects growth of bloom algae indirectly by complex formation. The extracellular polypeptides of the blue–green algae themselves have strong complex-forming properties and evidently promote the growth of the cells which produce them by modifying the ionic composition of the medium. That they are important in conditioning the medium in some way is suggested by the observation of Gerloff *et al.* (1950) that cultures of blue–green algae in artificial media could not be started from a single cell or filament but were successful if heavy inocula were used. More recently Eberley (1967) has reported quantitative studies with *Oscillatoria agardhii* which show that the duration of the lag phase of this alga in a mineral medium with ethylenediamine tetraacetic acid as chelator is inversely related to the size of the inoculum. The extra-cellular polypeptides of *Anabaena cylindrica* have been shown to be effective in reducing the toxicity of the cupric ion towards this alga (Fogg & Westlake 1955). Other organic substances besides extracellular products share these properties. Complex formation may maintain trace elements in solution under conditions in which they would otherwise be precipitated and be rendered unavailable to plankton algae and Wetzel (1965) has found evidence that this is important in maintaining algal growth in alkaline lakes. There may also be interactions between the various organic substances, as, for example, if one component neutralizes the antibiotic effect of another. Blue–green algae are generally sensitive to the polymixins but their extracellular products are markedly effective in reducing the toxicity of these antibiotics (Whitton 1967). There are indications that inhibitory organic substances play some part in controlling the development of myxophycean blooms in nature. Thus Vance (1965) found with impure cultures that *Euglena* sp.

and *Cryptomonas ovata* produce diffusible factors capable of suppressing the growth of *Microcystis aeruginosa*. It can be envisaged therefore that the apparent beneficial effect of dissolved organic matter might depend on some components, such as polypeptides, present in excess, neutralizing inhibitory substances produced by other organisms.

Another way in which organic matter might indirectly favour the growth of myxophycean blooms is by serving as a source of some nutrient element. For some blue-green algae such as *Gloeotrichia* even low concentrations of inorganic phosphate in artificial media appear to be inhibitory (Fogg, unpublished data). If such algae utilize organic phosphates or if the decomposition of organic matter yields a continuous supply of inorganic phosphate at a non-inhibitory concentration the apparent paradox that blooms develop when concentrations of inorganic phosphate are minimal might be explained.

In considering direct effects of organic substances on the growth of planktonic blue-green algae the possibility that they permit heterotrophic growth of the algae can be put aside as unlikely in the light of the evidence available at present. Hobbie & Wright (1965) have shown that planktonic algae compete poorly with bacteria for organic substrates in the absence of light. Appreciable uptake of glucose from concentrations as low as 3 $\mu g\ l^{-1}$ by *Oscillatoria agardhii* has been demonstrated in dark *in situ* experiments by a radioautographic technique (Dr G. W. Saunders, personal communication) so that further studies are desirable. Nevertheless, although reliable evidence from pure cultures is available for growth in the dark on organic substrates of a few species such as *Tolypothrix tenuis* (Kiyohara, Fujita, Hattori & Watanabe 1960), *Chlorogloea fritschii* (Fay 1965) and *Anabaenopsis circularis* (Watanabe & Yamamoto 1967), attempts to do this with most blue-green algae have been unsuccessful and the non-planktonic species *Anacystis nidulans*, *Anabaena cylindrica* and *Anabaena variabilis* seem from extensive evidence to be obligate phototrophs. Among planktonic forms *Oscillatoria rubescens* has been shown to be of this type (Staub 1961) and the probability is that other bloom-forming species are also. This inability to grow except in the light is a curious state of affairs and may arise from any one of

---

Description of plate 3

Figure 1. Planktonic blue-green algae from Esthwaite Water, English Lake District. A, *Microcystis aeruginosa;* B, *Aphanizomenon flos-aquae*; C, *Anabaena flos-aquae;* D, *Coelosphaerium naegelianum*. (×36.)

Figure 2. Planktonic blue-green algae: A, *Microcystis aeruginosa*, cells with gas-vacuoles; B, *Oscillatoria agardhii*, one filament with conspicuous gas-vacuoles. (×580.)

Figure 3. *Anabaena flos-aquae*, showing gas-vacuoles and wide mucilage sheath. (×470.)

Figure 4. Electron micrograph of gas cylinders extracted from *Anabaena flos-aquae* and negatively stained with phosphotungstic acid. (×64150.)

Figures 1, 2 and 3 by courtesy of Dr Hilda M. Canter; figure 4 by courtesy of Mr H. Eichelberger.

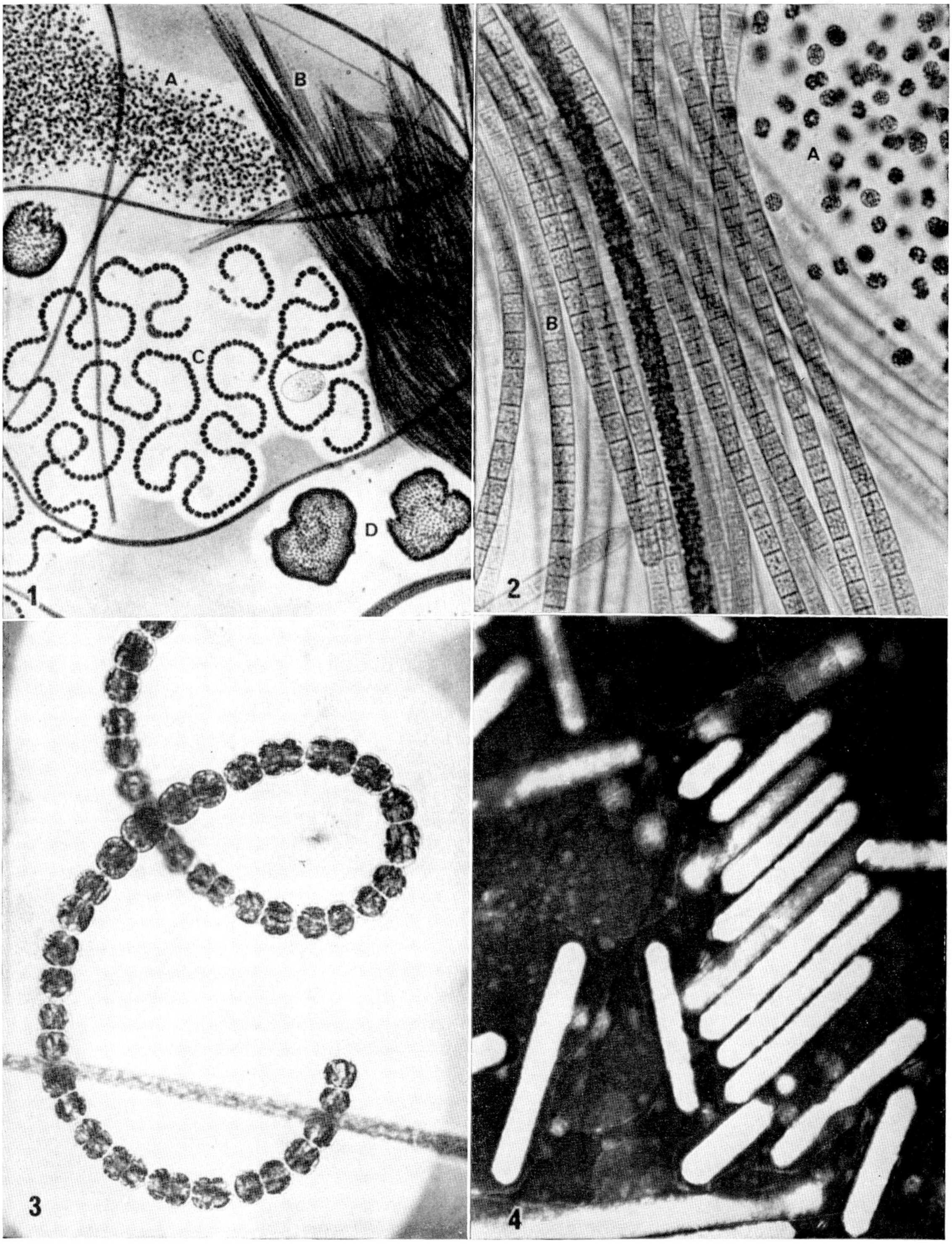

FIGURES 1–4. For legends see facing page.

PLATE 3

(*Facing p.* 96)

several different causes. With *Anabaena variabilis* it seems to be caused by lack of a complete tricarboxylic acid cycle. The evidence for this lies both in the pattern of incorporation of tracer into intermediates of this cycle following supply of ^{14}C-labelled substrates and in failure to demonstrate α-oxoglutarate dehydrogenase and succinyl coenzyme A synthetase activity in cell extracts (Smith, London & Stanier 1967; Pearce & Carr 1967*a*).

It does not appear therefore that an ability to grow at the expense of organic substrates in the depths of a lake where light is virtually absent can contribute to the success of planktonic blue–green algae. Nevertheless, given a low intensity of light, obligately phototrophic algae can assimilate organic substrates and this may be important. The photoassimilation (or photoheterotrophic assimilation) of substrates such as acetate and glucose by the green algae *Chlorella* spp. and *Chlamydomonas mundana* clearly takes place by use of the assimilatory power generated by the photochemical reactions of photosynthesis for direct incorporation of the substrate into cell components rather than by its breakdown in respiration followed by photosynthetic assimilation of the carbon dioxide produced (Bergmann 1955; Eppley, Gee & Saltman 1963; Marker & Whittingham 1966; Goulding & Merrett 1966). The situation is similar in the blue–green algae *Nostoc muscorum* (Allison, Skipper, Reid, Short & Hogan 1953) and *Tolypothrix tenuis* (Cheung & Gibbs 1966). Although *Anabaena variabilis* and *Anacystis nidulans* assimilate acetate in the light they fail to show enzyme adaptation towards this substrate, the activities of the acetate-activating enzymes, glyoxylate cycle enzymes and two tricarboxylic acid cycle enzymes remaining the same whether it was present or not (Pearce & Carr 1967*b*).

At light saturation photoassimilation of organic substances does not result in higher growth-rates than would be achieved with carbon dioxide alone. Thus Smith *et al.* (1967), who conducted their experiments at relatively high light intensities, found that supply of acetate, the organic substrate most readily assimilated by the blue–green algae which they studied, caused no increase in growth rate and contributed at the most 10% of the newly assimilated carbon in the cells. At light intensities which are definitely limiting, photoassimilation results in enhanced growth rates. This is most clearly illustrated by data for a planktonic strain of *Chlorella pyrenoidosa* supplied with 1 mg l^{-1} of glycollic acid at a light intensity of 500 lux, which was found to have a relative growth-rate double that of the alga in purely inorganic medium under otherwise similar conditions (Sen & Fogg 1966). With *Anabaena variabilis* at relatively low light intensities, photoassimilation resulted in 32% of its dry weight being derived from glucose when this was supplied and 18% from acetate (Barr & Pearce 1966). Glucose was found to have most effect on the growth-rate of *Tolypothrix tenuis* at about 500 lux (Kiyohara *et al.* 1962). The qualitative observations of Staub (1961) suggest that growth of the bloom alga *Oscillatoria rubescens* is improved if organic substrates are supplied at limiting light intensities.

The biological advantage of photoassimilation in dimly lit situations may be considerable in that by starting with an already reduced source of carbon the limited assimilatory power from the photochemical reactions may be used to provide for a much greater amount of growth than would otherwise be possible. Since substrates such as glucose, acetate (Hobbie & Wright 1965) and glycollate (Fogg, Eagle & Kinson 1969) are found in eutrophic waters in concentrations of the order of 0·05 mg l^{-1} it seems possible that photoassimilation could allow increased growth of plankton algae at the bottom of the photic zone.

The hypolimnion of a lake receiving substantial amounts of organic matter is usually partially or completely depleted of oxygen during the summer and it therefore seems significant that some blue–green algae not only tolerate low oxygen concentrations but actually are more active under such conditions. Both nitrogen fixation and photosynthesis in *Anabaena flos-aquae* proceed more rapidly the lower the concentration of oxygen, and growth of this planktonic blue–green alga is as rapid or more rapid in the presence of reducing agents such as hydrogen sulphide or thiosulphate as it is in ordinary media (Professor W. D. P. Stewart and Mr H. Pearson, personal communication). Free oxygen has an inhibitory effect on nitrogen fixation in cell-free preparations of *Anabaena cylindrica* (Fay & Cox 1967) and on photosynthetic carbon dioxide fixation in a variety of different organisms and isolated chloroplast preparations (Gibbs, Ellyard & Latzko 1968). Photoassimilation of glucose by *Tolypothrix tenuis* is as rapid in the absence as in the presence of oxygen (Kiyohara *et al.* 1962). Under natural conditions in a lake population maxima of *Oscillatoria agardhii* have been observed at depths at which reducing conditions prevail (Dr G. W. Saunders, personal communication).

Many blue–green algae possess the biologically advantageous property of being able to fix free nitrogen, their growth being thus independent of supplies of combined nitrogen. We have already seen that blooms tend to develop when concentrations of nitrate in the water are low and there is now ample evidence that they may carry out nitrogen fixation under these conditions. Fixation by samples of blooms *in situ* in lakes has been demonstrated using ^{15}N as a tracer (Neess, Dugdale, Dugdale & Goering 1962; Goering & Neess 1964; Billaud 1967; Fogg & Horne 1967) and by the acetylene reduction technique (Stewart, Fitzgerald & Burris 1967). Since the fixation is correlated with the abundance of particular species of blue–green algae and is light-dependent (Dugdale & Dugdale 1962; Fogg & Horne 1967; Stewart, Fitzgerald & Burris 1967) we may be fairly sure that it is actually the algae, and not associated bacteria, which are doing it. Fixation may even occur in the presence of a relatively high concentration of ammonium nitrogen (Goering & Neess 1964; Billaud 1967). This at first sight is surprising because ammonia is a potent inhibitor of nitrogenase formation, but since the algae form gelatinous masses (figures 1 and 3, plate 3) it may well be that the ammonia becomes locally exhausted so that the cells are forced to utilize the more abundant and rapidly diffusing N_2 molecule. Billaud (1967) showed that about one-seventh of the total nitrogen assimilated by the spring bloom of *Anabaena flos-aquae* in an Alaskan lake was provided by nitrogen fixation and Fogg &

Horne (1967) estimated the annual fixation of nitrogen by blue–green algae in Windermere as 10 metric tons, that is 0·67 g m^{-2}. Contributions of fixed nitrogen of this magnitude must appreciably aggravate eutrophication problems. However, not all planktonic blue–green algae are able to fix nitrogen. Species of *Anabaena*, *Aphanizomenon* and *Gloeotrichia* possess the property but those of *Microcystis*, *Coelosphaerium* and *Oscillatoria* do not (Stewart 1966; Stewart, Fitzgerald & Burris 1967, 1968). The latter forms do not possess the curious differentiated cells, heterocysts, which are now thought to be the site of the fixation process (Fay, Stewart, Walsby & Fogg 1968). The relative abundance of heterocysts, incidently, is an easily ascertained index of the nitrogen-fixing capacity of an algal bloom. Since some bloom-forming blue–green algae do not fix nitrogen the possession of this property cannot be the reason for their general abundance in eutrophic waters but it is probably an important factor determining the species which occur at any given time or place.

We must now consider a remarkable feature of the planktonic blue–green algae which certainly is a principal cause of their being a nuisance and which may also be largely responsible for their biological success, namely that their cells contain gas-vacuoles. Under the light microscope (figure 3, plate 3) gas-vacuoles appear as indefinite, irregular and highly refractive bodies. There can be no doubt that they really do contain gas; they are destroyed by sudden pressure and more gas can be extracted from algae which contain them than from otherwise similar material from which they are absent or in which they have been destroyed (Fogg 1941; Walsby 1969), furthermore, investigation by interference microscopy shows that the refractive index of the vacuoles corresponds to their contents being gas (Fuhs 1969). The occurrence of gas-filled cavities in protoplasm is most unusual. Otherwise, gaseous inclusions only occur in certain bacteria, such as *Halobacterium*, in which they seem to be of a similar nature to those in the blue–green algae (Larsen, Omang & Steensland 1967; Stoeckenius & Kunau 1968), and in the protozoan, *Arcella*, which produces intracellular bubbles of oxygen of a nature which seems entirely different (Cicak, McLaughlin & Wittenberg 1963).

Electron microscopy shows that gas-vacuoles are composed of three-dimensional packed arrays of electron-transparent cylinders (Jost & Matile 1966; Jost & Zehnder 1966; Smith & Peat 1967). The units, which have been called gas cylinders or *Hohlspindeln*, are similar in the different species of algae. They are of a uniform diameter of about 70 nm, but of variable length from 100 to 1300 nm, and have pointed ends (figure 4, plate 3). The wall has a thickness of 2 or 3 nm according to whether fixation is with osmic acid or potassium permanganate (Smith & Peat 1967) and has striations transverse to the long axis of the cylinder (Jost 1965; Walsby & Eichelberger 1968). Stoeckenius & Kunau (1968) found the gas-vacuole units of *Halobacterium halobium* to be essentially similar to this but shorter. Isolated gas-vacuole membranes from *H. halobium* were found to consist mainly of protein (70 % of the dry weight) and no lipide could be extracted from them with chloroform–methanol. Jost & Matile (1966) reported lipide as a major constituent of gas-vacuole membranes from *Oscillatoria rubescens* but Stoekenius and Kunau are critical of their preparative methods. The membranes from *Halobacterium* are thought to be constructed from a stack of hoops attached laterally to one another, although a helix cannot be ruled out. The relatively high buoyant density of the membranes (1·23) makes it unlikely that their structure is similar to that of other cellular membranes; that is, they do not represent half-unit membranes (Stoekenius & Kunau 1968). The fact that potassium permanganate, which is a good fixative for unit membranes, fixes the gas-vacuole membrane rather poorly also points in this direction (Smith & Peat 1967).

The work of Walsby (1969 and unpublished) makes it clear that the gas-vacuole membranes of *Anabaena flos-aquae*, and presumably those from other blue–green algae, are freely permeable to gases such as nitrogen, oxygen, carbon dioxide and argon. In an attempt to identify the vacuole gas by mass-spectrometry it was found that the amount of gas obtainable was dependent on the gaseous pressure under which the material had previously been kept, little or no gas being obtained if it had been in a vacuum. This result was confirmed using a modified Warburg respirometer, with which it could be shown that gas passes in and out of the vacuoles as the gas pressure over the algal suspension is raised or lowered. Gas vacuoles are only deflated by increased pressure if insufficient time is allowed for the gas to equilibrate between the outside and inside of the gas cylinders.

Nothing definite is known of the metabolic processes which result in the formation of gas-vacuoles, but they are actively produced in healthy cells (Pringsheim 1966; Smith & Peat 1967) and are not a pathological symptom as has sometimes been suggested (Fritsch 1945). Blue–green algae inhabiting anaerobic bottom deposits contain them as also do planktonic forms. In laboratory cultures of *Anabaena flos-aquae* Walsby (1969) has observed that they are produced more abundantly at low than at high light intensities, in agreement with the finding of Pfennig & Cohen-Bazire (1967) that the photosynthetic bacterium *Pelodictyon clathratiforme* produces gas-vacuoles most actively in very dim light. Pringsheim (1966) found that they were produced in *Oscillatoria agardhii* under conditions in which anaerobic fermentation was unlikely. Larsen *et al.* (1967) found that 2,4-dinitrophenol inhibits gas-vacuole formation in *Halobacterium* sp. These findings make it seem improbable that gas-vacuoles merely represent stabilized bubbles of gas accumulated as the result of intense photosynthetic or respiratory activity and the results of Walsby (1969) seem to preclude the possibility of accumulation of gas by an active transport mechanism as suggested by Larsen *et al.* (1967).

Walsby (1969) points out that it is unlikely that the highly organized structure of the gas cylinders could arise from a bubble, and this, together with the fact that the walls are freely permeable to gas, suggests that the gas-vacuole structure is self-erecting; that is, spaces are first created in the protoplasm and then fill with gas by diffusion. An assemblage of large molecules which, following some chemical transformation, undergo rearrangement so that a space is formed in the centre does not seem an impossibility. Such a process would be energy-requiring. At the least it begins to appear that gas-vacuoles are not the fortuitous and biologically valueless result of bubble formation in the special environment of myxophycean protoplasm but that they are specific structures with definite selective value.

Gas-vacuoles, if present in sufficient quantity, clearly render the alga buoyant. Klebahn (1922) determined the specific gravity of *Gloeotrichia echinulata* in which the gas-vacuoles had been destroyed as 1·0067 to 1·0085. To enable the alga just to float in water gas-vacuoles would then have to occupy 0·7 % of the cell volume and by direct observation of the diminution of volume following their destruction he found the value to be 0·8 % in his material. Smith & Peat (1967) found from measurements on electron micrographs that the percentage of cell volume occupied by gas-vacuoles was 22 for *Anabaena flos-aquae* and *G. echinulata* and 39 for *Oscillatoria agardhii*, the algae having been grown in culture. Because of their buoyancy blue–green algae tend to accumulate at the water surface in calm weather. As would be expected from hydromechanical theory this tendency is most marked for the forms, such as *Microcystis*, *Anabaena* and *Gloeotrichia*, having large colonies and least for species of *Oscillatoria* which occur as individual filaments of small diameter. Maximum populations of *Oscillatoria* spp., in fact, are characteristically found towards the bottom of the epilimnion and do not often accumulate at the surface (Eberley 1967). The observations of Talling (1957) on a Nile reservoir illustrate very well the accumulation of *Anabaena flos-aquae* at the water surface during the day-time thermal stratification and its uniform distribution after night-time isothermal mixing.

Putting together these various pieces of information a possible explanation of the success of blue–green algae in eutrophic waters begins to emerge. The main limitation on the growth of phytoplankton in general is that whereas the cells must remain near the surface in order to photosynthesize, the mineral nutrients they require, some of which are present in very low concentrations, are distributed through the bulk of the water and usually are more available towards the bottom. In a motionless medium the growth-rate of a cell is limited by the rate of diffusion up to it of the scarcer nutrient substances. As Munk & Riley (1952) pointed out, transfer of dissolved substances to the cell will be more rapid if the concentration gradient is kept steep by movement of the medium past the cell. Most phytoplankton species, having a specific gravity greater than that of water, achieve this 'forced convection' by sinking, the population being maintained near the surface by turbulence. Munk & Riley showed that division rates of various diatom species are in fact related to their calculated sinking rates (for further discussion of this see Hutchinson 1967). Planktonic blue–green algae show an upside-down relationship, achieving forced convection by rising through the water (Lund 1959) and thus becoming independent of turbulence to keep them in the photic zone.

The hypothesis that myxophycean blooms develop at the bottoms of lakes under anaerobic conditions has frequently been advanced (see Fritsch 1945). They do not appear to originate in or on the bottom deposit itself (Lund 1959) but the idea that the initial development of a bloom takes place in deep water has much to be said for it. At the bottom of the photic zone a more abundant supply of mineral nutrients will be available and the oxygen concentration will be more favourable for their growth. Photoassimilation of dissolved organic substances may enable high growth-rates to be achieved in spite of the low light intensity. Algae belonging to other groups might also find such conditions favourable but, unless they are actively motile, are constantly at risk, since turbulence is less in the depths than near the surface, their cells being liable to sink to depths at which the light intensity is below the compensation point. Gas-vacuoles being produced to the greatest extent at low light intensity this will not happen to the blue–green algae. It is conceivable that gas-vacuole formation may be related to light intensity in such a way that the positions of the cell in the water column is controlled (Walsby 1969) —as indeed seems to happen with blooms of *Oscillatoria* which characteristically maintain maximum populations towards the bottom of the photic zone (Lund 1959; Eberley 1967). It might even be that gas-vacuole formation in dim light causes a cell or colony to rise, inhibition of gas-vacuole formation and continued growth and cell division nearer the surface subsequently leading to it becoming heavier and sinking again. Movement of this sort up and down would make for maximum absorption of mineral nutrients. Near the surface light inhibition of photosynthesis may result in an inability to 'dilute out' the gas-vacuoles by growth so that the alga remains floating and becomes moribund. This is what appears to happen with the surface-bloom forms such as *Microcystis* and *Anabaena*. Fitzgerald (1967) has observed that such algae may readily be grown in culture if taken from deep-water samples whereas the same species from surface samples rarely survive if cultured. Reynolds (1967) has also commented that algae accumulated at the surface do not survive.

The correlation of myxophycean blooms with dissolved organic matter in fresh waters is thus seen to depend on a number of factors. Organic compounds of many different sorts may provide an ionic environment favourable for these algae by complex formation. Photoassimilation of organic substances may sustain growth at low light intensities and thus enable these algae to occupy the strata of water in which decomposition of organic matter has reduced the oxygen concentration to a level favourable for their growth. Anaerobic conditions in the hypolimnion of a lake usually develop in the summer at the time when the concentration of nitrate and phosphate in the surface waters is at a minimum. Organic decomposition in the deeper water gradually releases such inorganic nutrients in the large amounts required for the development of dense blooms. Blue–green algae appear to possess a means of regulating buoyancy which may enable them to draw on the mineral nutrients from larger volumes of water than other kinds of algae are able to exploit and thus to grow at times when the actual concentrations of these nutrients in the water is low. Even if a general explanation of the development of myxophycean blooms can be given along these lines it is certain that more complicating factors will have to be taken into account to obtain a complete picture of individual cases.

There is evidence that some of the algae concerned produce metabolites inhibitory to the growth of other algae (Vance 1965) and blue–green algae are themselves vulnerable to antibiotics (Vance 1965; Whitton 1967), fungal parasites (Canter 1963; Canter & Willoughby 1964) and viruses or cyanophages (Padan, Shilo & Kislev 1967).

It is blooms which accumulate at the water surface which are particularly a nuisance. Drifted by wind they may form scums of the consistency of paint and, decaying, these may lead to local exhaustion of oxygen in the water and death of fish by asphyxiation. However, fish may be killed when there is still an adequate oxygen supply and there is clear evidence that potent toxins are sometimes produced (Shilo 1964; Gorham 1964; Pankow 1964; Flint 1966). These may cause death of mammals and birds, as well as fish, very quickly—there is, for example, a record of a full-grown cow being killed within half an hour (Gorham 1964)—and they are therefore of great toxicological interest. Hughes, Gorham & Zehnder (1958) found toxins obtained from a strain of *Microcystis aeruginosa* grown in culture produced symptoms in mice similar to those produced by natural water-blooms and there is no doubt that the toxins are produced by the algae themselves. Distinct 'slow-death' and 'fast-death' factors were isolated and the latter, which kills livestock and other mammals but not water-fowl, has been identified as a cyclic polypeptide of molecular weight about 1200 and containing D-serine (Gorham 1964). A 'very fast-death' factor from *Anabaena flos-aquae*, which kills a variety of animals including water-fowl, is probably also a polypeptide. In addition, bacteria associated with the algae may produce toxins that cause less acute poisonings. The existence of toxic and non-toxic strains of algae is perhaps the main explanation of the very variable toxicity shown by myxophycean water-blooms. Peary & Gorham (1966) have shown that light intensity and temperature also have a considerable influence on toxin production by *Anabaena flos-aquae*. In addition the bloom must be concentrated if a lethal dose is to be absorbed and, with *Microcystis*, the toxin must be released from the cells by decomposition in order to be effective. Destruction or inactivation of the toxins may also affect toxicity under natural conditions (Gorham 1964). The investigations of Shilo (1967) on the ichthyotoxin produced by the flagellate *Prymnesium parvum* have shown how complex the interaction of factors affecting the production and activity of algal toxins may be.

Control of myxophycean water-blooms is difficult. Treatment with copper sulphate to give a concentration of about 0·2 mg l^{-1} in the water sometimes prevents development of blooms without damaging higher plants or fish but is not always effective, perhaps because extracellular polypeptides liberated by the algae produce a non-toxic complex with the copper ion (Fogg & Westlake 1955). In any case copper sulphate is only a temporary palliative and the wisdom of repeatedly adding a permanently poisonous substance to a lake or reservoir is doubtful (Hasler 1949). Some organic substances such as 2,3-dichloro-1,4-naphthoquinone (Fitzgerald & Skoog 1954) are selectively toxic towards blue–green algae, although here again the extracellular products of the algae seem to reduce the effect of the poison (Whitton & MacArthur 1967). Selective algicides of this sort do not appear to have been widely used against mycophycean water-blooms, possibly because of their high cost. Since, probably, the biological success and, certainly, most of the nuisance caused by these algae arises from their possession of gas-vacuoles, inhibition of the formation of these structures or their destruction by ultrasonic radiation or other means (Brook 1959) might be an effective means of control. Another possibility, if it is correct that active growth occurs in a circumscribed stratum of water in which light intensity, reducing conditions and nutrient supply are most favourable, may be artificial circulation to break down chemical stratification in the water. This method, introduced by the Metropolitan Water Board as a general means of algal control, does in fact appear to be particularly effective against blue–green algae.

In any case it should be remembered that blooms of blue–green algae are only a symptom of a condition of fresh water that arises from Man's disturbance of the ecological balance and which, even on the short-term view, is undesirable in other respects. If these and similar problems are to be solved it is important that microbiologists should follow Leeuwenhoek's example and, to a greater extent than they do, extend their studies into the natural environment.

I am most grateful to my colleagues Dr P. Fay, Mr H. Pearson, Professor W. D. P. Stewart and Mr A. E. Walsby, who greatly helped in the preparation of this lecture by their discussion and provision of unpublished results and illustrative material. My thanks are also due to Dr H. M. Canter (Mrs Lund) for her kindness in allowing me to use her photomicrographs of blue–green algae.

References

Allen, M. B. 1952 The cultivation of Myxophyceae. *Arch. Mikrobiol.* **17**, 34–53.

Allison, R. K., Skipper, H. E., Reid, M. R., Short, W. A. & Hogan, G. L. 1953 Studies on the photosynthetic reaction. I. The assimilation of acetate by *Nostoc muscorum*. *J. biol. Chem.* **204**, 197–205.

Bergmann, L. 1955 Stoffwechsel und Mineralsalzernährung einzelliger Grünalgen. II. *Flora, Jena* **142**, 493–539.

Billaud (Dugdale), V. M. 1967 Aspects of the nitrogen nutrition of some naturally occurring populations of blue–green algae. In *Environmental requirements of blue–green algae*, pp. 35–53. U.S. Department of the Interior, Pacific Northwest Water Laboratory, Corvallis, Oregon.

Brook, A. J. 1959 The waterbloom problem. *Proc. Soc. Water Treat. Exam.* **8**, 133–137.

Canter, H. M. 1963 Concerning *Chytridium cornutum* Braun. *Trans. Br. mycol. Soc.* **46**, 208–212.

Canter, H. M. & Willoughby, L. G. 1964 A parasitic *Blastocladiella* from Windermere plankton. *Jl R. microsc. Soc.* **83**, 365–372.

Carr, N. G. & Pearce, J. 1966 Photoheterotrophism in blue–green algae. *Biochem. J.* **99**, 28–29*P*.

Cheung, W. Y. & Gibbs, M. 1966 Dark and photometabolism of sugars by a blue–green alga: *Tolypothrix tenuis*. *Pl. Physiol., Lancaster*, **41**, 731–737.

Cicak, A., McLaughlin, J. J. A. & Wittenberg, J. B. 1963 Oxygen in the gas-vacuole of the rhizopod protozoan, *Arcella*. *Nature, Lond.* **199**, 983–985.

Dugdale, V. A. & Dugdale, R. C. 1962 Nitrogen metabolism in lakes. II. Role of nitrogen fixation in Sanctuary Lake, Pennsylvania. *Limnol. Oceanogr.* **7**, 170–177.

Eberly, W. R. 1967 Problems in the laboratory culture of planktonic blue–green algae. In *Environmental requirements of blue–green algae*, pp. 7–34. U.S. Department of the Interior, Pacific Northwest Water Laboratory, Corvallis, Oregon.

Eppley, R. W., Gee, R. & Saltman, P. 1963 Photometabolism of acetate by *Chlamydomonas mundana*. *Physiologia Pl.* **16**, 777–792.

Fay, P. 1965 Heterotrophy and nitrogen fixation in *Chlorogloea fritschii*. *J. gen. Microbiol.* **39**, 11–20.

Fay, P. & Cox, R. M. 1967 Oxygen inhibition of nitrogen fixation in cell-free preparations of blue–green algae. *Biochim. biophys. Acta* **143**, 562–569.

Fay, P., Stewart, W. D. P., Walsby, A. E. & Fogg, G. E. 1968 Is the heterocyst the site of nitrogen fixation in blue–green algae? *Nature, Lond.* **220**, 810–812.

Fitzgerald, G. P. 1967 Discussion. In *Environmental requirements of blue–green algae*, pp. 97–102. U.S. Department of the Interior, Pacific Northwest Water Laboratory, Corvallis, Oregon.

Fitzgerald, G. P. & Skoog, F. 1954 Control of blue–green algae blooms with 2,3-dichloronaphthoquinone. *Sewage ind. Wastes* **26**, 1136–1140.

Flint, E. A. 1966 Toxic algae in some New Zealand freshwater ponds. *N.Z. Vet. J.* **14**, 181–185.

Fogg, G. E. 1941 The gas-vacuoles of the Myxophyceae (Cyanophyceae). *Biol. Rev.* **16**, 205–217.

Fogg, G. E. 1952 The production of extracellular-nitrogenous substances by a blue–green alga. *Proc. Roy. Soc.* B **139**, 372–397.

Fogg, G. E. 1966 The extracellular products of algae. *Oceanogr. Mar. Biol. Ann. Rev.* **4**, 195–212.

Fogg, G. E., Eagle, D. J. & Kinson, M. E. 1969 The occurrence of glycollic acid in natural waters. *Verh. int. Ver. theor. angew. Limnol.* **17**, (in the Press).

Fogg, G. E. & Horne, A. J. 1967 The determination of nitrogen fixation in aquatic environments. In *Chemical environment in the aquatic habitat*, pp. 115–120. Amsterdam.

Fogg, G. E. & Westlake, D. F. 1955 The importance of extracellular products of algae in freshwater. *Verh. int. Ver. theor. angew. Limnol.* **12**, 219–232.

Fritsch, F. E. 1945 *The structure and reproduction of the algae*, vol. 2, Cambridge University Press.

Fuhs, G. W. 1969 *Verh. int. Ver. theor. angew. Limnol.* **17**, (in the Press).

Gerloff, G. C., Fitzgerald, G. P. & Skoog, F. 1950 The isolation, purification, and culture of blue–green algae. *Am. J. Bot.* **37**, 216–218.

Gibbs, M., Ellyard, P. W. & Latzko, E. 1968 Warburg effect: control of photosynthesis by oxygen. In *Comparative biochemistry and biophysics of photosynthesis*, pp. 387–399. University of Tokyo Press and University Park Press.

Goering, J. J. & Neess, J. C. 1964 Nitrogen fixation in two Wisconsin lakes. *Limnol. Oceanogr.* **9**, 530–539.

Gorham, P. R. 1964 Toxic algae. In *Algae and Man* (ed. Jackson, D. F.), pp. 307–336. New York: Plenum Press.

Goulding, K. H. & Merrett, M. J. 1966 The photometabolism of acetate by *Chlorella pyrenoidosa*. *J. exp. Bot.* **17**, 678–689.

Griffiths, B. M. 1939 Early references to waterblooms in British lakes. *Proc. Linn. Soc. Lond.*, **151**, 12–19.

Hasler, A. D. 1949 Antibiotic aspects of copper treatment of lakes. *Trans. Wis. Acad. Sci. Arts Lett.* **39**, 97–103.

Hobbie, J. E. & Wright, R. T. 1965 Competition between planktonic bacteria and algae for organic solutes. *Mem. Ist. Ital. Idrobiol.* **18** (Suppl.), 175–185.

Hughes, E. O., Gorham, P. R. & Zehnder, A. 1958 Toxicity of a unialgal culture of *Microcystis aeruginosa*. *Can. J. Microbiol.* **4**, 225–236.

Hutchinson, G. E. 1967 *A treatise on limnology*, vol. II. New York: Wiley.

Jost, M. 1965 Die Ultrastruktur von *Oscillatoria rubescens* D.C. *Arch. Mikrobiol.* **50**, 211–245.

Jost, M. & Matile, P. 1966 Zur Characterisierung der Gasvacuolen der Blaualge *Oscillatoria rubescens*. *Arch. Mikrobiol.* **53**, 50–58.

Jost, M. & Zehnder, A. 1966 Die Gasvakuolen der Blaualge *Microcystis aeruginosa*. *Schweiz. Z. Hydrol.* **28**, 1–3.

Klebahn, H. 1922 Neue Untersuchungen über die Gasvakuolen. *Jb. wiss. Bot.* **61**, 535–589.

Kyohara, T., Fujita, Y., Hattori, A. & Watanabe, A. 1960 Heterotrophic culture of a blue-green alga, *Tolypothrix tenuis* I. *J. gen. appl. Microbiol., Tokyo* **6**, 176–182.

Kyohara, T., Fujita, Y., Hattori, A. & Watanabe, A. 1962 Effect of light on glucose assimilation in *Tolypothrix tenuis*. *J. gen. appl. Microbiol., Tokyo* **8**, 165–168.

Larsen, H., Omang, S. & Steensland, H. 1967 On the gas vacuoles of the Halobacteria. *Arch. Mikrobiol.* **59**, 197–203.

Lund, J. W. G. 1959 Buoyancy in relation to the ecology of the freshwater phytoplankton. *Br. phyc. Bull.* no. 7, 1–17.

Lund, J. W. G. 1967 Eutrophication. *Nature, Lond.* **214**, 557–558.

Marker, A. F. H. & Whittingham, C. P. 1966 The photoassimilation of glucose in *Chlorella* with reference to the role of glycollic acid. *Proc. Roy. Soc.* B **165**, 437–485.

Munk, W. H. & Riley, G. A. 1952 Absorption of nutrients by aquatic plants. *J. mar. Res.* **11**, 215–240.

Neess, J. C., Dugdale, R. C., Dugdale, V. A. & Goering, J. J. 1962 Nitrogen metabolism in lakes. I. Measurement of nitrogen fixation with N^{15}. *Limnol. Oceanogr.* **7**, 163–169.

Padan, E., Shilo, M. & Kislev, N. 1967 Isolation of 'cyanophages' from freshwater ponds and their interaction with *Plectonema boryanum*. *Virology* **32**, 234–246.

Pankow, H. 1964 Bemerkung über die Schädlichkeit von Blaualgenwasserbluten für Tiere. *Naturwissenschaften*, **51**, 146–147.

Pearce, J. & Carr, N. G. 1967*a* An incomplete tricarboxylic acid cycle in the blue–green alga *Anabaena variabilis*. *Biochem. J.* **105**, 45*P*.

Pearce, J. & Carr, N. G. 1967*b* The metabolism of acetate by the blue–green algae, *Anabaena variabilis* and *Anacystis nidulans*. *J. gen. Microbiol.* **49**, 301–313.

Pearsall, W. H. 1932 Phytoplankton in the English Lakes. 2. The composition of the phytoplankton in relation to dissolved substances. *J. Ecol.* **20**, 241–262.

Peary, J. A. & Gorham, P. R. 1966 Influence of light and temperature on growth and toxin production by *Anabaena flos-aquae* NRC-44-1. *J. Phycol.* **2** (Suppl.), 3–4.

Peelen, R. 1969 *Verh. int. Ver. theor. angew. Limnol.* **17**, (in the Press).

Pfennig, N. & Cohen-Bazire, G. 1967 Some properties of the green bacterium *Pelodictyon clathratiforme*. *Arch. Mikrobiol.* **59**, 226–236.

Pintner, I. J. & Provasoli, L. 1958 Artificial cultivation of a red-pigmented marine blue–green alga, *Phormidium persicinum*. *J. gen. Microbiol.* **18**, 190–197.

Prescott, G. W. 1951 Algae of the western Great Lakes area. *Cranbrook Institute of Science Bull.* no. 30

Pringsheim, E G. 1966 The nature of pseudovacuoles in Cyanophyceae. *Nature, Lond.* **210**, 549–550.

Provasoli, L. 1966 In *Marine Biology*, vol. II (ed. C. H. Oppenheimer), pp. 93–98. New Academy of Sciences.

Reynolds, C. 1967 The breaking of the Shropshire meres; some recent investigations. *Shropshire Conservation Trust Bull.* no. 10, pp. 9–14.

Schwimmer, D. & Schwimmer, M. 1964 Algae and medicine. In *Algae and Man* (ed. D. F. Jackson), pp. 368–412. New York: Plenum Press

Sen, N. & Fogg, G. E. 1966 Effects of glycollate on the growth of a planktonic *Chlorella*. *J. exp. Bot.* **17**, 417–425.

Shilo, M. 1964 Review on toxigenic algae. *Verh. int. Ver. theor. angew. Limnol.* **15**, 782–795.

Shilo, M. 1967 Formation and mode of action of algal toxins. *Bact. Rev.* **31**, 180–193.

Singh, R. N. 1955 Limnological relations of Indian inland waters with special reference to waterblooms. *Verh. int. Ver. theor. angew. Limnol.* **12**, 831–836.

Smith, A. J., London, J. & Stanier, R. Y. 1967 Biochemical basis of obligate autotrophy in blue–green algae and thiobacilli. *J. Bact.* **94**, 972–983.

Smith, R. V. & Peat, A. 1967 Comparative structure of the gas-vacuoles of blue-green algae. *Arch. Mikrobiol.* **57**, 111–122.

Staub, R. 1961 Ernährungsphysiologische-autökologische Untersuchungen an der planktischen Blaualge *Oscillatoria rubescens*. DC. *Schweiz. Z. Hydrol.* **23**, 82–198*a*.

Stewart, K. M. & Rohlich, G. A. 1967 Eutrophication—a review. *State of California, State Water Quality Control Board, publication* no. 34.

Stewart, W. D. P. 1966 *Nitrogen fixation in plants*. Athlone Press of the University of London.

Stewart, W. D. P., Fitzgerald, G. P. & Burris, R. H. 1967 *In situ* studies on N_2 fixation using the acetylene reduction technique. *Proc. natn. Acad. Sci. U.S.A.* **58**, 2071–2078.

Stewart, W. D. P., Fitzgerald, G. P. & Burris, R. H. 1968 Acetylene reduction by nitrogen-fixing blue-green algae. *Arch. Mikrobiol.* **62**, 336–348.

Stoekenius, W. & Kunau, W. H. 1968 Further characterization of particulate fractions from lysed cell envelopes of *Halobacterium halobium* and isolation of gas vacuole membranes. *J. Cell Biol.* **38**, 337–357.

Talling, J. F. 1957 Diurnal changes in stratification and photosynthesis in some tropical African waters. *Proc. Roy. Soc.* B **147**, 57–83.

Tischer, R. G. 1965 Pure culture of *Anabaena flos-aquae* A-37. *Nature, Lond.* **205**, 419–420.

Vance, B. D. 1965 Composition and succession of cyanophycean water blooms. *J. Phycol.* **1**, 81–96.

Walsby, A. E. 1969 The permeability of blue-green algal gas-vacuole membranes to gas. *Proc. R. Soc.* B **173**, 235.

Walsby, A. E. & Eichelberger, H. H. 1968 The fine structure of gas-vacuoles released from cells of the blue-green algae *Anabaena flos-aquae*. *Arch. Mikrobiol.* **60**, 76–83.

Watanabe, A. & Yamamoto, Y. 1967 Heterotrophic nitrogen fixation by the blue-green alga *Anabaenopsis circularis*. *Nature, Lond.* **214**, 738.

Wetzel, R. G. 1965 Nutritional aspects of algal productivity in marl lakes with particular reference to enrichment bioassays and their interpretation. *Mem. Ist. Ital. Idrobiol.* **18**, (Suppl.), 137–157.

Whitton, B. A. 1967 Studies on the toxicity of polymixin B to blue-green algae. *Can. J. Microbiol.* **13**, 987–993.

Whitton, B. A. & MacArthur, K. 1967 The action of two toxic quinones on *Anacystis nidulans*. *Arch. Mikrobiol.* **57**, 147–154.

Review Lecture

Experiment and the crust of the Earth

By W. S. Fyfe

Royal Society Research Professor, Manchester University

(*Lecture delivered* 15 *February* 1968—*MS. received* 20 *May* 1968)

Proceedings of the Royal Society, A, *volume* 307, *pp*. 373–389, 1968

1. Introduction

In this review I should like to consider our progress in unravelling some problems in Earth science, particularly problems where experiment has led to a much deeper understanding of observations concerning the formation of rocks now found at the surface of the Earth. While geophysical probes into the Earth's interior become increasingly sophisticated, I think it is still true that most of our models of Earth properties and materials at depth, are based on observations and experiments on materials seen at the surface if we include meteorites in the sample. It is also true that crustal phenomena provide most of the evidence concerning the dynamics and evolution of the entire Earth.

2. The scale of the geochemical problem

When the Earth is discussed, it is usual to consider its division into core, mantle, crust and atmosphere (figure 1). If we are concerned with the total chemical composition of the Earth, it must be remembered that about 99·5 % of the mass is contained in the mantle and core and is hence not directly observed. Our position is not simple, for unlike the chemist or physicist interested in atoms and nucleii, we cannot

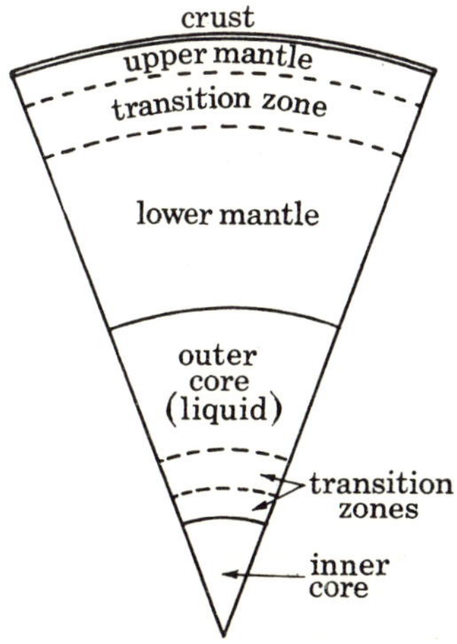

Figure 1. Division of the Earth's interior, true scale (after Birch 1965).

split the Earth and examine its properties. Any statement about the chemistry of the Earth must thus be based on limited observations which reflect this chemistry. These observations, which fortunately are quite restrictive, include:

(1) The composition must be suitable to explain the mean density and density distribution known from observations on gravitational and seismic properties.

(2) The composition must explain the observed heat production of the Earth in so far as radioactive elements contribute.

(3) The composition must be fitting to the layered structure of the Earth.

(4) The chemistry of the core must be such that the magnetic field can be generated.

(5) The composition probably should not be too unique in comparison to solar and cosmic abundances of the elements.

(6) Observed crustal abundances should not conflict with the principles of chemical equilibrium in a gravitational field.

(7) The composition of the upper mantle must be such that certain volcanic products can be generated from it by partial fusion.

When all such limitations are considered it is found that a rather good model of Earth composition can be derived by analogy with the composition of the chondritic meteorites (table 1), the most common meteorites seen to fall. Further, the

TABLE 1. BULK PERCENTAGE COMPOSITION OF THE EARTH BASED ON A CHONDRITIC MODEL

Fe	34·63
Ni	2·39
Co	0·13
S	1·93
O	29·53
Si	15·20
Mg	12·70
Ca	1·13
Al	1·09
Na	0·57
Cr	0·26
Mn	0·22
P	0·10
K	0·07
Ti	0·05
total	100·00

formation of the Earth, its crust and atmosphere, fit well into a model based on the metamorphism, partial fusion, and gravitational separation of chondritic materials (Birch 1965; Clark & Ringwood 1964). Thus in a gross fashion, the Earth as we see it today is the result of a vast distribution experiment. Metallic phases dominated by nickel and iron, melt and sink to form the liquid and solid core. This metal phase will extract and bury most of the inert heavy elements such as gold and platinum and possibly much of the sulphides. The mantle will be dominated by silicates and oxides in the system Mg–Fe–Si–O. The crust is the low density, low melting fraction of the chondritic material enriched in materials such as the alkalis and any elements with large oxygen affinities. Brewer (1951) has shown that element distribution in the Earth is in reasonable agreement with that to be expected from chemical equilibrium based on the general relation:

$$d\mu_i = 0 = \left(\frac{\partial \mu_i}{\partial T}\right)_{P,h,n} dT + \left(\frac{\partial \mu_i}{\partial P}\right)_{T,h,n} dP + \left(\frac{\partial \mu_i}{\partial h}\right)_{P,T,n} dh + \Sigma \left(\frac{\partial \mu_i}{\partial n_j}\right)_{P,T,h,n_{j-1}} dn_j$$

(μ_i is the chemical potential of component i, P the pressure, T the temperature and h the position in the gravitation field and n_i the number of moles of component i).

When we consider the variables in this equation we become alerted to differences between the Earth system and laboratory systems which so strongly influence our chemical intuition. Temperature variation within the Earth probably does not exceed about 3000 °C while pressure variation is about $3{\cdot}5 \times 10^6$ atm. If we consider possible reactions in solid systems, entropy and volume changes tend to have sympathetic values and similar signs. Averages from a large number of examples show that if ΔV is -1 cm³ mol⁻¹, ΔS is of the order of $-0{\cdot}5$ cal mol⁻¹ deg⁻¹. If we consider the possible Earth range, and the above figures, then as

$$\left(\frac{\partial \Delta G}{\partial T}\right)_P = -\Delta S \quad \text{and} \quad \left(\frac{\partial \Delta G}{\partial P}\right)_T = \Delta V,$$

the 3000 °C temperature range produces a change in free energies of about 1500 cal mol^{-1} while the pressure leads to a change of the order of 10^5 cal mol^{-1}. Thus to a first approximation, pressure effects will be more important than temperature effects. As the magnitude of pressure effects is large and comparable with bond energies, one might expect to find some quite unfamiliar compounds or unfamiliar structural states. An important example of such a change is provided by the transformation of SiO_2 in the quartz structure to silica in the rutile structure at pressures near 10^5 atm with a density change of 2·65 to 4·29, accompanying the change from 4 to 6 coordination. It is not difficult to show that such changes must occur in the Earth by application of the simplest treatment of lattice energies. Another interesting case is related to the magnetic properties of transition metal oxides in the mantle. In most crustal materials these occur in the normal high-spin states but as pairing leads to a large decrease in radius of the ion, mantle states will tend to be low spin. In the case of ferrous iron, crustal materials are paramagnetic or anti-ferromagnetic but in the deep mantle, ferrous iron compounds will become diamagnetic. Thus consideration of materials in the deep Earth may lead us to consider quite new structural states.

3. Mineralogy

The ultimate object of Earth science is to understand or describe the evolution of the Earth since its origin some $4{\cdot}5 \times 10^9$ years ago. The existence of a stable crust is known since about $3{\cdot}5 \times 10^9$ years and much of the history of the Earth must be extracted from the early fragments of crust and the rocks which have been added to it throughout geologic time. The inorganic history (and the organic for that matter) must largely be won from a study of the minerals and mineral assemblages of this crust. For example, the spin orientation of atomic magnetic moments in minerals such as magnetite (Fe_3O_4) records part of the history of the Earth's magnetic field with all the resultant implications in terms of crustal and ocean floor mobility and stability; the state of ordering of the Al–Si atoms in a mineral such as $KAlSi_3O_8$ may record thermal history; the $^{16}O/^{18}O$ fractionation between a pair of minerals may record quite exact temperatures; the content of a given component in a solid solution or the presence of a given polymorph may leave a record of pressure; the tiny fluid inclusions trapped in a crystal may preserve a sample of a high temperature fluid which transported rare ore materials. The same remarks apply equally to the mineralogy of meteorites and the nature of their parent bodies. Thus, to understand the evolution of the Earth we must be able to extract from minerals many types of information often requiring quite sophisticated laboratory techniques.

Many of the properties of minerals which we need to understand are equilibrium properties. But many processes are rate controlled; phases stable only at great depth in the crust may reach the surface preserved, and to appreciate the significance of their preservation we may require data on reaction rates. Some such cases preserve data on the geothermal gradient during tectonic processes. When solid mineral systems are considered, most processes involving diffusion are so slow that they occur only to a trivial extent at crustal temperatures. Most crustal reactions require catalysts and the common catalyst of the crust is an aqueous phase. We have much to learn about the migration of fluids at depth but it appears that rapid diffusion only occurs when fluid pressures are similar to the rock pressure. In many cases, it appears that large crustal masses are almost floating on such fluids and the ease with which enormous gravity slides or thrusts occur (as in many mountain building processes) can only be explained on such a model.

4. Apparatus

Today, tools are available to describe the materials of the Earth in immense detail. With the electron microprobe we can obtain an adequate chemical analysis of minute amounts of materials (*ca.* 10^{-11} g) or compositional zones in a small crystal. Activation analysis has allowed rather exact estimation of element abundances in the part/10^9 range and much problematical data of the past have been eliminated. Analysis may often include a range of stable and unstable isotopes. In minerals, structural analysis by many techniques from normal X-ray diffraction to Mössbauer spectra is providing more exact data on site populations and a beginning is being made on similar studies with silicate liquids. In the latter case, we are still a very long way from having the data needed to apply thermodynamics adequately.

As we are much concerned with equilibrium processes, we require thermodynamic properties of minerals over a wide range of conditions. Calorimetric studies of silicates (e.g. the determination of ΔH data) is a difficult problem because of their refractory nature and the very high precision required. Formerly, the only suitable solvent was aqueous hydrofluoric acid, but use of fused salt solvents has led to a tenfold increase in precision. In most cases direct study of reactions is still required.

As the maximum thickness of the crust is about 70 km, pressures at the base will rarely exceed 2×10^4 atm. Temperatures in this region rarely exceed the melting temperatures of granitic rocks, about 800 °C. Apparatus is available for static, hydrostatic pressure conditions, for all this range. Some crustal rocks have an origin considerably deeper (100–200 km) and for such cases were require apparatus for 10^5 atm and temperatures of over a 1000 °C. To a limited extent these conditions can be met, but beyond $2{\cdot}5 \times 10^5$ atm shock wave techniques must be used. While shock pressures and temperatures can be produced to simulate core conditions, the micro-second reaction times are inadequate for processes where major diffusional steps are required. There is a constant search for new techniques and materials to generate larger and more sustained high pressures and temperatures.

5. The variables

Chemical processes which occur in the crust and upper mantle involve solid, liquid and gas phases. The important variables in the description of equilibrium states include pressure (including non-hydrostatic terms), temperature and the partial pressures of significant gaseous components. One of the objectives of experimental studies is to describe the environment of formation of any given rock in terms of such variables. Non-hydrostatic pressure terms present problems, for at present, while we know that such terms are significant, we have no way of estimating their magnitude during many processes. We are also a long way from being able to predict their thermodynamic consequences. Experiments on this aspect of the steady state are difficult and as yet there is little useful information.

6. Specific problems

I now wish to discuss some specific examples of mineral equilibria to indicate some of the problems and the approaches used. Many of these have been of particular concern to the experimental group at Manchester University.

(*a*) *Polymorphism of the* Al_2SiO_5 *minerals*

Three polymorphic modifications of Al_2SiO_5 are widely distributed in metamorphic rocks of the crust. They are kyanite

$$(S^0 = 20{\cdot}02\,\text{cal mol}^{-1}\,\text{degC}^{-1}, \quad V^0 = 44{\cdot}11\,\text{cm}^3\,\text{mol}^{-1}),$$

andalusite ($S^0 = 22{\cdot}28$, $V^0 = 51{\cdot}54$) and sillimanite ($S^0 = 22{\cdot}97$, V^0 49·91). Geologists have recognized that kyanite-bearing rocks appear to be those most deeply buried while sillimanite is the form stable in regions of high temperature. On the basis of the entropy-volume relations, and assuming that all have regions of stable existence (a reasonable assumption in the context of geologic time for a metamorphic cycle involves times of the order of 10^8 years), the equilibrium phase diagram must have a form as shown in figure 2. Because the minerals are so widespread (they form from the dehydration of clay rich sediments) it is important that we can place numbers on this phase diagram for the occurrence of a given form in a rock will place limits on the P–T conditions of its formation. Providing the phases are iso-chemical, their stability will depend only on the variables P–T–P_{stress}.

The experimental problem is difficult for a number of reasons. The fact that all modifications appear at the surface indicates that reaction rates in the solid state are very slow. In addition, many rocks which have passed through a transition region, frequently contain two polymorphs over a wide transition zone. This again indicates sluggish kinetics. Consideration of the entropy data shows that we are dealing with very small free energies of reaction. Near the andalusite–sillimanite equilibrium curve, the ΔS of reaction is of the order of $0 \cdot 3 \pm 0 \cdot 3$ cal mol^{-1} degC^{-1} and this means that even 100 °C from the phase boundary, the ΔG of reaction will only be about 30 cal mol^{-1}. Such a small value (not uncommon for solid-solid reactions in silicates) indicates that the supersaturation required for nucleation and growth of a new phase will be built up slowly as we cross a phase boundary.

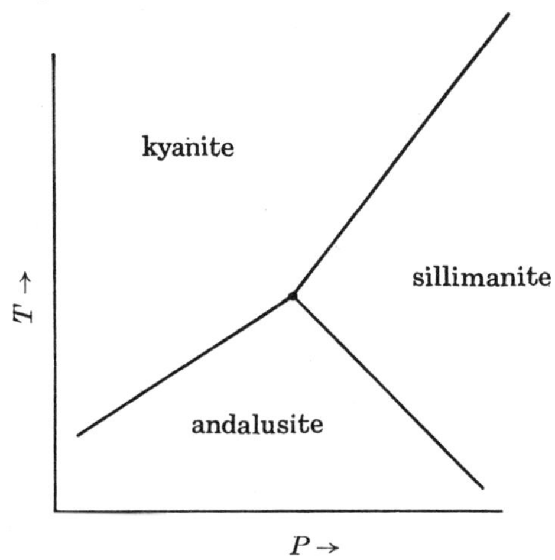

FIGURE 2. Form of the Al_2SiO_5 phase diagram required by entropy and volume data.

The small entropy changes imply that thermochemical data required for estimating the phase diagram, will need to be of very high precision. Transition temperatures occur near 1000 °K and errors in the free energy of reaction arising from uncertainties in ΔS alone may be as large as $\pm$ 300 cal mol^{-1} and such errors cannot lead to a geologically significant prediction. To use calorimetric data, at least a tenfold improvement in present techniques is required. But if we can directly determine the phase boundaries to within a few degrees, we can obtain very precise thermodynamic data. This has been previously demonstrated from studies of polymorphism of the silica minerals. Direct studies provide free energies accurate to within a few calories while calorimetric data is accurate to within a few hundred calories.

There have been numerous attempts to find the phase boundaries of figure 2. The boundaries and triple-point locations reported by at least a dozen workers cover such a range that the geologist is left with values ranging from upper crust to upper mantle pressures. Inspection of the methods used makes this apparent chaos less surprising. To avoid the frustrations of slow reaction rates, many results are based on synthetic methods. In such methods, the crystallization of reactive materials such as a mixture of amorphous oxides, is studied as a function of P–T conditions. It is optimistically assumed that phases formed will reflect thermodynamic stability and that the synthesis diagram will be similar to the equilibrium phase diagram. But there is no reason why nucleation and growth rates should reflect the trivial free energy changes and such an approach will be reasonable only if we know that rates of conversion of metastable products are rapid. As this is not the case, we can dismiss this group of experiments from serious consideration.

A second group of results have been obtained in opposed anvil pressure devices which provide a simple method of obtaining very high pressures. A thin film of material is compressed between pistons of very high compressive strength and in such devices pressures of 10^5 atm are easily attained. As there is a large shearing stress component, reactions in solids are often more rapid than in hydrostatic pressure devices. The problem with such devices is the great uncertainty in the pressure profile across the piston faces. Pressures are normally obtained from a simple force-area argument but careful calibration studies have shown that departures by as much as 50 % are common. Thus we can dismiss this group of results if accurate data are required.

Recent studies have shown that at temperatures above 700 °C, direct reactions are possible in an aqueous hydrostatic environment. Mixtures of two phases are used to avoid the nucleation barrier and reversible growth between the pairs kyanite–sillimanite and kyanite–andalusite has been achieved (see Fyfe 1967*a*). Conversions with the pair andalusite–sillimanite have not yet been satisfactorily demonstrated.

Rates of solutions of minerals are invariably faster than rates of growth. Thus an additional approach involves measurement of solubilities of each polymorph in an identical solvent. Ease of metastable persistence of forms is a necessary condition to allow such experiments. Weill (1966) has measured the incongruent solubilities of all polymorphs in liquid cryolite (Na_3AlF_6). The solution equilibrium involves the reaction:

$$Al_2SiO_{5\,\mathrm{solid}} + Na_3AlF_6 \rightarrow Al_2O_{3\,\mathrm{solid}} + SiO_{2\,\mathrm{solution}}.$$

If the silica concentrations of solutions in equilibrium with two forms can be measured at the same temperature, the free energy difference between the two polymorphs will be a function of these silica concentrations. To evaluate this free energy difference one must know how to express the chemical potential of silica in the solution. Weill used the Temkin model of fused salts and his derived data are in reasonable agreement with the direct studies.

Although the reliable experiments are few, a geologically significant phase diagram can now be constructed using existing entropy and volume data. As will be discussed below, the same information provides us with a suitable method for examining the molecular chemistry of silica–water solutions at high pressures and temperatures.

An additional problem should be noted. If entropy and free energy differences between polymorphs are very small, then a phase diagram is likely to be unique only if it is certain that the polymorphs have identical chemistry. Because rocks are chemically complex, solid solutions are ubiquitous in minerals. It is well known that the polymorphs under discussion frequently contain minor amounts of oxides such as Fe_2O_3 and Mn_2O_3 in solid solution and it is also known that certain polymorphs favour solutions more than others. Andalusite has one site well suited to an ion such as $Mn^{3+}(d^4)$ which exhibits a large Jahn-Teller distortion. Thus we may have a reaction such

$$\text{andalusite} \rightarrow \text{sillimanite}$$

$$(AlAl_{1-x}Mn_x)SiO_5 \rightarrow 1-\tfrac{1}{2}x\,Al_2SiO_5+\tfrac{1}{2}x\,Mn_2O_3+\tfrac{1}{2}x\,SiO_2.$$

If ideal mixtures are considered, x values as small as 0·03 could make the entropy of transition near zero and larger values could change the sign resulting in totally different phase relations. This aspect of the problem is still being investigated.

(b) *Metamorphism of the Pacific margin*

Recent geological and geophysical research has provided such convincing evidence that the process of continental drift can now be considered fact and not hypothesis. In addition, data on ocean floor spreading and heat flow show that this process involves much more than a thin crustal shell. If continental drift occurs, there is good reason to suspect that continental margins may be sites of interesting phenomena. Certainly they are regions where much earthquake and volcanic activity are concentrated today. Over the past decade it has also become clear that these are regions where unique metamorphic processes occur, and that by the study of these rocks we can learn more about the response of the crust to the forces producing drift. We may even be able to map fossil margins where the same pattern of metamorphism is now contained far within continents. This is another example of how rocks and minerals preserve history of Earth evolution.

In long, narrow belts around much of the Pacific margin, we find Triassic–Jurassic–Cretaceous sediments which have been deeply depressed into the crust. This depression has led to the formation of a unique assemblage of minerals. The sediments are dominated by volcanic debris and some masses are easily identified as products of submarine volcanism. Roughly parallel to the sedimentary troughs are some of the world's most spectacular fault systems such as the San Andreas fault of California. These faults are spectacular for their long period of activity and large lateral displacements. Associated with the sediments and their metamorphic equivalents are ultrabasic rocks and eclogites, rocks of possible mantle origin.

In the buried sediments a number of interesting mineralogical changes are observed. The changes have led to the formation of dense minerals with large average atomic coordination numbers. Some unique reactions affecting the volcanic debris include:

$$\text{plagioclase} + H_2O \rightarrow \text{lawsonite} + \text{albite}$$

$$(NaAlSi_3O_8\text{–}CaAl_2Si_2O_8)_{(\text{solid soln.})} + H_2O \rightarrow CaAl_2Si_2O_7(OH)_2 . H_2O + NaAlSi_3O_8$$

$$(\text{note: } CaAl_2Si_2O_8 + 2H_2O \rightarrow CaAl_2Si_2O_7(OH)_2 . H_2O$$

$$100{\cdot}7\ \text{cm}^3 + 36\ \text{cm}^3 \rightarrow 101{\cdot}3\ \text{cm}^3).$$

This reaction is followed by

$$\text{calcite} \rightarrow \text{aragonite}$$

$$CaCO_3 \rightarrow CaCO_3$$

$$\Delta V = -2{\cdot}7\ \text{cm}^3$$

and finally

$$\text{albite} \rightarrow \text{jadeite} + \text{quartz}$$

$$NaAlSi_3O_8 \rightarrow NaAlSi_2O_6 + SiO_2$$

$$\Delta V = -16{\cdot}5\ \text{cm}^3.$$

The final metamorphosed volcanic sediment may contain some selection of the minerals jadeite–quartz–lawsonite–aragonite–glaucophane–muscovite–chlorite. Rocks with these assemblages are almost cntirely restricted to long, narrow regions and as well as those in the Pacific margin, other series are found through the European Alps–Greece–Turkey–N. India and another much older series in the Urals. The density of the product rock is remarkably high for a water and silica-rich material.

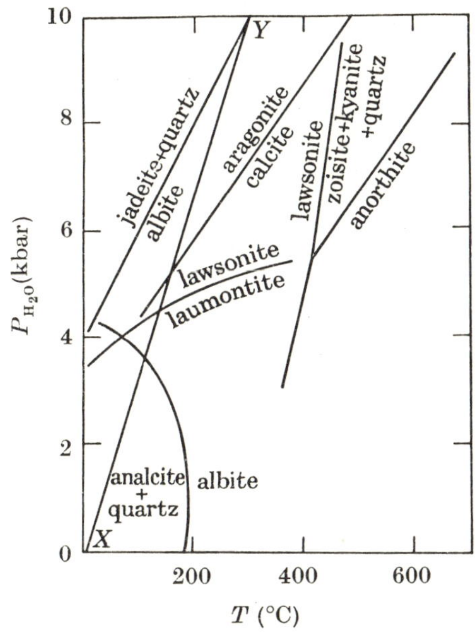

FIGURE 3. Phase transitions involving the minerals albite, jadeite, lawsonite, aragonite.

There has been considerable interest in the conditions necessary to form the mineral assemblages. The rocks are associated with eclogites which could have an origin in the upper mantle. Early experimental work indicated that jadeite could not be synthesized at low pressures. Further, as many of these regions have a complex structural history, it is a matter of extreme difficulty to obtain more than a guess about possible depths of burial from mapping techniques.

Over a number of years data have been accumulated on the conditions of their formation. Some of these data are summarized in figure 3 (Fyfe 1967*b*). These data are recorded in terms of the variables P_{H_2O}–T and there is good reason to consider that total pressure and water pressure in these regions are similar. Data on the stability of zeolites (Campbell & Fyfe, 1965) applied to very thick sedimentary sections such as that described by Coombs (1954) in Southern New Zealand, provide convincing evidence for this equality as do observations on fluid pressures in very deep drill holes. Experiments in progress at Manchester on the diffusion coefficients of water in confined rocks also indicates that diffusion is only rapid when P_{H_2O} approaches P_{total}.

The phase boundaries of figure 3 indicate that regions where lawsonite may be stable are restricted to $T < 400$ °C and $P > 4000$ atm. As these data apply to a system of lawsonite composition, we know that in real rocks with additional species such as Fe_2O_3–FeO–MgO etc., this field will be further restricted.

The calcite-aragonite equilibrium has been measured by several workers and results are in reasonable agreement. Although aragonite is a high-pressure phase, it is commonly formed as a metastable phase in the skeletons of marine organisms. But such aragonite is rapidly transformed to stable calcite under conditions of shallow burial and calcite is the normal polymorph found in all but the most recent sediments. There is little doubt that the aragonite in these metamorphic rocks was derived from calcite. The highest pressure transition found in these rocks involves the transformation of albite into the nearly pure pyroxene jadeite. It thus seems that to account for the observations we must pass along a burial path such as XY in figure 3 where the geothermal gradient is about 10 °C/km or less, and reach conditions where pressures of about 10^4 atm are attained at temperatures of 300 °C or less. Pressures of this order of magnitude suggest burial to depths of about 35 km or average crustal thickness.

Other lines of evidence point to the same cool-deep metamorphic environment. The fractionation of oxygen isotopes between mineral pairs such as quartz and muscovite, or quartz and calcite, is a sensitive function of temperature and rather insensitive to pressure. Results of Epstein & Taylor (1967) indicate temperatures of formation in the range 250–300 °C. The ordering of the Al–Si atoms in the silicate framework of a mineral such as albite ($NaAlSi_3O_8$) is also temperature sensitive. It is not surprising that when albites are examined from rocks formed over a wide range of conditions, those present in these rocks are the most ordered known.

Studies of reaction rates also confirm the general conclusions. As mentioned above, the interconversions of aragonite and calcite are quite rapid. Detailed studies of reaction rates at various P–T conditions have been carried out. Such studies show that if a free fluid (pore fluid) was present at the close of metamorphism and during unloading, high pressure aragonite would never be preserved while returning to the surface unless the rate of uplift was catastrophic. There is no evidence for such catastrophy in the geological record. Even in the absence of an aqueous catalyst, reaction rates in the solid state are quite rapid and to bring aragonite back to the Earth's surface without conversion to calcite, the geothermal gradient must be 10 °C/km or less. Gradients such as 15 °C/km are quite impossible. As shown in figure 3 a gradient of 10 °C/km indicates that aragonite enters the area of calcite stability at temperatures of about 150 °C, and at such temperatures rates are sufficiently slow that survival for millions of years is possible.

In the discussion above, attention has been focused on rather simple reactions. But solid solutions are the rule in mineral systems and while at first sight they may appear to complicate the interpretation of experimental data, if the interpretation is possible, they add new dimensions in detail. The pyroxenes of these rocks provide an excellent example. While almost pure jadeite is sometimes found, many pyroxenes are solid solutions of three pyroxenes, $NaAlSi_2O_6$, $CaMgSi_2O_6$ and $NaFeSi_2O_6$ (jadeite, diopside and acmite). If such a pyroxene solid solution is in equilibrium with albite and quartz as frequently happens, then for a given temperature, the jadeite content of the solid solution is a function of pressure.

Consider a simple system, $NaAlSi_3O_8$–$NaFeSi_2O_6$ (albite–acmite). At 300 °C, acmite is stable at one atmosphere while albite forms jadeite at 10^4 atm. If the solid solutions were ideal then:

$$\mu_{\mathrm{jd(s.s.)}} = \mu^0_{\mathrm{jd}} + RT \ln x_{\mathrm{jd(s.s.)}}$$

where $\mu_{\mathrm{jd(s.s.)}}$ is the chemical potential of $NaAlSi_2O_6$ in the solid solution and $x_{\mathrm{jd(s.s.)}}$ is the mole fraction in the solid solution. If albite and quartz are present then

$$\mu_{\mathrm{jd(s.s.)}} + \mu_{\mathrm{qtz}} = \mu_{\mathrm{albite}}$$

and hence

$$\begin{aligned} RT \ln x_{\mathrm{jd(s.s.)}} &= \mu_{\mathrm{albite}} - \mu_{\mathrm{qtz}} - \mu^0_{\mathrm{jd}} \\ &= -\Delta G \text{ of the reaction, albite} = \text{jadeite} + \text{quartz}. \end{aligned}$$

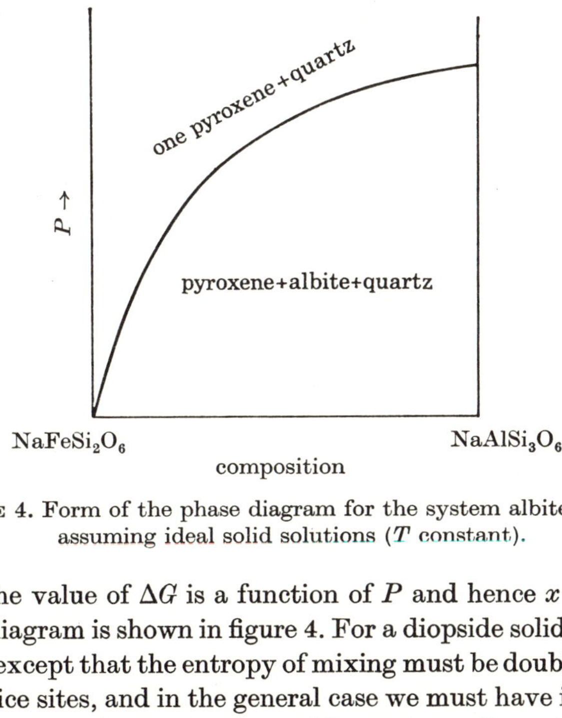

FIGURE 4. Form of the phase diagram for the system albite–acmite, assuming ideal solid solutions (T constant).

For a given T, the value of ΔG is a function of P and hence x can be calculated. The ideal phase diagram is shown in figure 4. For a diopside solid solution the treatment is identical except that the entropy of mixing must be doubled because mixing involves two lattice sites, and in the general case we must have information on site populations and site preference in the solid solutions. Experimental calibration of such systems is still incomplete but a close approach to ideality is indicated by petrological observations (Essene & Fyfe 1967).

The value of such solid-solution equilibria is apparent. From observations on isotope equilibria, it seems possible to obtain a rather close estimation of temperature of crystallization. From data on pure phases or polymorphic transitions, we can assign only limits to P unless the rock is very close to a region of a phase transition.

But with a solid-solution equilibrium of the pyroxene type, we can assign a definite pressure. Eventually this should allow sensitive mapping of P–T gradients over large crustal regions. Further, if crystals are chemically zoned, a single crystal preserves a record of a large series of events over a long period of time. Slow rates of solid diffusion at low temperatures thus work in our favour.

Many problems remain. Several new minerals have been recently found and some are high-pressure phases the significance of which is not yet known. We do know, however, that these metamorphosed sediments require greater pressures for their formation than eclogites at low temperatures. They thus provide important information about trends in mantle materials. The problems remaining move into the areas of geophysics and the dynamics of the crust on the leading margin of a drifting continent. It also appears, that our guess that the similar but more ancient rocks in the Urals represent an old drifting margin, may be substantiated by paleomagnetic studies of older rocks.

(c) *Element fractionation during magmatic crystallization*

One of the ultimate problems of geochemistry, and one which involves the basis of economic geology, is to understand how elements are distributed in the Earth and how they are redistributed during various processes. Almost all crustal rocks begin their history as igneous rocks then pass through a sedimentary or metamorphic stage. We are concerned with all stages but obviously the primary igneous stage is of general importance.

Many of the ideas on distribution which have influenced geochemical thought for several decades were based on the classical studies of V. M. Goldschmidt. These ideas were summarized in his Liversidge lecture to the Chemical Society in 1937. The views which Goldschmidt then held, and which have been only slightly modified by later workers, were based on the impressive correlation he observed between the size of ions (Goldschmidt first determined many of these values) and geochemical behaviour. In particular he noted:

(1) If a crystal is forming from a melt, ions may enter solid solutions only if they are of similar size.

(2) If two ions have similar size and identical charge, the smaller ion is concentrated in the first fraction to crystallize.

(3) If two ions have similar size and differ in charge, the more highly charged species is concentrated in early fractions to crystallize. The rules suggest a correlation between geochemical behaviour and lattice energy. Compounds with large lattice energies tend to have high melting points (but there are numerous exceptions).

Numerous exceptions to the rules have been found and later ideas introducing partial covalent behaviour have failed to explain most of these. A simple case will be found with any binary system where a solid solution series exhibits a minimum or maximum melting point.

One much discussed example indicates why such exceptions should be anticipated. Basaltic rocks contain minor amounts of nickel. When these rocks crystallize, most of this nickel is concentrated in solid solution in one of the early formed minerals, olivine, $(MgFe)_2SiO_4$. The systems Mg_2SiO_4–Fe_2SiO_4 and Mg_2SiO_4–Ni_2SiO_4, have been studied experimentally. Both show similar melting behaviour of the form shown in figure 5. On the basis of such diagrams we would expect both iron and nickel enrichment in late fractions. In basaltic rocks this is true for iron but the reverse trend is seen for nickel. The radii of Ni^{2+} and Mg^{2+} are almost identical and no rules will explain the different behaviour in the two systems involving an orthosilicate or a basalt melt. It is obvious that the reasons for the different patterns must be related to the different melt chemistry or differences in the liquid structures.

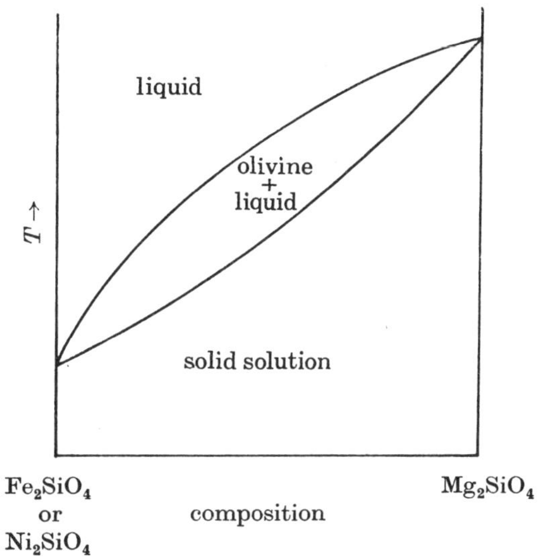

FIGURE 5. Form of the melting diagram for solid solutions in the systems Mg_2SiO_4–Fe_2SiO_4 or Mg_2SiO_4–Ni_2SiO_4.

Williams (1959) suggested that the geochemical behaviour of transition metals appeared to reflect their octahedral crystal-field parameters. He observed that ions with large crystal-field splitting values (Δ) in solids, were those which concentrated in solids which crystallized at any early stage. Burns & Fyfe (1967) studied the spectra of a number of silicate minerals and derived liquids (glasses) and concluded that it was not the Δ values which were important, for these were very similar in both liquid and solid states. This had been previously noted for many aqueous solutions of transition metal compounds. But it was noted that with an ion such as nickel, as alkali silicates were added to a melt, the liquid structure changed and nickel ions were found in tetrahedral as well as octahedral sites. Associated with the new geometry is a much smaller crystal-field stabilization energy. McClure (1957) demonstrated that a very strong site preference energy results from these effects, and used this term to explain the peculiar patterns of site preference observed in transition metal compounds in the spinel structure, patterns which make no sense in relation to ion sizes.

It now seems that we can reasonably explain the change in behaviour. In the simple system, Mg_2SiO_4–Ni_2SiO_4, Δ values and site geometry, are similar in both liquid and solid states. Thus fractionation behaviour simply reflects the difference in melting points of the two pure substances. The nickel compound actually has a larger lattice energy than the magnesium compound, but melts at a lower temperature because the heat of fusion is lower. When we pass to more silica and alkali-rich liquids such as a basalt liquid, tetrahedral and octahedral liquid sites are occupied by the ions and site preference energy in effect raises the heat of fusion of the nickel silicate relative to the magnesium silicate.

If we compare the thermodynamics of dissolving Mg_2SiO_4 and Ni_2SiO_4 in a liquid with a definite ratio of octahedral and tetrahedral sites, the entropy of solution of the Mg^{2+} ion will be more positive than that of Ni^{2+}, for its preference for octahedral holes will allow less spreading over all available sites. The heat of solution terms will also reflect this site preference and if Ni^{2+} ions occupy tetrahedral sites, this will lead to a more positive value. Both terms combined, will lead to a more positive free energy of solution for the nickel compound and a lower solubility. Tables of site preference energies are available and the correlation with the behaviour of various transition metals is convincing.

The example considered above illustrates the necessity of accumulating data on the structures and site distributions in silicate liquids just as we have such data for solids. The problem is difficult and techniques are limited but various spectroscopic methods appear useful, in particular Mössbauer spectroscopy. Given such data, the thermodynamics of melting can be approached fundamentally. Progress is being made, and calculations such as those of Blander (1968) using computers are impressive. As the number of silicate systems of interest in petrology is near infinite, the number of necessary experiments must be reduced by sound theoretical approaches.

(d) *Transport properties and high-temperature solutions*

Almost all mineralogical reactions in the Earth's crust involve chemical transport. In common reactions, movement may only involve distances related to grain size, a centimetre or less. In other cases tens of metres of rock may be converted to a new chemistry. When rare or highly dispersed elements are concentrated to form useful ore bodies, migration may even be over distances of kilometres. Not many years ago, it was considered by many, that diffusion in solids could account for some large-scale phenomena but when the relevant diffusion coefficients are examined, it is obvious, that except in rare cases, solid diffusion is not the mechanism given crustal temperatures and even geologic times. But there is good evidence that the dominant transporting media are aqueous solutions. While aqueous transport has been recognized for a very long time (for example in Lyell's, *Principles of Geology*, 1872) data on the nature of these solutions has accumulated slowly. That minerals can be very soluble in water is well demonstrated by the work of Kennedy, Wasserburg, Heard & Newton (1962). Over much of the crustal range, saturated quartz solutions are about 0·1 M. At temperatures of 1080 °C and pressures of 10^4 bars, gas phase solubilities of 60 wt. % have been determined. It is frequently supposed that silica in solution is present as the molecule $Si(OH)_4$ but work of Weill & Fyfe (1964) indicates that there is little evidence for this conclusion.

At present we are trying to resolve this problem using data concerning the stability of the Al_2SiO_5 polymorphs. These react with water according to the reaction:

$$Al_2SiO_{5\,(solid)} + H_2O \rightarrow Al_2O_{3\,(solid)} + SiO_{2\,(soln.)}.$$

If such equilibria are measured for two polymorphic modifications of known stability we have:

$$\mu_{Al_2SiO_5\,(1)} = \mu_{Al_2O_3} + \mu^1_{SiO_2\,(soln.)}$$

$$\mu_{Al_2SiO_5\,(2)} = \mu_{Al_2O_3} + \mu^2_{SiO_2\,(soln.)}$$

$$\Delta G_{(transition)} = \mu^1_{SiO_2\,(soln.)} - \mu^2_{SiO_2\,(soln.)}.$$

If ΔG is known from a reliable phase diagram, we can set limits on polymerization of the silica solutions by finding what concentration function will describe the free energy difference. The problem is simplified by knowledge that the solutions are non-ionic. Our present data has not yet reached the stage to produce unique answers but this will soon be the case. Moreover, this system is one which can be used over a very wide range of crustal conditions because of the slow rate of interconversion of these polymorphs.

High-temperature aqueous inorganic solutions are very different from their low-temperature counterparts which so strongly influence our normal experience. Inspection of entropies of solution of salts clearly indicates the changes to be expected. Reactions such as

$$MX + H_2O \rightarrow M^+_{(soln.)} + Y^-_{(soln.)}$$

tend to be associated with negative entropy changes. If separate ions are formed, water molecules must solvate them and lose freedom. As the aqueous solvent decreases in density with increased temperature, the entropy changes tend to become even more unfavourable in terms of ionization. On the other hand, reactions such as

$$M^{z+} + Y^- \rightarrow MY^{(z-1)+}$$

become increasingly favoured as T increases. The final result should be the formation of molecular solutions at high temperatures. Measurements of the conductivities of solutions of simple species such as NaCl, KCl, HCl, NaOH confirm this trend. Above 400 °C at most crustal pressures, molecules dominate the solutions. Further, as temperatures increase, mixtures tend to become more ideal and even concentrated solutions in the system $NaOH–H_2O$ begin to behave like simple mixtures of molecules.

While these trends lead to a rather simple treatment of such solutions, another factor leads to complexity. If inorganic solids mix with gases to a considerable extent, normally strong molecular interaction must compensate for the strong solid binding forces. If solute–solvent interaction tends to decrease at increased temperature, this must be replaced by solute–solute interaction. Thus Brewer (1956) showed that the vapour phase in equilibrium with solids (the normal mineralogical situation) becomes more polymeric as temperature increases. This must be the case for the 50:50, $SiO_2–H_2O$ gas mixture referred to above. In summary, solutions at depth in the Earth's crust must be largely non-ionic polymer solutions and the geochemist must be as much an organic as inorganic chemist. Data on such systems are still meagre.

Recently we have seen a very old transport problem at least partially resolved. This is the problem of the mechanism of transport and deposition of insoluble sulphides. At low temperatures the aqueous solubility of sulphides of metals such as Fe, Zn, Pb and Cu, are very slight. To transport hundreds of tons of these on the basis of solubility product data, would require vast volumes of water for which there is little geological evidence. Studies of oxygen isotope ratios indicate quite modest volumes of water. Hemley *et al.* (1967) appear to have provided the answer. A reaction such as:

$$\underset{\text{muscovite}}{KAl_2(AlSi_3O_{10})(OH)_2} + \underset{+\,\text{quartz}\,+\,\text{KCl}}{6SiO_2 + 2KCl_{(aq.)}} \rightarrow \underset{\text{orthoclase}}{3KAlSi_3O_8} + 2HCl + H_2O$$

produces a significant partial pressure of HCl (molecules) at moderate temperatures. A sulphide such as PbS (galena) will react according to the relation:

$$PbS + 2HCl_{(aq.)} \rightarrow PbCl_{2(aq.)} + H_2S_{(aq.)}.$$

The chemistry is molecular. Solubilities as high as 1 mol. $l.^{-1}$ have been measured at 500 °C and $P = 1$ kbar. At lower temperatures the equilibrium is reversed. There is nothing exotic about the reagents as small amounts of alkali halides are always found in deep water samples. Further, the reactions indicate a correlation between silicate and sulphide reactions, a correlation long noted by economic geologists. Many fascinating problems remain in describing the mechanisms of transport and deposition of rare elements. But an understanding of solution chemistry and crystal nucleation, is necessary for this description.

7. Concluding remarks

We are approaching the time when we will be able to examine a rock in the laboratory and place definite limits on the variables necessary for its formation. In a general way, the thermodynamics of mixing or the thermodynamics of solutions provides the key to geochemical processes whether it be the origin of granites or the origin of a gold deposit. Once such data can be provided, geological, geochemical and geophysical approaches will reinforce each other in our understanding of the evolution of the Earth.

References

Birch, F. 1965 *Bull. Geol. Soc. Am.* **76**, 133.

Blander, M. 1968 *Chem. Geol.* **3**, 33.

Brewer, L. 1951 *J. Geol.* **59**, 490.

Brewer, L. 1956 *J. Chem. Educ.* **35**, 153.

Burns, R. G. & Fyfe, W. S. 1967 *Researches in geochemistry* **2** (Abelson, P. H., ed.), p. 259. New York: Wiley.

Campbell, A. S. & Fyfe, W. S. 1965 *Am. J. Sci.* **263**, 807.

Clark, S. P. & Ringwood, E. A. 1964 *Rev. Geophys.* **2**, 35.

Coombs, D. S. 1954 *Trans. R. Soc. N.Z.* **82**, 65.

Epstein, S. & Taylor, H. P. 1967 *Researches in geochemistry* **2** (Abelson, P. H., ed.), p. 29. New York: Wiley.

Essene, E. J. & Fyfe, W. S. 1967 *Contr. Miner. Petrol.* **15**, 1.

Fyfe, W. S. 1967*a* *Chem. Geol.* **2**, 67.
Fyfe, W. S. 1967*b* *Trans. Leicester lit. phil. Soc.* **61**, 36.
Goldschmidt, V. M. 1937 *J. chem. Soc.* p. 655.
Hemley, J. J., Meyer, C., Hodgson, C. J. & Thatcher, A. B. 1967 *Science, N.Y.* **158**, 1580.
Kennedy, G. C., Wasserburg, G. J., Heard, H. C. & Newton, R. C. 1962 *Am. J. Sci.* **260**, 501.
McClure, D. S. 1957 *J. Phys. Chem. Solids* **3**, 311.
Weill, D. F. 1966 *Geochim. cosmochim. Acta* **30**, 223.
Weill, D. F. & Fyfe, W. S. 1964 *Geochim. cosmochim. Acta* **28**, 1243.
Williams, R. J. P. 1959 *Nature, Lond.* **184**, 44.

Review Lecture

Two atoms—the basis of the steel industry

By Sir Charles Goodeve, F.R.S.

BISRA, The Inter-Group Laboratories of the British Steel Corporation, 24 Buckingham Gate, London, S.W. 1

(*Delivered* 25 *April* 1968—*Received* 28 *June* 1968)

Proceedings of the Royal Society, A, *volume* 308, *pp.* 437–446, 1969

For 2000 years or more steel and its parent material, iron, have dominated the civilized world and it appears likely that they will continue to do so for a long time ahead.

Steel has achieved this position because it is unique in possessing in combination the three characteristics of cheapness, strength and versatility. I propose to review the reasons behind these characteristics and to describe some of the ways in which modern knowledge in numerous branches of science has helped the steady advance in the technology of steelmaking. It will be shown that ferrous materials could not have achieved their position without a partnership with carbon, a material which is unique in other ways. Naturally numerous simplifications will be necessary.

The cheapness of steel results first of all from the prevalence of iron ore which is to be found in tremendous quantities all over the surface of the Earth and probably throughout the whole universe. Indeed, it is likely that the centre of the Earth consists, as do most meteors, primarily of iron, perhaps with some nickel. This plentiful supply fulfils one of the requirements for cheapness and also means that the price of iron does not fluctuate as does that of other metals.

This prevalence of the element iron is probably associated with the unique characteristic of its nucleus. In figure 1 the important elements are arranged in the order of their atomic masses (Wehr & Richards 1960), and placed vertically according to the average mass of their nuclear components, i.e. protons and neutrons. The elements on the extreme right of the diagram can split up into smaller elements, lose mass and release energy as in a nuclear power station. Those on the extreme left can, at least from an energy point of view and sometimes in practice, fuse together, as hydrogen does in the sun, to form heavier elements releasing very large quantities of energy. The iron nucleus is at the lowest point of this energy-mass diagram and this, coupled with other structural reasons, makes it likely that iron is the 'ashes' of nuclear reactions at some stage in the evolution of the universe.

Strength needs to be looked upon from three different but related points of view, rigidity, high yield point and resistance to fracture. Steel is not challenged seriously by any other material in these respects and indeed it retains its strength up to very high temperatures.

By versatility we mean that steel can be made very hard, as in ball bearings, or soft if so required for shaping, or a balance between the two. Steel piano wire can retain its strength and hence remain in tune indefinitely and yet can be wound around a small tuning pin without breaking. Steel can be made magnetic or non-magnetic and stainless if required.

These characteristics of steel result from the arrangement of the iron atom's 26 electrons and in particular the outer 8. The electrons of all atoms move about the nucleus according to a definite series of orbitals or patterns. The less energetic electrons move in patterns close to the nucleus and others with more energy in patterns further out. Their movement is similar to that in a Scottish Reel.

The letters *s* (spherical), *p* (polar) and *d* (directional) are used to describe the shapes of the patterns and the numbers 1, 2 and 3, etc., the energy levels. Approximate representations of these three types of patterns are shown in figure 2, plate 4 (Hume-Rothery 1960).

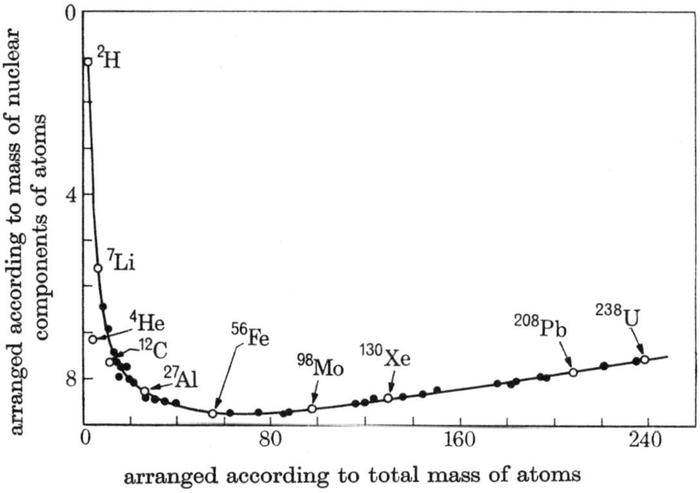

FIGURE 1. Potential energy per nucleon relative to the state of complete separation.

Let me first draw your attention to the striking difference between the patterns of movement for the outer electrons of iron and copper. I should remind you that interchange of electrons between neighbouring atoms is the basis of most chemical forces.

At least seven of the outer eight electrons of iron spend most of their time in a 3*d* pattern as shown and the remaining one of the eight in an *s* pattern. There is room for ten 3*d* electrons and therefore up to three electrons can come in from neighbouring atoms. In these two facts, the *d* pattern and the empty spaces therein, lies steel's secret. Its versatility, and especially the magnetism of one of its crystal forms, are due to this electron structure which turns out to be unusual and very fortunate. The high rigidity of ferrous materials follows from the directional character of the *d* patterns and also from the close packing of the atoms which these patterns bring about.

Copper has only one outer electron moving in a spherical or non-directional pattern as do those of most other metals and consequently the metal is easy to bend and to shape. However, copper's *s* electron and the *s* space leads to high mobility and therefore high electrical conductivity.

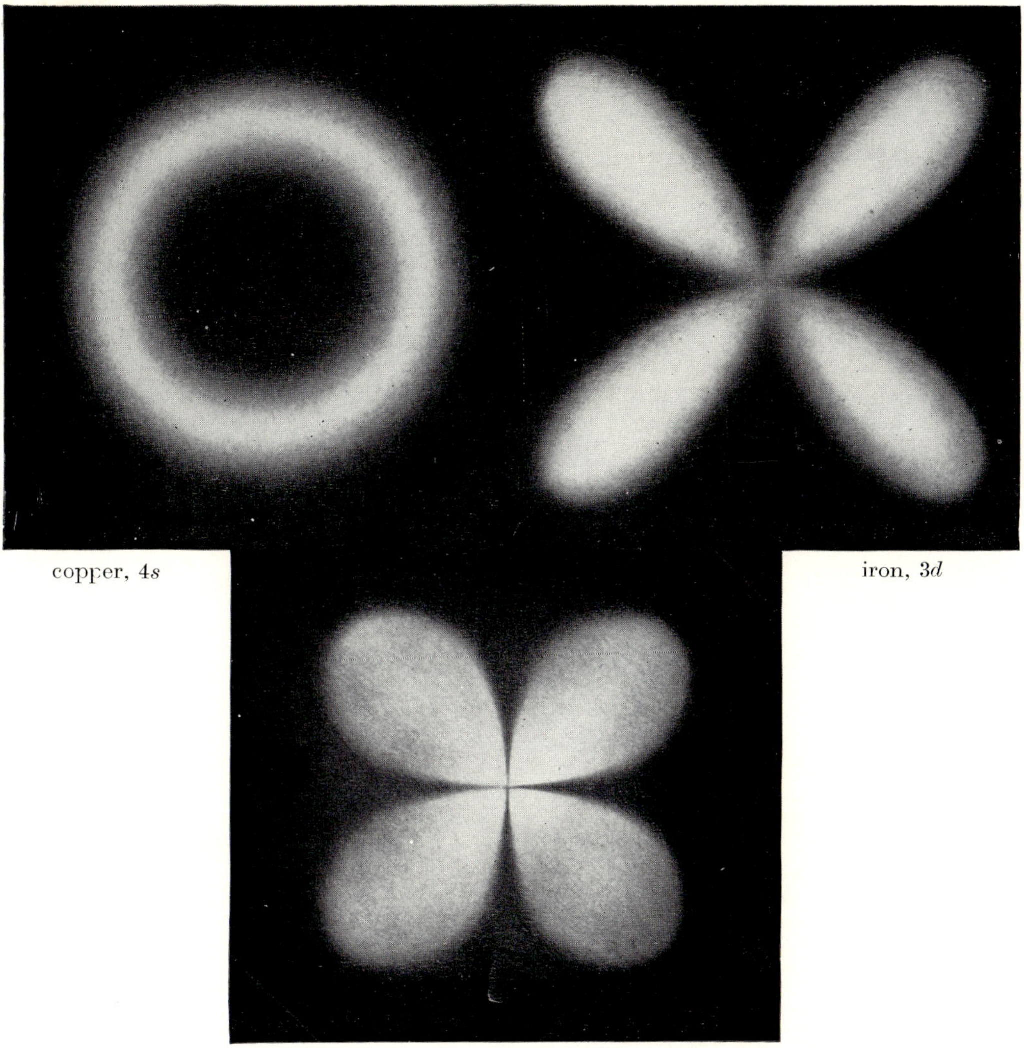

copper, 4s

iron, 3d

carbon, 2p

FIGURE 2. Patterns of movement of outer electrons of copper, iron and carbon.

PLATE 4

(*Facing p.* 128)

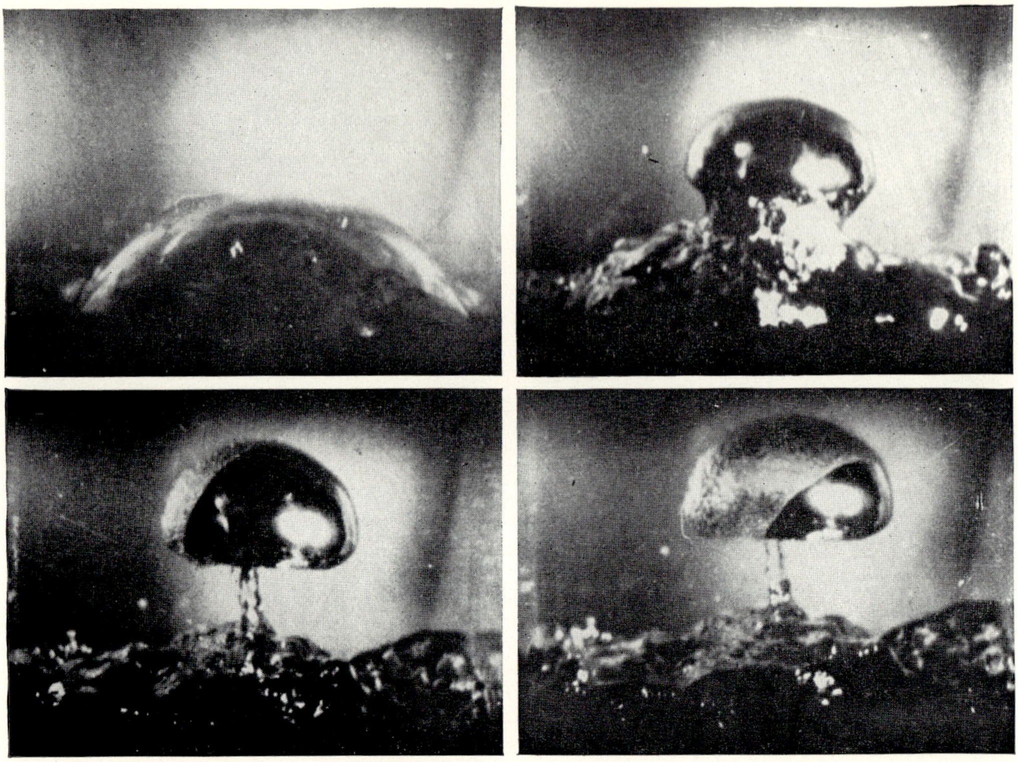

1 2 3 4

FIGURE 7. Mass transfer by bubbles.

FIGURE 6. The boil in an open-hearth furnace (reproduced by courtesy of Union Carbide Corporation).

FIGURE 10. The cementation furnace in Sheffield.

PLATE 5

Some other metals besides iron have outer *d* electrons and some spaces, but in practically all the cases the pattern is either nearly full or the electrons are mostly paired off in a manner which renders them less able to interchange from atom to atom and therefore they are less able to contribute to the chemical or physical properties of the element. It is interesting to note that only two other, but rare, elements, ruthenium and osmium, have a *d* structure similar to that of iron (at higher energy numbers) and both have a comparable strength to iron.

For a long time, however, it was a mystery as to why steel could be made soft with a yield strength of 5 Kg/mm^2 compared with a theoretical strength one hundred times higher, but the work of Sir Geoffrey Taylor and Dr A. H. Cottrell and others showed how this was due to the movement of dislocations in the metal lattices. The bending of a steel bar can only take place through the movement of millions of dislocations running through the bar and allowing the layers of atoms to slip over one another.

However, carbon now begins to play its first major role. The atoms of this element, being very small, can, at intermediate temperatures, worm their way between the layers of the iron lattice until they come to dislocations which they lock, thus preventing slippage. For this reason we can make steel fifty times stronger than soft steel or up to half the theoretical flow strength.

When we consider the processes used in the production of iron and steel, we come to the real impact of the carbon atom. Its four $2p$ polar electrons give carbon, in all its forms, diamond, graphite and coke, a high strength and an ability to maintain its strength up to extremely high temperatures. These two factors are vital to the primary process, the production of metallic iron from iron ore in the blast furnace. The cross-sectional diagram in figure 3 shows that a column of coke resting on the bottom of the hearth supports about 1000 tons of ore and coke at temperatures well over 1500 °C and keeps the whole mass open so that gas can percolate. Furthermore, carbon provides a very satisfactory refractory for the walls of the hearth where it not only withstands the high temperature, but is attacked neither by the slag produced in the furnace nor by the iron itself, because the latter is saturated with carbon from the coke.

Of course, carbon and its compounds, being cheap, play a fourth role in that, used as a fuel, they supply the energy required for metallurgical processes. Carbon is cheap and prevalent for reasons similar to those applying to iron: low specific mass and a favourable structure of its nucleus (see figure 1).

Most of the common elements exist in nature in combination with oxygen, and their commercial production depends on coke and carbon in some form or other for their release. This fact earns for carbon the well deserved title of 'the key to metallurgy'.

The reason why carbon is in this remarkable position can be found from chemical thermodynamics. The universal second law of thermodynamics states that before any system can pass by itself from one state to another connected state the second must be at a lower potential than the first.

Thus chemists can define and measure an oxygen potential based on the partial pressure of oxygen; if oxygen is to pass from a state of combination with any metal

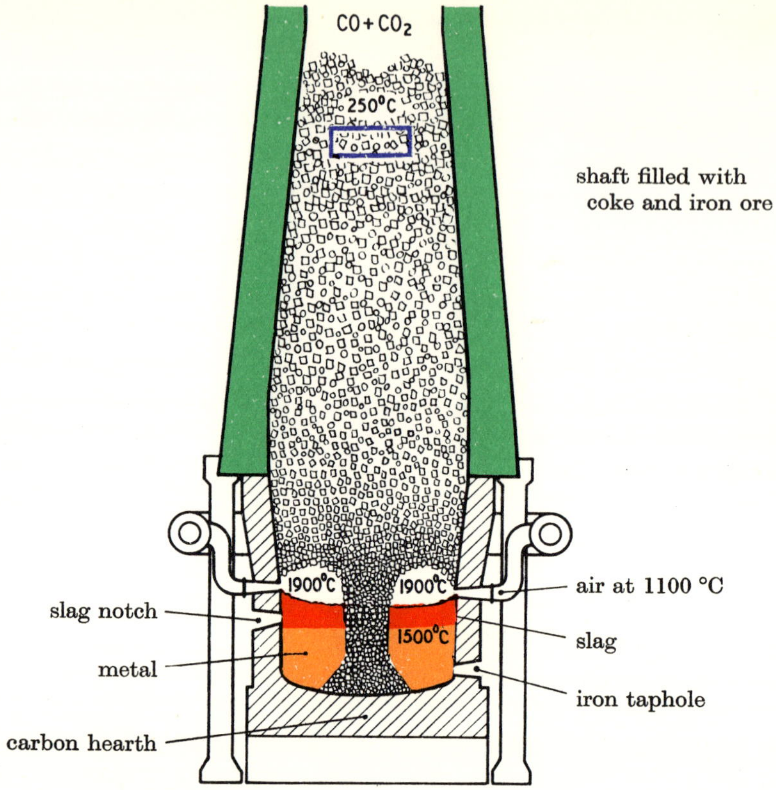

FIGURE 3. Carbon in iron making.

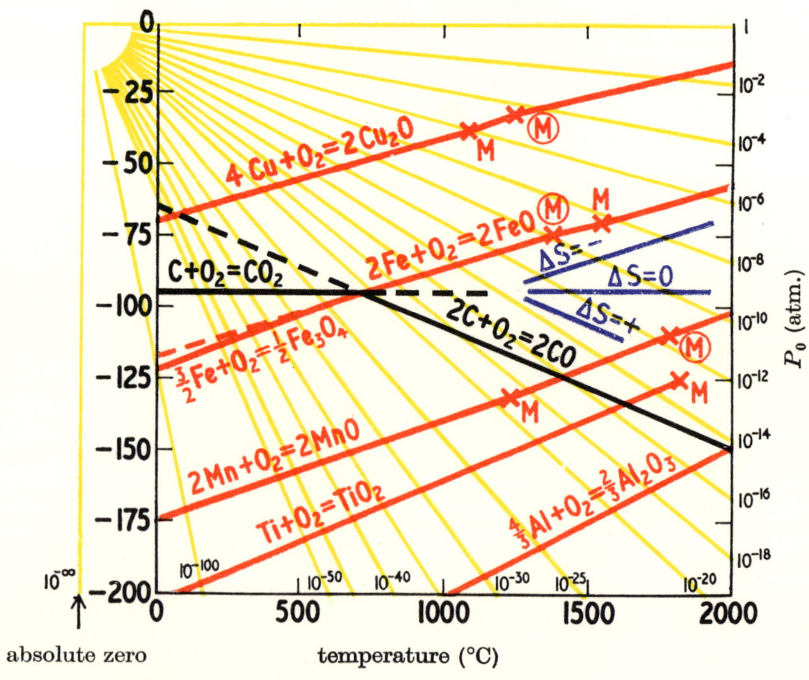

FIGURE 4. Oxygen potentials in metallurgical processes.

such as iron to join with carbon, the oxygen potential in the second state must be lower than in the first.

In figure 4 is shown an example of an oxygen potential-temperature diagram of a type which now forms the basis of extraction metallurgy (Richardson & Jeffes 1948). To avoid numerous complications the oxygen potential is not defined as the partial pressure, but rather as $RT \times$ logarithm (partial pressure). The actual pressure is shown by the yellow lines with the corresponding scale on the right and along the bottom.

The first thing that will be noticed from this diagram is that the potential of oxygen in equilibrium with iron and iron oxide is less than that with copper and its oxide and the potential in the presence of aluminium and aluminium oxide is still lower. Therefore oxygen will readily leave copper oxide to combine with iron or in turn it will leave iron oxide to combine with aluminium. This latter chemical reaction is, of course, the basis of the Thermite process by which steel is produced by the reaction of aluminium powder on powdered iron ore.

A striking point shown in this diagram is that, whereas the oxygen potential rises with temperature in the case of all the metal oxides, it falls in the case of carbon monoxide. The important consequence of this is that, if a high enough temperature is reached, the carbon–carbon monoxide line will be lower than the metal–metal oxide line. In simple language, as the temperature rises, oxygen's preference turns from the metal to carbon.

It may be seen that the slopes, i.e. the rise in the oxygen potential with temperature, are effectively the same for every metal and its oxide. It can also be seen that the slope of the carbon–carbon monoxide line is almost exactly the same, except that the sign is the opposite, i.e. it is negative.

How do these slopes come about? First of all, each of the metal–metal oxide reactions as written involves the disappearance of one molecule of gas, i.e. oxygen. It is well known from chemical thermodynamics that the oxygen potential is equal to the heat absorbed in a reaction, ΔH, minus the change of 'entropy', ΔS, multiplied by the temperature:

$$\text{oxygen potential} = RT \log p_{O_2} = \Delta H - \Delta S\, T.$$

Entropy is a measure of disorder and, therefore, the disappearance of one molecule of gaseous oxygen to enter a much more ordered state, such as in an oxide of a metal, means that the entropy has decreased or its change has a negative sign. Therefore, the higher the temperature the greater is the tendency for the reaction to be reversed, i.e. the partial pressure of oxygen (its potential) increases as the temperature rises.

Furthermore, the entropy change corresponding to the disappearance of one molecule of diatomic gas of any kind is almost always exactly the same, and can be precisely calculated. This entropy change provides the main contribution to the slope of the chemical potential/temperature line and its value is shown by the upper blue line on the right of the diagram. Practically all the metal–metal oxides have a positive slope corresponding closely to this blue line. There are deviations at the melting points, as indicated by the letter M. These deviations indeed throw light on the nature of the metal and its oxide.

Oxidation of carbon to carbon monoxide, results in an increase of one in the total number of gaseous molecules, i.e. one oxygen molecule disappears and two carbon monoxide molecules appear. Therefore, the entropy increases in this case by exactly the same amount as it decreased in the metal oxide cases; the entropy change therefore is positive and the slope of the oxygen potential line negative. Of interest but of less importance is that, for the oxidation of carbon to its dioxide, the slope is zero as one would expect.

Carbon is therefore the key to metallurgy, because it produces on oxidation two molecules of a gaseous monoxide; this product has a higher entropy than the reactants and, consequently, high temperature favours the reaction.

It is sometimes difficult to get the feel of entropy, but in process metallurgy it comes strongly to life and dominates the entire reaction. Normally chemical reactions proceed in such a direction so as to give out heat, but in metallurgical processes involving carbon, heat is absorbed, the driving force of the reaction being the attraction of higher entropy.

The position of carbon in metallurgy arises from the unique character of carbon monoxide. Its chemical bond formed by only two of carbon's four outer electrons is the strongest known—256 kcal/mole—exceeding even that of nitrogen (225 kcal/mole). It has the same total number of electrons as has the nitrogen molecule, and the patterns of its valency electrons are probably very similar. It differs from nitrogen in the asymmetry of its two nuclei, but this asymmetry is not shown externally by the molecule; indeed, carbon monoxide is non-polar and hence is a gas even at very low temperatures.

The molecules of metal oxides, on the other hand, are highly polar and readily perturb one another to form strong chemical bonds and thus build up crystal structures. Metal oxides are rarely gaseous except at very high temperatures.

I have digressed somewhat from the blast furnace in order to bring out part of the science behind this primary piece of ironmaking equipment. On the basis of modern knowledge such furnaces can produce over 5000 tons/day at a cost of only a few pounds per ton above that of the materials it uses. Speed and scale, of course, do not follow from thermodynamics, but from the kinetics of the reactions and from mass and heat transfer. The first, the kinetics, is only of importance in a few cases, because temperatures are high, but the transfer rates are dominant in controlling the speed and hence the economics.

Until recently the blast furnace was a typical black box and there were difficulties in determining exactly what went on inside. Among the techniques used to advance our knowledge is one known as SCICE (Klemantaski & Wild 1962; Johnson 1968). A horizontal section of the furnace charge, such as that outlined in blue in figure 3, will move from the top down to near the hearth in about 5 h as a result of the reactions with the hot gases coming up and through the section. In SCICE such a section can be reproduced exactly in the laboratory and, while held stationary, can be taken through the 5 h cycle. This is a rather tricky operation, but it makes possible a precise study of what goes on in this black box.

Liquid iron from the blast furnace is still some distance from steel in its many forms and compositions in that it contains, not only 15 atomic per cent of carbon, but also other deleterious elements such as sulphur, phosphorous, silicon and manganese which are present in most ores.

The next stage, that of steelmaking itself, has the object of bringing the carbon down to its desired level, that is somewhere between 0·1 and 2 atomic per cent. This is done by controlled oxidation. At the same time the other elements are oxidized and transferred to a molten slag. Here one could wander into a very complex branch of physical chemistry based on the thermodynamics of slags. In steelmaking new parameters, such as the basicity, the oxygen potential and the temperature, and gradients of the latter two, can now be varied and must be controlled.

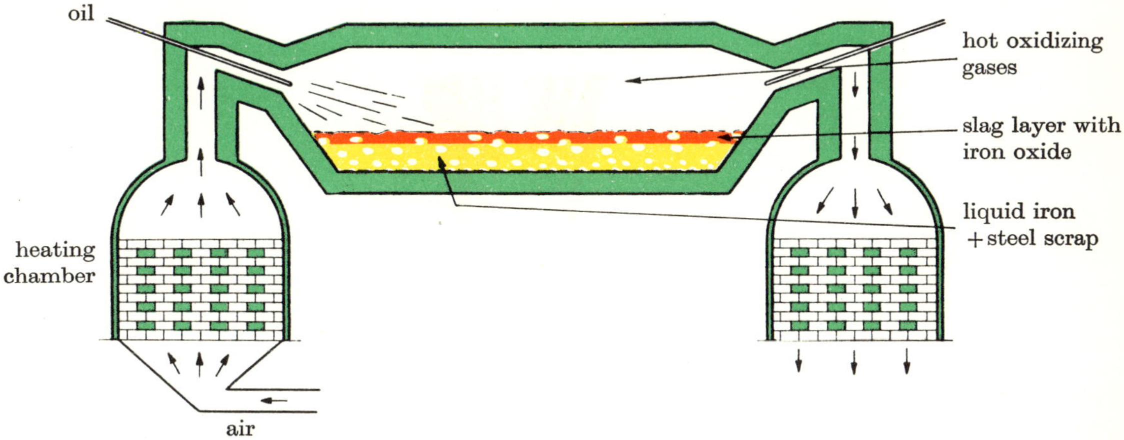

FIGURE 5. The open hearth furnace process. The direction of the flow is reversed about every 20 min.

I do not propose to explore these areas, but to concentrate on two major economic factors, speed and control. The main process used over this last hundred years to convert liquid iron to steel is that of the open hearth furnace, figure 5, where the oxidation is carried out rather gently by hot carbon dioxide, air and iron ore. Varying quantities of steel scrap are also put in with the liquid iron. Control here is by sampling and analysis and as the process takes 5 to 12 h, there is no special difficulty.

One problem, however, is how to get rapid transfer of phosphorus, sulphur, silicon, manganese, etc. into the slag. At the temperature of 1600 °C it is impossible to stir the bath directly, but carbon again comes to the rescue. The formation of carbon monoxide gas is difficult to nucleate in a homogeneous molten metal, but it is readily nucleated by the refractory bottom of the furnace. The small bubbles on rising grow very rapidly and stir the metal and then the slag, as seen in figure 6, plate 5.

Equally important, however, is the fact that the bubbles in passing through the metal to the slag carry metal with them, as shown in figure 7, plate 5, by a sequence of stills from a high-speed film. It is to be noted that a bubble of gas, having come to the surface of mercury, carries with it a complete film of the metal into the viscous aqueous liquid above, which film on breaking finishes up as fine droplets. This illustrates a sixth role that carbon plays in this partnership with iron; without these rising bubbles there would be effectively very little mass transfer and very little chemical reaction. Indeed this provides a hazard in steelmaking because a bath can sometimes lie quiet while it develops a strong oxygen potential gradient. When the bubbles start the whole bath boils up in a most spectacular and rather dangerous way.

The open hearth furnace is, however, very slow and now relatively costly. The Bessemer process, also a century old is extremely fast and indeed is the fastest commercial process existing today. In this process air is blown through the bottom of a vessel containing liquid iron. The enormous surface area of gas to liquid overcomes all the nucleating problems and also provides ample mixing of the slag and metal. The process is still in use, but is in the decline, largely because of the fact that it cannot use concentrated oxygen; nitrogen from the air plays an essential part in the Bessemer process, but it dissolves in the liquid steel rendering it of limited use.

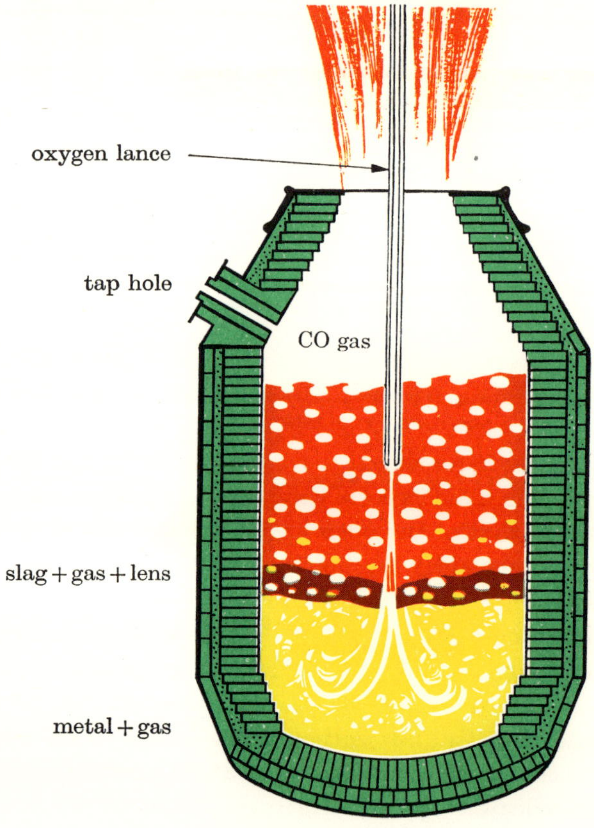

FIGURE 8. The L.D. process.

The introduction of cheap tonnage oxygen twenty or thirty years ago has led to the development in Austria of the L.D. process, which is similar to the Bessemer in some ways, but gets over its handicaps. In this process a high velocity jet of oxygen penetrates into the surface of the slag and metal, as can be seen in figure 8. The degree of turbulence produced is remarkably high, partly as a result of the energy of the jet, but more because of the rapid generation of carbon monoxide throughout the whole bulk. Indeed, a sample of the slag taken at any time during the height of the reaction shows large quantities of metal temporarily suspended in the slag in the form of fine droplets. Here again we have large surface areas and excellent mixing of the slag and metal as well as ample opportunity for nucleation of carbon monoxide.

This process has been expanding continuously over the last fifteen years and is rapidly overtaking the open hearth process. It has now become sufficiently consistent to respond to control by computers.

In sight, however, is a new and exciting process, called 'spray steelmaking' (Rhydderch 1967). It has the merits of being extremely fast, is semi-continuous and the equipment required (see figure 9) is simple and cheap.

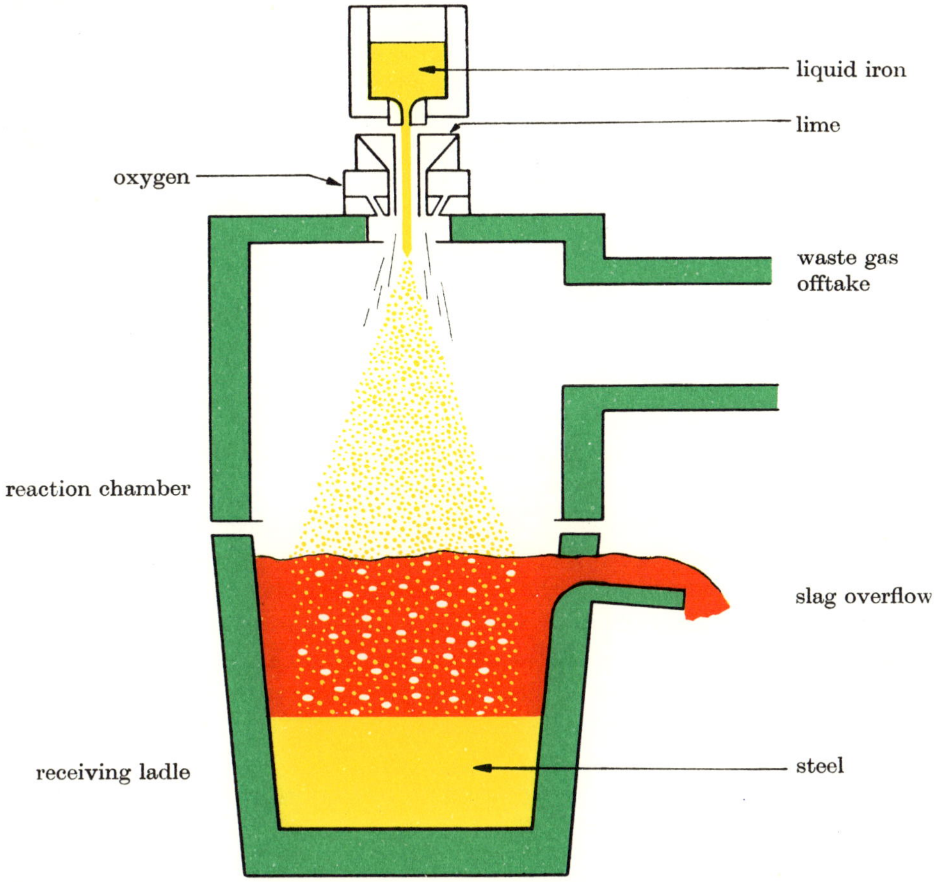

FIGURE 9. The spray steelmaking process.

The essentials of the process are that a stream of liquid iron is poured at a constant rate through a refractory nozzle and, on emerging, is atomized by high velocity jets of oxygen gas. Lime and iron oxide are added as a curtain around the stream. The spray process takes place in a refractory lined chamber and the resulting steel is collected in a receiving vessel placed underneath.

The droplets formed by the spray are further broken up by the violence of the reaction, as can readily be seen if one merely allows a drop of liquid iron to fall in a column of oxygen; the iron quickly becomes super-saturated with carbon monoxide and effectively explodes.

The heat generated in this process is very high because the carbon is fully burnt to CO_2. This means that scrap iron or steel must be used as a coolant, but this has good economic consequences. Better still, we can use pre-reduced pellets of iron ore which are coming on to the market.

At the end of the run the whole is allowed to settle. The receiving vessel is drawn out from under the equipment and the metal poured into a ladle where it receives such additions as are required for the final product.

In designing this process not only do we use all our knowledge of chemical thermodynamics, of reaction kinetics and of mass and heat transfer, but we must also bring in another branch of science, that of fluid dynamics to ensure uniform and rapid mixing and better control.

Maintaining control of such a high-speed reaction is obviously something which requires careful study. However, a speedy semi-continuous process such as this has some advantages over a batch process in modern control technology, largely because the correcting feed-backs can be quicker.

Trials of this process so far have met with no insuperable difficulties. The spray steelmaking process has today reached the stage where the L.D. process was about ten or so years ago. The greater background of scientific knowledge which we have today should allow it to move much faster and it might well overtake the L.D. process before very long.

I have picked out the central part of the steel industry because it shows up the numerous roles of carbon. But I would remind you that there are other processes tailored to specific raw materials or products. These, however, introduce few new phenomena of interest.

I have spoken of 2000 years of steel. In our laboratories in Sheffield can be seen a cementation furnace (see figure 10, plate 5) which uses a process developed by the Egyptians. It ceased production less than 20 years ago. In this process the conversion of iron to steel required two to three months, while in the adjacent building, the pilot plant for spray steelmaking carries out the conversion almost instantaneously. This is an extreme example, but it illustrates the fact that most of the technical progress about which I have spoken has occurred during the last twenty-five years.

(The lecture was followed by a short film lent by Professor F. D. Richardson of Imperial College. See Davenport, Bradshaw & Richardson 1967.)

References

Davenport, W. G., Bradshaw, A. V. & Richardson, F. D. 1967 *J. Iron Steel Inst.* **205**, 1034.

Hume-Rothery, W. 1960 *Atomic theory for students of metallurgy*. London: Institute of metals.

Johnson, A. 1968 *J. Iron Steel Inst.* **206**, 671.

Klemantaski, S. & Wild, R. 1962 *Zéme Journées int. Siderurgie Luxembourg.*

Rhydderch, M. J. 1967 *J. Iron Steel Inst.* **205**, 814.

Richardson, F. D. & Jeffes, J. H. E. 1948 *J. Iron Steel Inst.* **160**, 261.

Wehr, M. R. & Richards, J. A. 1960 *The physics of the atom*, p. 297. Reading, Mass. Addison-Wesley.

THE BAKERIAN LECTURE, 1968

Review of recent developments in cosmology

BY F. HOYLE, F.R.S.

(*Delivered* 13 *June* 1968—*Received* 10 *July* 1968)

Proceedings of the Royal Society, A, *volume* 308, *pp*. 1–17, 1968

It is hard to believe today that most scientists in the year 1920 believed that our Galaxy was all there is to the whole Universe—the 'island Universe' as it was then called. The difference between our present day view, with all its subtle complexities, and such a primitive notion has been brought about very largely by the accumulation of observational data. Almost immediately following the year 1920 Hubble disposed of the 'island Universe' concept and for the past fifty years astronomers have worked on the basis that our Galaxy is but one among thousands of millions strewn more or less uniformly throughout space.

Although observation plays the key role in determining which ideas survive and which are rejected, ideas themselves frequently come from theoretical studies. Already in 1922 Friedmann discovered the theoretical models of the Universe which are now often described as the 'big bang' cosmologies. In this lecture I shall not be much concerned with these models, for the personal reason that I happen to be not very interested in them. But it is of relevance that I should explain to you why this is so.

Most of the phenomena studied by astronomers have evolutionary lifetimes that are very long compared to everyday happenings. The shortest evolutionary lifetimes of stars exceed 10^6 years, while significant changes in galaxies require more than 10^8 years. Indeed to discuss the general problem of the origin and evolution of galaxies one must work on a time scale approaching 10^{10} years, and this is close to the time that has elapsed since the 'big bang' in the cosmologies in question. So the natural time scale of the galaxies and of the Universe are comparable with each other on this point of view.

This means that there is not much scope for the operation of delicate controlling processes, which one might suppose to have been operative, since galaxies everywhere seem strikingly similar to each other. The situation is a little like trying to produce the human species all in a single generation. The difficulty is compounded by the expansion of the Universe which makes any coherent communication over large distances impossible once the Universe has expanded appreciably away from its initial state. The similarity of widely separated parts of the Universe then makes it necessary to suppose that initial conditions were responsible for the state of the world as we now find it. So the further study of big bang cosmologies, beyond the stage already achieved by Friedmann some fifty-six years ago, consists in the investigation of the effects of initial boundary conditions on the subsequent development of the Universe. To those with a taste for parameter-fitting exercises this may seem a satisfactory way of spending one's time, but for myself I much prefer the direct relationship between theory and observation, such as we have had in the past fifteen years in the case of the steady-state cosmology—I much prefer it even if at times the confrontation may be somewhat hectic.

Then there is a technical argument against such a boundary condition philosophy. Initial fluctuations tend to die away rapidly in an adiabatically expanding system. Since in big bang cosmologies the Universe has expanded adiabatically through a very large density range, boundary condition investigations always show that the initial conditions make little difference unless the initial fluctuations are very large indeed. This explains why the modern trend is to investigate major deviations from the homogeneity and isotropy that was always assumed until very recently. This trend seems to me to fly in the face of the greatest triumph of theoretical cosmology.

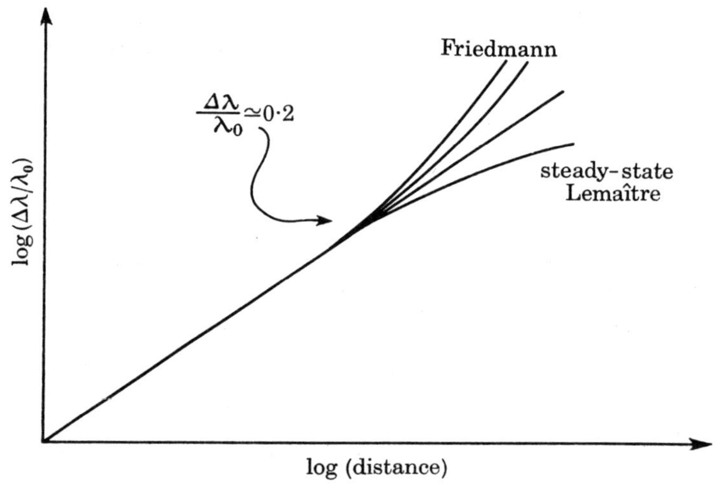

FIGURE 1

It used to be believed that the Earth was at the centre of the solar system. As recently as the beginning of the present century astronomers believed the solar system lay at the centre of the Galaxy, and that the Galaxy was the Universe. Thrice bitten by these mistakes modern cosmologists have never sought to place the solar-system in a privileged position with respect to the Universe. Instead, in the half-century 1915 to 1965, they went to the opposite extreme—by postulating that there was no such thing as a privileged position. The postulate of homogeneity requires the Universe to appear the same whatever your spatial position, while the postulate of isotropy requires that whatever direction you look out in things are also the same. Now one can formulate an explicit mathematical problem. Granted these two powerful symmetry postulates—homogeneity and isotropy—what motions for the Universe are still permitted? The answer is either an expansion or a contraction obeying certain rules. The expansion, for example, obeys a rule of the kind shown in figure 1.

The ordinate is the rate of expansion expressed in terms of the red-shift displacement $\Delta\lambda$ of a spectrum line of laboratory wavelength λ_0—the ratio $z = \Delta\lambda/\lambda_0$ is the same for all lines. No significance is to be attached to the linear position of the curve passing through the origin, since the distance unit can always be chosen so that this is so. The relevant features are the linear part of the curve and the departures from linearity which begin to show themselves for $\Delta\lambda/\lambda_0 > \sim 0{\cdot}2$.

The Friedmann cosmologies show the upward curvature of figure 1 because of the finite travel time of light. The values of z which we calculate for galaxies at different distances therefore belong to different moments of time—the greater the distance the earlier the time. Now if we take account of gravitation in an expanding system we expect the expansion rate to slow up to some degree as time proceeds, the extent of the slowing up depending on the initial conditions. The effect therefore is that the $\Delta\lambda/\lambda_0$ values belonging to earlier moments are enhanced. This explains the upward curvature. The degree of curvature depends on the initial conditions, so the Friedmann cosmologies have an infinite family of possibilities—but they all show upward curvature. (This assumes that 'distances' are appropriately defined. The appropriate definition is somewhat different from the astronomers operational determination of distance, but a known relation exists between the operational distance and the 'distance' of figure 1.)

Also marked in figure 1 are the expected curves for Lemaître's exploding universal atom and for the steady-state cosmology. Why do these cases show an opposite curvature to the right? Because the physical laws used in these cosmologies are different. In Lemaître's case, gravitation has been altered by the inclusion of the cosmical constant in Einstein's gravitation equations. In the steady state a new scalar field appears in Einstein's equations which affects the red-shift-distance relation in a similar way to the cosmical constant.

How about the facts? In figure 2 we have essentially the same plot between red-shift and distance for a number of nearby galaxies of all types. The linear relationship is apparent, although the intrinsic variations between one galaxy and another produces some dispersion. But this is already sufficient to establish the success of the theoretical prediction of a linear relationship provided the distances are not too large. I referred to this success before as the greatest triumph of cosmology. You recall that the success was based on the symmetry requirements of homogeneity and isotropy. It is this success which makes me reluctant to abandon these symmetry requirements.

Hubble found it was possible to improve the observed linearity between distance and $\Delta\lambda/\lambda_0$ provided care was taken to ensure reasonable intrinsic similarity in the chosen sample of galaxies. His method was to choose the third brightest member of fairly rich clusters of elliptical galaxies. This meant that a uniform class of galaxies was used, and that statistical scatter was reduced by the use of well-populated clusters.

The third brightest criterion was also thought to reduce statistical scatter, but Sandage now says it is actually better to use the brightest member.

In figure 3 we have the situation as it existed in 1956 (Humason, Mayall & Sandage). The red-shift values extend up to $\Delta\lambda/\lambda_0 \simeq 0{\cdot}2$, covering pretty well the full range of the expected linear dependence. You can also see that one might suspect the beginning of a curvature to the left, in agreement with the Friedmann

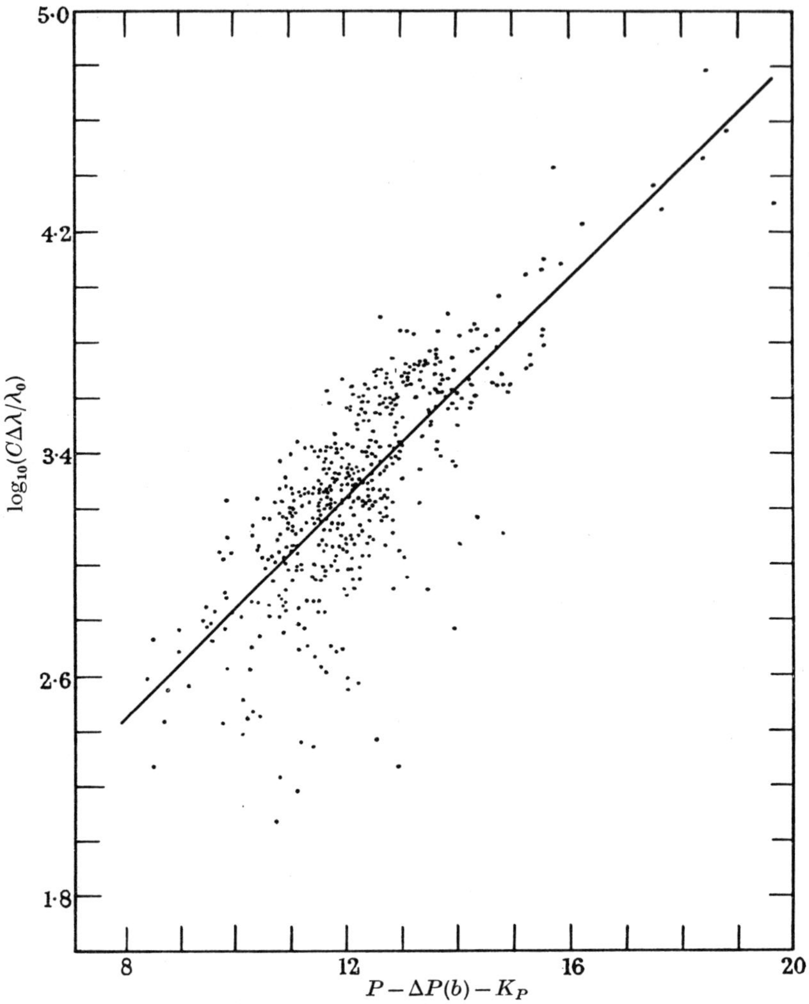

FIGURE 2. The red-shift magnitude relation for 474 field nebulae of all nebular types.

cosmologies and in contradiction to Lemaître's use of the cosmical constant and also in disagreement with the steady-state theory.

This was the third crisis for the steady-state theory in only eight years. The first crisis concerned the origin of the chemical elements. As an outcome of the work of Gamow, and of his students Alpher and Herman, it was thought around 1950 that only in a big bang cosmology did nuclear conditions exist which could lead to the synthesis of the heavy elements. It was pointed out that it would need a remarkable range of stellar conditions if the full range of the chemical elements owed their origin to processes inside stars. But the astronomical observations of the preceding 50 years had shown a profusion of types of star, and we now know that this profusion is directly related to the requisite range of nuclear conditions. The richness of stellar evolution is a product of nuclear physics—this indeed is what the modern subject of stellar evolution is all about. By 1956 this was becoming clear. Indeed by then observation had shown that stars differ so much in chemical composition that the concept of a universal big-bang synthesis could not be correct. It was still possible that particular processes, for example, H→He, might occur in a big-bang, but not the more complicated aspects of the building of the chemical elements from light nuclei. This whole problem has been worked over again during the past two or three years The use of high speed computers has made more ambitious calculations possible, but the conclusion remains the same as before—one can have H→He in a big bang, but the chemical structure of the world around us seems stellar in its origin.

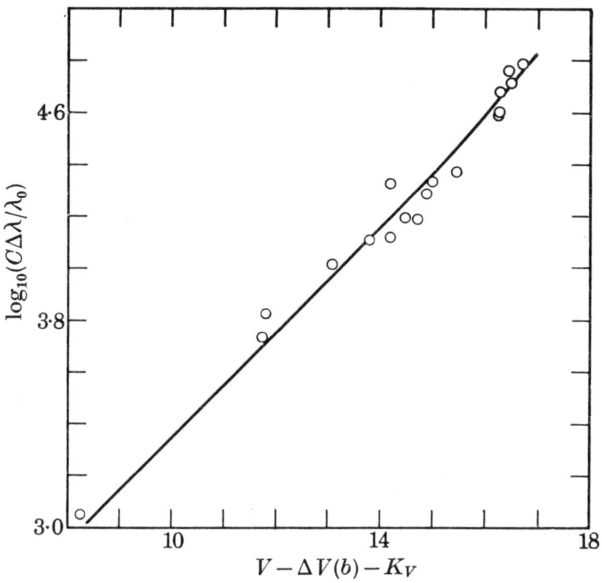

FIGURE 3

The second crisis concerned what became known as the Stebbins–Whitford effect. Because of the red-shift, distant galaxies should naturally appear redder than nearby galaxies. But in the early 1950s Stebbins & Whitford found an excess reddening that seemed to increase with distance. This excess was attributed to the kind of evolutionary process which is difficult to understand within the steady-state cosmology. However, within a year or two both Stebbins and Whitford began to suspect that errors of calibration might have been responsible for these first results. So in 1956 a question mark had to be attached to the Stebbins & Whitford effect. The issue has remained unresolved for more than 10 years, but recent work by Sandage (private communication) appears at last to show that there is no effect. Distant galaxies do not have excess reddening, so on this score, at least, the steady-state theory can breathe again.

But how of the suggested curvature to the left which seems to show up in figure 3? By pursuing this topic I will arrive eventually at the quasi-stellar objects and the problems and mysteries which surround them. Since this is a large subject in itself, I would prefer to interpolate a discussion of a still more serious threat to the steady-state theory which was raised in 1957 by Ryle.

If one counts the number of radiosources N brighter than a given flux-level S it is possible to plot a curve of N as a function of S, as in the curves of figure 4. No importance is to be attached to the absolute value of $N(S)$, only to the functional dependence—i.e. to the shape of the curve. For intrinsically similar objects in a Euclidean Universe the curve should be a line in a logarithmic plot with slope $-1\cdot5$. For the steady-state cosmology, apart from possible fluctuations at high values of S due to local departures from uniformity, the curve should begin as in a Euclidean Universe but then should fall over more and more to the left as the red-shift becomes more and more dominant. What was observed was an approximately straight line

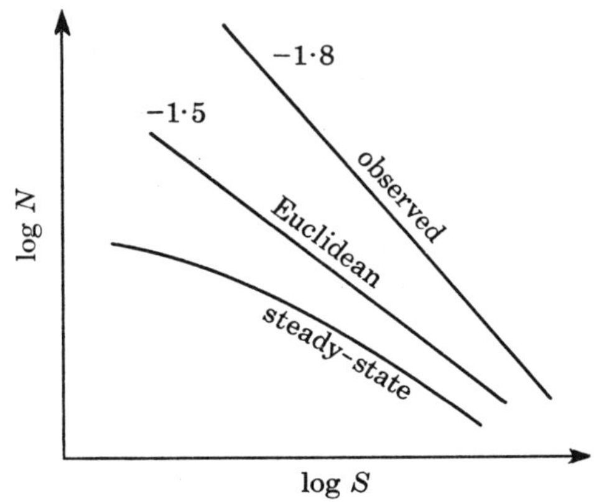

FIGURE 4

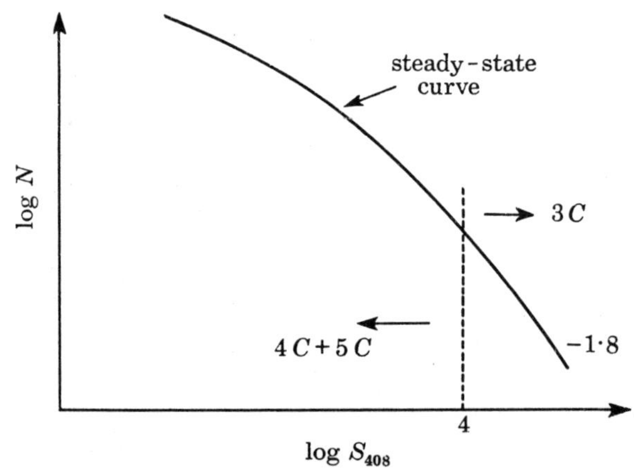

FIGURE 5

that rose with a slope of about $-1\cdot8$. This observation seemed to require radio-sources at great distance (and therefore with low flux values) to have a greater spatial density than local sources. Remembering that we see the distant sources at an earlier time, because of the finite speed of light, we naturally interpret the observed counts as a time effect rather than as a spatial effect. We agree that at any particular moment the spatial density of sources is everywhere the same, but we require the density to have been higher in the past than at present. Local fluctuations apart, this is the kind of behaviour which it is hard to reconcile with the steady-state theory.

The situation would seem to me to have been appreciably changed by the recent work of Professor Ryle and his colleagues,† as well as by recent work by Shimmins, Bolton & Wall (1968). Counts to much lower flux values have been made with the result shown schematically in figure 5.

If the curve is divided into two parts at $S_{408} = 4 \times 10^{-26}\,\text{w.m.}^{-2}\,\text{Hz}^{-1}$, the part to the left agrees very well with the steady-state curve. Narlikar and I have considered the steady-state theory for a number of possible luminosity-function distributions of the sources. Within wide limits we find the same general behaviour. The following table gives results for a typical case:

TABLE 1

$$P_{\text{I}} = 10P_{\text{II}} = 10^2 P_{\text{III}}, \quad \Sigma_{\text{III}} = 30\Sigma_{\text{II}} = 3 \cdot 10^3 \Sigma_{\text{I}}$$

S_{408}	N	z_{II}
5	56	0·083
0·5	10^3	0·23
0·05	$1·3 \times 10^4$	0·57
0·005	10^5	1·26

The sources are taken to be of three types I, II, and III, with I ten times more powerful intrinsically than II, i.e. $P_{\text{I}} = 10P_{\text{II}}$, and II ten times more powerful than III. The spatial densities Σ were taken such that the density of sources of type II was 10^2 times the density of sources of type I, while III had 30 times the spatial density of II. This distribution would have the effect in a Euclidean situation of giving a count

$$N(S) = N_I + N_{II} + N_{III}$$
$$\sqrt{10}\,N_I = N_{II} = \tfrac{1}{3}\sqrt{10}\,N_{III},$$

i.e. sources II and III would contribute about equally, while type I would contribute about one-third of II. Also in a Euclidean situation we would have the following relations between the respective red-shifts,

$$z_I/\sqrt{10} = z_{II} = \sqrt{10}\,z_{III},$$

these equations holding for sources giving the same observed flux density S.

The only free parameter to be applied to the numbers of table 1 is a scale factor in N. This corresponds to a free parameter in the spatial densities Σ. The free parameter has been chosen so that the value of N for $S_{408} = 5$ agrees with the number of radiogalaxies per steradian in the 3CR survey. This we took to be 56. The remaining values of N are then fixed and are in close agreement with the recent Cambridge results at low flux values (Pooley & Ryle 1968). Thus the recent work gives $N \simeq 8 \times 10^4\,\text{sr}^{-1}$ at $S_{408} = 0·01$. At $S = 0·005$ the appropriate value read off from a curve drawn through the observational points is $N \simeq 1·4 \times 10^5\,\text{sr}^{-1}$. But this refers to all sources. If, as is usually supposed, radio-galaxies comprise about two-thirds of the total, then the appropriate number of radio-galaxies is close to $10^5\,\text{sr}^{-1}$ at

† This is a personal view, as also are many other opinions expressed in this lecture. Ryle has expressed his disagreement with my arguments here—in particular with the division of 3CR sources into categories (tables 2 and 3).

$S = 0{\cdot}005$, in almost exact agreement with the steady-state calculation. I have preferred to work in terms of radio-galaxies here because there is no ambiguity in the interpretation of the counts for radio-galaxies, as there is for quasi-stellar objects.

It is important that no Friedmann cosmology seems capable of fitting the counts at low-flux values, except by introducing special evolutionary assumptions. The steady-state theory appears to be the only cosmology that will fit the observations without recourse to a parameter-fitting exercise.

I also wish to emphasize that the red-shift values of table 1 are determinate. The results of table 1 therefore require that for radio-galaxies with $S_{408} > 5$ units we must have $z_{\mathrm{I}} < \sim\sqrt{10} \times 0{\cdot}083 = 0{\cdot}26$, $z_{\mathrm{II}} < \sim 0{\cdot}083$, $z_{\mathrm{III}} < 0{\cdot}083/\sqrt{10} = 0{\cdot}026$. In short, we expect the average red-shift to be $\sim z_{\mathrm{II}}$, i.e. somewhat less than 0·1, and the red-shifts of the intrinsically brighter sources to extend up to $\sim 0{\cdot}25$. A considerable fraction of radiogalaxies in the 3CR survey—i.e. with $S_{408} > \sim 5$ flux units have had their red-shifts determined. Out of 86 such radiogalaxies, 28 have $z \leqslant 0{\cdot}026$, 33 have $0{\cdot}026 < z \leqslant 0{\cdot}083$, 23 have $0{\cdot}083 < z \leqslant 0{\cdot}26$, and only 2 have $z > 0{\cdot}26$.

You will recall that it is the sources of the 3CR survey with $S_{408} > \sim 5$ that give the troublesome steep slope of $-1{\cdot}8$ in the $\log N$–$\log S$ plot. Fortunately these sources have been analysed in detail, so that many optical identifications and red-shifts are available for them. Véron (1966) has given the following breakdown, while Jauncey (1967) had given a statistical analysis of the slopes of the $\log N$–$\log S$ curve for various categories that appear in the breakdown. Results are shown in table 2.

TABLE 2. BREAKDOWN OF 3CR SOURCES INTO CATEGORIES

r.g. ≡ radiogalaxies. r.g.? ≡ doubtful radiogalaxies
q.s.s. ≡ quasi-stellar sources. q.s.s.? ≡ doubtful q.s.s.
e.f. ≡ apparently empty fields
e.f.? ≡ fields in which no satisfactory identification can be made.

total number of sources:		328
galactic	32	
heavily observed	42	74
	74	
sources available for analysis		254

category	number	slope of $\log N$–$\log S$ curve	
r.g.	100	$-1{\cdot}26 \pm 0{\cdot}13$	
	+		
r.g.?	44		$-1{\cdot}58^{+0{\cdot}14}_{-0{\cdot}13}$
q.s.s.	39	$-1{\cdot}56^{+0{\cdot}27}_{-0{\cdot}25}$	
	+		
q.s.s.?	21		$-2{\cdot}00^{+0{\cdot}30}_{-0{\cdot}27}$
e.f.	36		
	+		
e.f.?	14		$-2{\cdot}50^{+0{\cdot}46}_{-0{\cdot}43}$

It is immediately interesting that the well-identified r.g. and q.s.s. categories, comprising 139 objects out of the total of 254 available for study, show no significant deviation of the $\log N$–$\log S$ slope from what is expected. It is also interesting that the doubtful identifications r.g.? and q.s.s.? both lead to increased steepness, r.g.? when added to r.g., q.s.s.? when added to q.s.s. This would suggest that at least a proportion of r.g.? and q.s.s.? represent incorrect identifications. This in fact has turned out to be correct for most objects in the q.s.s.? category. Since Véron made his classification, spectra of these objects have been studied and it is clear they are not quasi-stellar. Probably most of q.s.s.? should be added to e.f.? I think it is possible that an appreciable fraction of r.g.? should also join e.f.? If one adds a half of r.g.? to r.g. and a half to e.f.? one arrives at the following results:

TABLE 3

category	number	slope of $\log N$–$\log S$ curve
r.g.	122	$-1{\cdot}4 \pm 0{\cdot}1$
q.s.s.	39	$-1{\cdot}56 {+0{\cdot}27 \atop -0{\cdot}25}$
e.f. + e.f.?	93	$-2{\cdot}50 \pm 0{\cdot}3$

The issue then turns on the 93 sources in the e.f. + e.f.? category. The situation is that one-third or so of sources which the optical astronomers find difficult to identify turn out to be just the sources showing the anomalously steep rise in the $\log N$–$\log S$ curve. There now remains only a total of about 93 sources, distributed over a solid angle of 3 sr. The difficulty therefore concerns some 30 sources/sr.

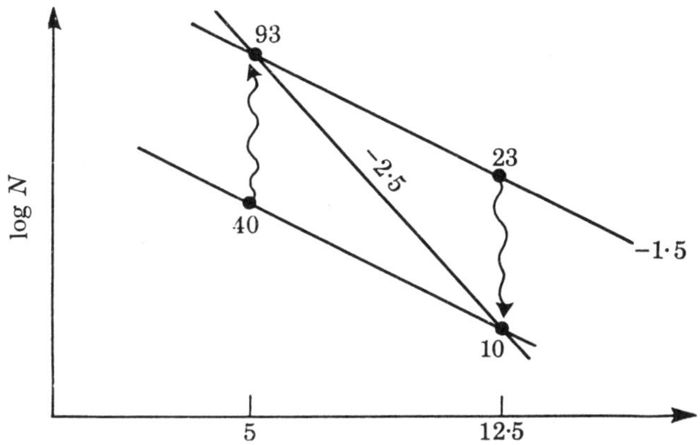

FIGURE 6

There are two ways of looking at the remaining discrepancy. In figure 6 we have the $\log N$–$\log S$ curve for the troublesome 93 sources. Because of the steep $-2{\cdot}5$ slope the count falls from 93 to 10 when S_{408} increases from 5 to 12·5. Suppose we draw two parallel lines each of slope $-1{\cdot}5$, one through $N = 93$, $S_{408} = 5$, the other through $N = 10$, $S_{408} = 12{\cdot}5$. The first of these passes through $N = 23$, $S_{408} = 12{\cdot}5$, while the second passes through $N = 40$, $S_{408} = 5$. We can either describe the curve of slope $-2{\cdot}5$ as having an excess of 53 sources at $S_{408} = 5$, or a deficit of only 13 sources at $S_{408} = 12{\cdot}5$. There is no way from the observations of distinguishing between these two possibilities. However, if a theory is under attack it is only reasonable to choose the least awkward possibility. This amounts as I have shown to a deficit of only 13 sources over about 3 sr, i.e. about 4 to 5 sources/sr. It does not seem at all difficult to suppose that such a deficit might be due to a local cosmological fluctuation, one confined say to the nearest 50 mpc. In relation to this possibility it will be of interest to see how new counts how being made in the southern hemisphere turn out. If the effect is due to as little as 5 sources/sr there could be a difference between counts made in the north and in the south.

I wish now to return to figure 3, and to the points near $\Delta\lambda/\lambda_0 \simeq 0{\cdot}2$, the points which seemed to indicate curvature to the left. It was soon found that these points had to be moved over towards the right, not far enough to reach the correct position for the steady-state theory but far enough to include the steady-state curve within the probable error associated with the data. The change was due to a change from a photographic to a photoelectric technique—the latter being essentially linear in behaviour in place of the awkward non-linear behaviour of the photographic plate.

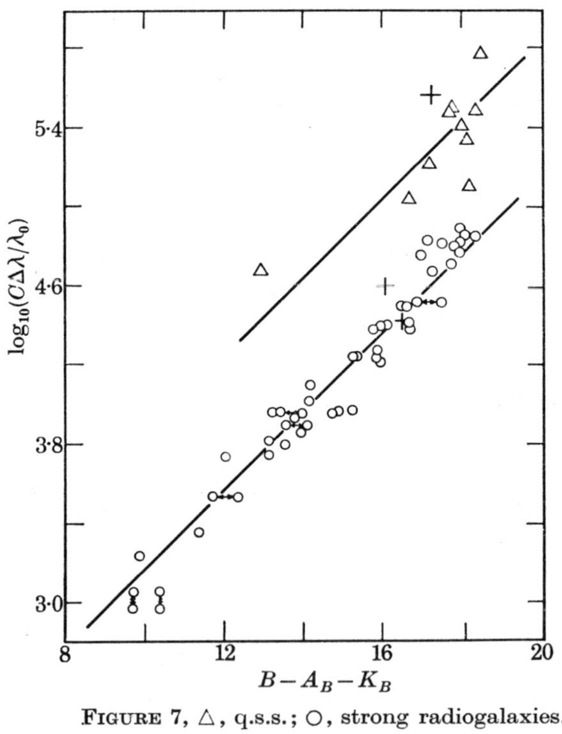

FIGURE 7, △, q.s.s.; ○, strong radiogalaxies.

The situation as it was in 1964 is shown in figure 7 (Sandage 1965) where the circles are for galaxies—actually radiogalaxies. Once again the linear behaviour for smaller red-shifts is well shown, but once again any curvature to the left or to the right at $\Delta\lambda/\lambda_0 \simeq 0{\cdot}2$ is masked by the scatter of the data. However, in figure 7 we also have a linear relationship for the first batch of quasi-stellar objects (q.s.o.). You remember that the ordinate scale depends on the intrinsic properties of the objects—so the fact that the line for the q.s.o. lies above that for galaxies only means that the q.s.o. are intrinsically brighter. What is important in the present connexion is the shape of the line drawn through the observed points. In 1964 this was thought to be a straight line, not a line with curvature to the right, as the steady-state theory would require.

(Here it should be noted that the magnitude scale of figure 7 is not the same as the distance scale of figure 1. There is a known relationship between the two. If figure 7 was represented in the form of figure 1 the straight line through the q.s.o. of figure 7 would become a line with curvature to the left, in agreement with the Friedmann cosmologies.)

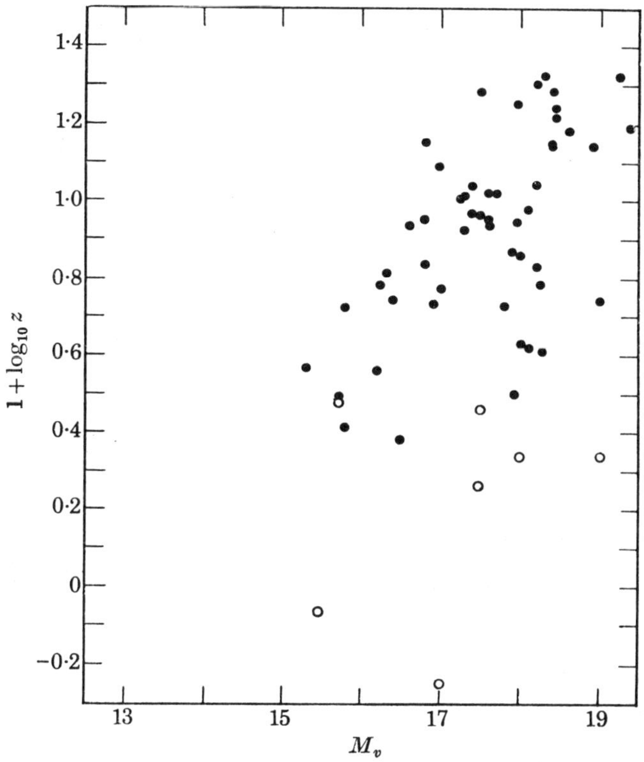

FIGURE 8. Red-shift plotted against optical magnitude (M_v) for all q.s.o. and N-type galaxies for which data are available.

However, on figure 7 you will notice three points marked not by open triangles △ but by crosses +. If these are taken account of, one might place a very steeply rising curve through the points. While this would not agree with the steady-state cosmology, it would not agree with any cosmology—so we should be obliged to discuss the data either as being incomplete or as not being of cosmological significance at all.

In figure 8 we have the situation now that more than 100 q.s.o. have been observed (Burbidge & Burbidge 1967). The steep rise is confirmed, and once more the steady-state theory survives. But what does the steep rise of figure 8 mean? Either that the data has been affected by peculiar selection effects or that the observed very large red-shifts are not due to the expansion of the Universe.

It is known that q.s.o. must consist of very large masses in small volumes—small volumes astronomically speaking. The source 3C273B has been shown by radio interferometric methods to have a diameter not greater than about 10 light years. On the other hand, the energy output of 3C273B in the form of cosmic rays has been estimated to be not less than 10^{57} erg/year. Even for the very short lifetime of 1000 years this would require an energy production of 10^{60} erg, equivalent to the total release of the rest mass energy of 10^6 solar masses. Hence at least 10^6 solar masses must be confined within a volume having a diameter less than 10 light years.

A still stronger requirement comes from recent infrared measurements of galaxies. The extreme central parts of many galaxies are similar to q.s.o. in that very high mass concentrations must be involved. Values as high as 10^{46} erg/s at infrared wavelengths up to 25 μm have recently been reported by Low (1968). The lifetimes of the galaxies in question cannot be much less than 10^9 years, otherwise it would not be possible to explain why so many galaxies show these high infrared emission rates. Since 10^{46} erg/s taken over 10^9 years amounts to 3×10^{62} erg we see that the mass involved must rise appreciably to *ca.* 10^9 solar masses. The concept insistently forced on us is of a mass of galactic order confined within a volume having a radius no more than a few light years. This implies unprecedentedly strong gravitational fields.

The question therefore arises as to whether the red-shifts of q.s.o. could not be due to gravitation, instead of to the expansion of the Universe. If this were so the observed apparent magnitudes of the q.s.o. would be uncorrelated with distance. The only correlation with red-shift should be the comparatively minor one that quanta lose energy in climbing out of a gravitational field. If this effect is allowed for, we would expect the observed optical magnitudes to become uncorrelated with red-shift. This was found to be indeed the case (Hoyle & Burbidge 1966).

The same thing holds for radio magnitudes. In figure 9 we have radio magnitudes (i.e. flux values S on the logarithmic magnitude scale $M_R = -2{\cdot}5 \log S + \text{constant}$) plotted against $z = \Delta\lambda/\lambda_0$. Once again there is a scatter diagram. If we consider this scatter diagram together with the $\log N$–$\log S$ slope of about $-1{\cdot}5$ for the q.s.o. we have a puzzling situation. We have seen that such a slope is characteristic of intrinsically similar objects distributed uniformally in a Euclidean Space—this would mean objects not very far away from us.

In figure 10 we have the $\log N$–$\log S$ count for the 39 q.s.o. of table 3. The number is rather small but the slope appears to be settling down to $-1{\cdot}5$. Moreover, a similar result is required by the counts of Shimmins, Bolton & Wall (1968) which involve a considerably greater number of q.s.o. The conclusion seems inescapable that either the $\log N$–$\log S$ count for the q.s.o. must be interpreted in some new way or that the red-shifts are not due to the expansion of the Universe.

Everything would be straightforward if the spectra of q.s.o. were characteristic of the dense gas conditions which we would expect to be associated with a strong gravitational field. But they are not, they are spectra characteristic of a diffuse gas. The spectra strongly suggest that the gas lies well outside any strong gravitational field and hence that the red-shifts of the spectrum lines cannot be gravitational in origin.

It is natural that faced with a dilemma astronomers fall into two groups; those who seek to dismiss the non-spectral data as due to selection effects and those who

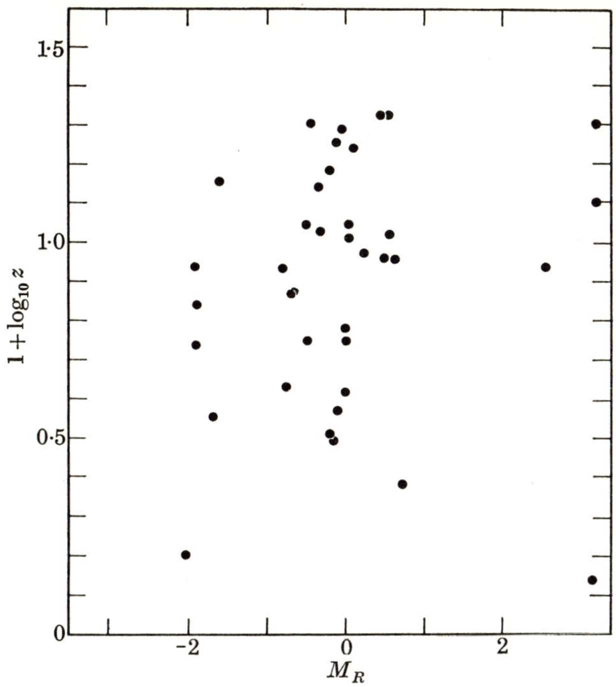

FIGURE 9. Red-shift plotted against radio magnitude for quasi-stellar objects.

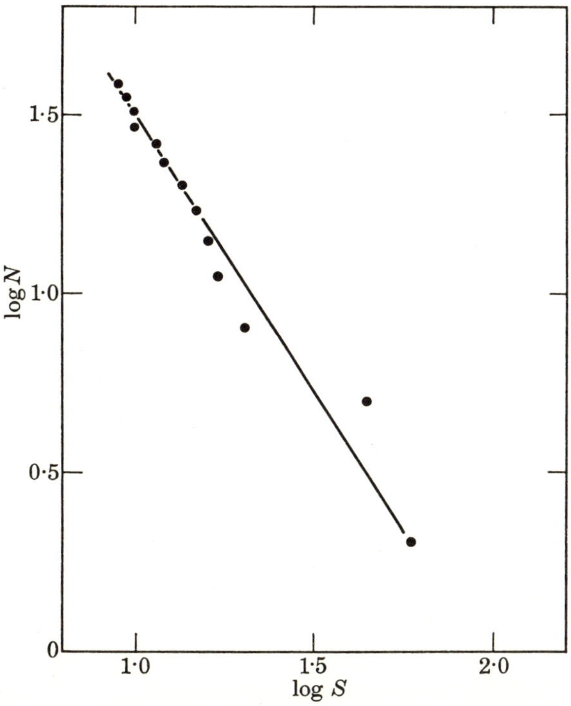

FIGURE 10. Plot of log N against log S for the 39 q.s.o. of table 3.

seek somehow to reconcile the spectra with strong gravitational fields, and it is natural that each side finds the arguments of the other extremely irritating, since each side emphasizes those aspects of the data which the other side would like to ignore.

For the past two years I have preferred the gravitational side because strong gravitational fields are a new phenomenon. Such fields cannot be studied in the laboratory so it would be extremely valuable if they were accessible to investigation by the astronomer. Those of us on the gravitational side have argued that the diffuse gas characteristics are not the only peculiar feature of the spectra. Red-shifts up to $z = \Delta\lambda/\lambda_0 = 2$ are readily found, but shifts significantly above 2 are not found. There appears to be a concentration of red-shift values of $z = 1{\cdot}955$, as can be seen from the data set out in table 4.

TABLE 4. QUASI-STELLAR OBJECTS WITH EMISSION LINE RED-SHIFTS GREATER THAN 1·9

object	$z_{em.}$	$z_{abs.}$	object	$z_{em.}$	$z_{abs.}$
4C 29·50	1·927	—	3C 9	2·012	—
PHL 61	1·93	—	Ton 1530	2·046	1·9362, 1·9798, 2·0553
3C 191	*1·956*	*1·947*	PHL 1305	2·064	—
PKS 0119 − 04	*1·955*	*1·965*	PKS 0229 + 13	2·07	—
PHL 938	*1·956*	1·9064, 0·6128	B 189	2·075	—
BSO 6	*1·956*	—	BSO 11	2·084	—
PHL 5200	1·98	1·90–1·98, *1·9502*	PKS 0106 + 01	2·107	—
PKS 1148 − 00	1·892	—	PKS 1116 + 12	2·118	*1·947*
PHL 1127	1·990	*1·95* (one line)	PKS 0237 − 23	2·223	2·2017, *1·9556*, *1·9517* 1·6744, 1·6715, 1·6564 1·5958, 1·5132, 1·3646
LB 8755	2·010	—			

About one-half of the q.s.o. of large red-shift possess absorption lines as well as emission lines. The nature of the absorption lines can be identified without undue difficulty in some objects—for example, 3C191. But identification has proved exceedingly difficult in other cases; in particular PKS 0237 − 23. Recently Burbidge & Lynds (1968) on the one hand and Bahcall, Greenstein & Sargent (1968) on the other have identified many of the lines in the very rich absorption spectrum of PKS 0237 − 23. Both groups agree that many different red-shifts are present in the same object. Both groups give values ranging from $z = \Delta\lambda/\lambda_0 = 1{\cdot}36$ up to 2·20 While it would be possible to explain one or two discrepant absorption red-shifts in terms of intergalactic clouds intervening between ourselves and the object in question, I think it strains credulity to imagine that so many different red-shifts as seem to be present in PKS 0237 − 23 could be explained in this way. Why should PKS 0237 − 23 have so many absorption red-shifts while the majority of q.s.o. have none?

PHL 938 has been found by Burbidge & Lynds (1968) to have two widely different absorption red-shifts, one at $z = 0{\cdot}61$, the other at $z = 1{\cdot}91$, whereas the emission line red-shift is $z = 1{\cdot}955$. There is quite strong evidence that lines have changed in this object in only a year, almost certainly requiring that the lines be produced in the object itself not in an intergalactic cloud.

Several red-shifts in the same object suggests variable positioning in a strong gravitational field. But this is not the only explanation. The object could have fired out a cloud of gas at a speed approaching light. If such a cloud now intervenes between us and the object, absorption at a widely different red-shift is possible. This point of view permits the cosmological interpretation of the emission line red-shift to be retained. But then why are the absorption lines almost always very narrow? Velocity variations in a rapidly expanding cloud would more usually lead to diffuse absorption features.

For myself, I think it is best to go back to the unequivocal position that q.s.o. must be associated with very large masses in small volumes and to ask what one might expect in such circumstances. The inevitable answer according to the usual theory of gravitation is that one would expect gravitational collapse inside the so-called Schwarzschild radius, leading to the development of singularities—one would expect this to occur for at least some proportion of the material of the object. The situation is entirely analagous to the big-bang origin of the Universe according to the Friedmann cosmologies—but in reverse; we now have retreat into a singularity instead of emergence out of a singularity. The great importance of the q.s.o. is therefore that they bring us face to face with the key question 'Do singularities exist in the real world?'

At an earlier stage I spoke of the steady-state as requiring the presence of a new scalar field in Einstein's equations. The need for this new field is the penalty one must pay for escaping from the mathematical singularity of the Friedmann cosmologies. But of course the reduction of physics to metaphysics which is what a singularity implies (this is not the comparatively straightforward task of tracing a known mathematical solution through a singularity) is a severe—and I believe an overriding—penalty of the usual theory. Some field of the steady-state type is essential if metaphysical singularities are to be removed. The same is true for the q.s.o. as for the Universe. Singularities do not exist within the q.s.o. if the steady-state type of scalar field exists.

This relates back to the paradox of the spectra of the q.s.o. I do not believe it makes any sense to discuss these spectra on the basis of the usual ideas of the properties of a strong gravitational field. Such ideas lead to a quick retreat from the external world not to the violent explosive outbursts which we actually observe.

I might add that some theoreticians have sought to avoid this difficulty by reversing the argument, by suggesting that the q.s.o. represent emergence from singularities, not retreat. But the q.s.o. have a wide range of characteristics similar to the nuclei of galaxies and the great ages of the galaxies seem to dispose of this idea.

The latest challenge to the steady-state theory is the discovery of a universal microwave background. For wavelengths greater than about 1 cm the observed radiation can be interpreted as the Rayleigh–Jeans tail of a black-body radiation field of temperature 2·7 °K. It is not yet known whether the black-body characteristics are maintained at the higher frequencies—wavelengths $\sim$ 1 mm. This radiation field is believed by many astronomers to be a fossil relic of a Friedmann cosmology, and it is widely quoted as strong evidence against the steady-state theory.

Why is it so quoted? Because the critics feel it is hard to understand the origin of the radiation in terms of known processes, in terms of the known activity of galaxies and radiosources. But of course we do not know everything concerning galaxies and radiosources, although we always think we do. When the microwave background was first discovered it was not realized that the frequency spectra of sources could turn upward at high frequency. Then an upturn was found for one or two sources. But it was still argued that such behaviour is exceptional. Now we know this upturn is characteristic of an appreciable fraction of all sources. Next, the extraordinary high infrared emission of galaxies was found. It seems as if about 1 % of all galaxies may emit upwards of 10^{46} erg. s^{-1} at frequencies $\sim 10^{12}$ Hz, not much different from the maximum of the observed background. Indeed we have the curious situation shown in table 5 concerning energy densities.

TABLE 5

phenomenon	energy density (erg cm^{-3})
infrared from galaxies	$> 10^{-13}$
microwave background	4×10^{-13}
H $\rightarrow$ He conversion in galaxies	6×10^{-13}
starlight in our galaxy	8×10^{-13}
cosmic-rays	3×10^{-12}

According to conventional astronomical views the similarity of the numbers are coincidental. Such views seem to me to arise out of ignorance. I simply cannot accept so many coincidences. There may be one accident—one criminal in the list—but it is unlikely there is more than one. The criminal could of course be the microwave background, which could be a fossil relic of a big-bang cosmology. But I do not think we are driven to this view today with any real force. Moreover, the case today is significantly weaker than it was only two or three years ago when the background was first discovered. At that time the theory could only survive by predicting that galaxies and radiosources were much stronger emitters of radiation in the wavelength range 10 cm down to 1 mm or less than was formerly thought possible. This prediction has turned out to be correct. Whether all the details can be made to fit remains to be seen.

To summarize then: over the past 20 years the steady-state theory has had to face six different crises. Five in my view have been wholly or largely surmounted, the sixth—the microwave background—remains to be resolved. I think it fair to say that the theory has demonstrated strong survival qualities, which is what one should properly look for in a theory. There is a close parallel between theory and observation on the one hand and mutations and natural selection on the other. Theory supplies the mutations, observation provides the natural selection. Theories are never proved right. The best a theory can do is to survive.

References

Bahcall, J. N., Greenstein, J. L. & Sargent, W. L. W. 1968 *Astrophys. J.* **153**, 689.
Burbidge, G. R. & Burbidge, E. M. 1967 *Quasi-stellar objects.* London and San Francisco: Freeman.
Burbidge, E. M. & Lynds, C. R. 1968 *Astrophys. J.* (in the Press).
Hoyle, F. & Burbidge, G. R. 1966 *Nature, Lond.* **210**, 1346.
Humason, M. L., Mayall, N. U. & Sandage, A. R. 1956 *Astr. J.* **61**, 97.
Jauncey, D. L. 1967 *Nature, Lond.* **216**, 877.
Low, F. J. 1968 Conference on Seyfert galaxies and related objects. *Astr. J.* (in the Press).
Pooley, G. G. & Ryle, M. 1968 *Mon. Not. R. astr. Soc.* **139**, 515.
Sandage, A. R. 1965 *Astrophys. J.* **141**, 1560.
Shimmins, A. J., Bolton, J. G. & Wall, J. C. 1968 *Nature, Lond.* **217**, 818.
Véron, P. 1966 *Nature, Lond.* **211**, 724.

Review Lecture

Hydrostatic extrusion

By H. Kronberger, F.R.S.

(*Lecture delivered* 21 *March* 1968—*MS. received* 4 *November* 1968)

Proceedings of the Royal Society, A, *volume* 311, *pp*. 331–347, 1969

Introduction

The object of this lecture is to present an account of a technological development which has taken place within the last twenty years, from very simple beginnings founded largely on the pioneer high-pressure work of the late P. W. Bridgman, to a metal-forming process of considerable industrial interest.

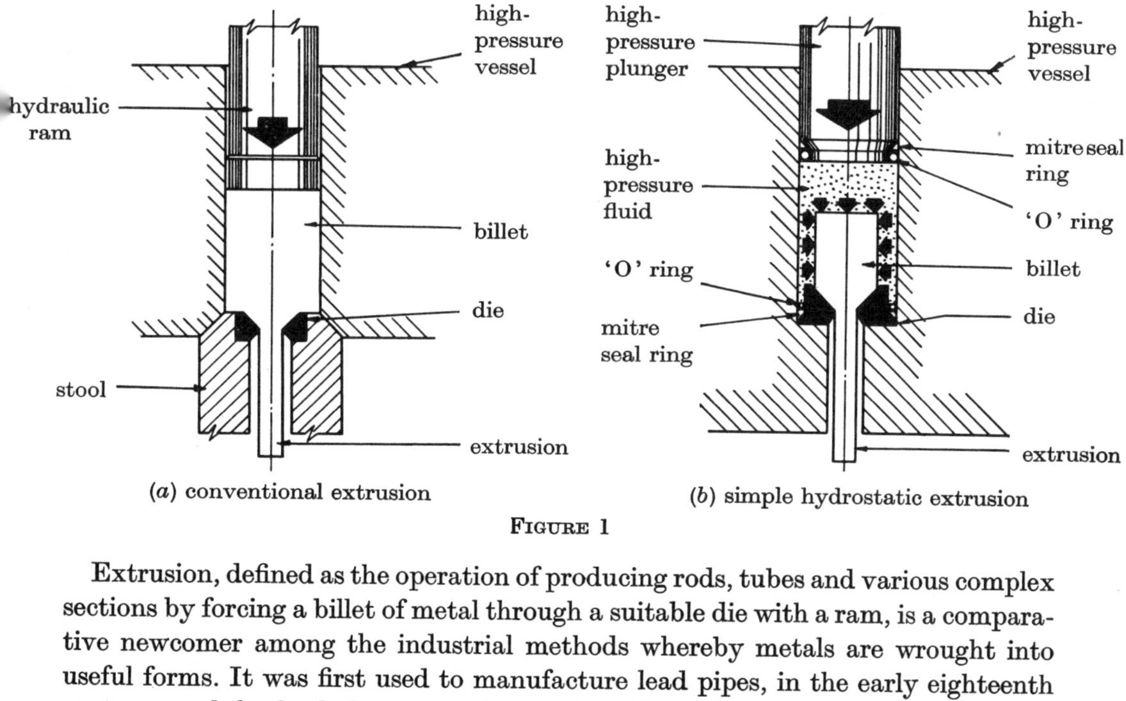

(*a*) conventional extrusion (*b*) simple hydrostatic extrusion

Figure 1

Extrusion, defined as the operation of producing rods, tubes and various complex sections by forcing a billet of metal through a suitable die with a ram, is a comparative newcomer among the industrial methods whereby metals are wrought into useful forms. It was first used to manufacture lead pipes, in the early eighteenth century, and for lead sheathing of electric cables about seventy years ago. The inventive genius of Alexander Dick, whose first patent was taken out in 1894, led to the use of hot extrusion for copper and its alloys and to the design of large machines with high throughput. Extrusion of steel has attained significance only within the past fifteen years.

In conventional direct extrusion (figure 1*a*) the metal billet, which is usually heated to reduce the yield strength of the metal, is compressed in a container by a ram, and is forced through the orifice of the die in the direction of the ram travel.

Frictional forces between the billet and die and between billet and container may often be more than 20% of the force required for extrusion; these forces increase rapidly with the length of the billet, and thus restrict the length of billet that can be extruded from a given extrusion container.

In hydrostatic extrusion (figure 1*b*) the compressive stress is not applied directly to the metal billet, but to a fluid surrounding the billet in the extrusion container. The pressure generated in the fluid induces a state of hydrostatic stress throughout the metal, except near the die aperture, where the imbalance in the stress system exceeds the yield strength of the metal, so that plastic deformation occurs, and metal is extruded through the die.

James Robertson patented the use of hydrostatic pressure for extrusion in 1893 and anticipated many recent developments in the art, but the idea was not taken up commercially at that time. Robertson was limited to cast iron pressure vessels and leather seals, and in general did not have available the necessary techniques for containing high pressures to exploit his invention.

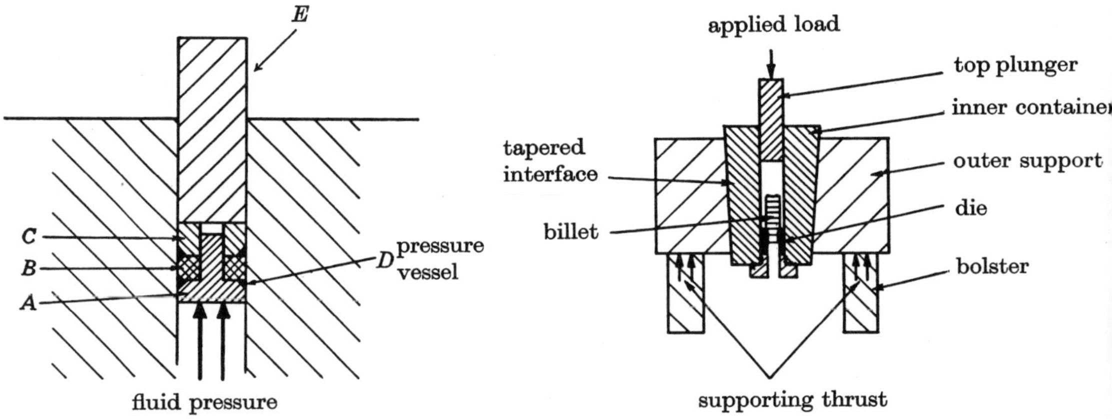

(*a*) unsupported area seal (Bridgman) (*b*) tapered duobloc pressure vessel

FIGURE 2

Such techniques were developed by Bridgman, who began with columns of mercury to obtain pressures up to 12 kbar and had by 1912 obtained pressures of 30 kbar in piston cylinder apparatus; in 1935–7 he extended the pressure range to 50 kbar with two pistons and an anvil and his Bakerian lecture to the Royal Society (Bridgman 1950) illustrated how a very small vessel, mounted inside a 'conventional' 30 kbar apparatus, could produce and contain pressures up to 100 kbar.

Key inventions assisting in this progressive extension of the pressure range were the manganin-resistance pressure gauge, the 'unsupported area' seal (figure 2*a*) and the use of the tapered multibloc pressure vessels (figure 2*b*).

The unsupported area seal consists of a hard steel plug A supported on an annular neoprene packing B, resting on a hard steel sleeve C and a plunger E. As the area of the neoprene is smaller than that of the face of the plug A, pressure in the neoprene is higher than in the fluid and leakage of fluid past the plug is impossible. Extrusion of the neoprene is prevented by the mitre rings *D*. Other seals have been devised to give advantages such as allowing air to escape from the pressure vessel on the return stroke of the plunger but the principle of the Bridgman seal has been retained.

By using an inner container with a small taper (typically 1 or 2°) on its outer diameter, supported by an outer collar with a matching bore, the maximum liquid pressure that can be obtained in the fluid without failure of the pressure vessel can be substantially increased. The downward force exerted by the top plunger and the upward reaction on the outer collar act so that increasing pressure forces the container further down the taper into the collar, and generates increasing support at the interface. Bridgman had carried out some wartime experiments at Watertown Arsenal on the extrusion of metals under hydrostatic pressure, and mentions his results in his book *Large plastic flow and fracture* (1952), but apparently considered the subject to be of limited interest.

Since about 1957, Beresnev & Vereschagin in Moscow and Pugh and his colleagues at the National Engineering Laboratories in the U.K. have been actively engaged in the development of hydrostatic extrusion as a metal forming process. The term 'hydrostatic extrusion' was first used by Pugh & Ashcroft (1962). Work for the Commisariat a l'Energie Atomique in France had begun about the same time, using large amounts of lubricant for conventional hot extrusion of steel and extrusion of uranium fuel rods for nuclear reactors, but the results have only recently been published (Sauve 1965).

Further impetus was given at a meeting of the Institute of Metals in London in March 1961 (Institute of Metals 1962), devoted to cold extrusion of metals with particular reference to lubrication.

It was clear that, by 1962, a new method of cold extrusion existed for metals and alloys, including some high strength materials that were only normally extruded at elevated temperatures where the yield strength of the material was reduced. Now the possibility of designing practical machines for exploiting the process began to be examined.

A small machine of 200 tons load working at a modest pressure of 7 to 8 kbar was installed (Green 1965) in the Reactor Fuel Laboratories of the U.K.A.E.A. There followed an increasingly widespread interest in the possibilities of applying the process as a production tool and equipment manufacturers began at this time to play an important role.

Effects of hydrostatic pressure on the properties of materials

It was observed comparatively early that the fracture of metal test samples by tension under high hydrostatic pressure was accompanied by much greater ductility than in the same material under atmospheric pressure. Experiments on the effect of hydrostatic pressure on the total strain to fracture, initiated by Bridgman and continued in other laboratories, suggested that the effect of pressure is to prevent the formation of cavities and of internal damage which appear in tensile testing.

The subject, which has an extensive literature of its own, was reviewed recently (Pugh 1965) in considerable detail. Two points particularly relevant to hydrostatic extrusion are:

(1) Pressure has no significant effect on the shape of the stress strain curve for either mild steel or copper, other than to extend it. In general, metals yield at the same values of stress difference under high pressure as at atmospheric pressure.

(2) Although the strain to fracture under pressure increases for all metals tested, the way in which it increases may range from a gradual change as, for example, in mild steel to a dramatic increase in ductility as in zinc.

(a)

(b)

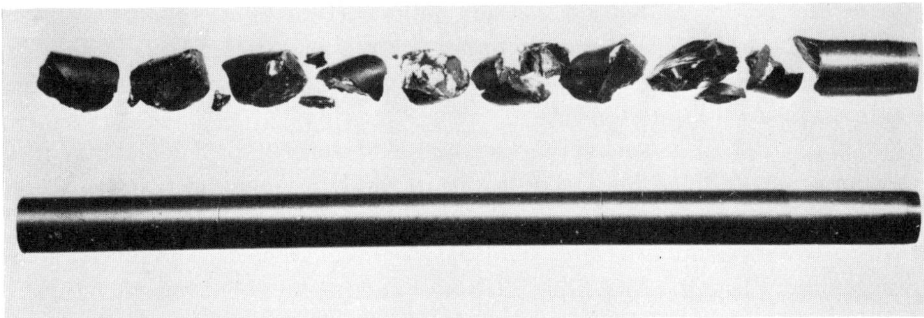

FIGURE 3. Extrusion of a brittle silver alloy: (a) into the atmosphere; (b) into a pressurized receiver.

When brittle materials such as bismuth are extruded through a die, the product develops circumferential cracks, or may even break into fragments. It was first demonstrated by Pugh & Green (1958) and independently by Beresnev *et al.* (1958), that cracking could be prevented by extruding from a fluid at very high hydrostatic pressure into a receiver containing fluid at a rather lower pressure. Figure 3 shows the two kinds of extrusion obtained with a brittle silver alloy. This technique of differential pressure extrusion has been studied by the Pressure Technology Corporation Ltd. (Bobrowsky & Stack 1965), where good results have been obtained with molybdenum. The reason for its success is believed to be that hydrostatic pressure suppresses the formation of cracks and cavities, during and after extrusion.

THE MECHANICS OF HYDROSTATIC EXTRUSION

The parameter of greatest importance is the extrusion ratio—the ratio of the cross-sectional area of the initial billet to that of the extruded product. It is a measure of the work that must be done by the compressive stresses against the resistance to deformation of the metal, mainly by shear.

The relation, $P = a \ln (A_0/A) + b$, has been verified for many materials where P is the hydrostatic pressure necessary for extrusion, A_0 is the original cross-sectional area of the billet, A is the cross-sectional area of the extruded product, a is a constant characteristic of the material, and b is a small term mainly dependent on friction and die configuration. Johnson (1965) had found an expression of this form applicable to well lubricated conventional extrusion of aluminium and lead.

Another empirical relation (Pugh 1964) exists between extrusion pressure, extrusion ratio and hardness of the material H on the diamond pyramid scale. For 45° dies, the published results for a wide range of materials are in good agreement with $P = 0{\cdot}056H \ln(A_0/A)$. This provides a useful guide to the extrusion ratio attainable by a given hydrostatic pressure for materials of known hardness.

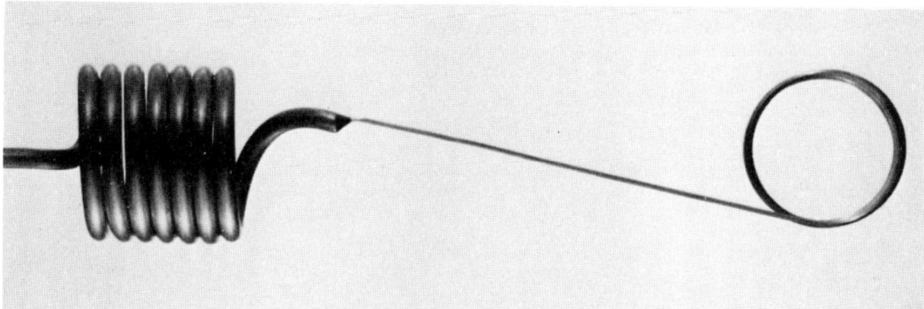

FIGURE 4. Extrusion of a coiled billet into wire.

Because there is no contact between the billet and the container, the shape and length of the billet to be extruded are limited only by the size of the pressure vessel, and not by frictional forces. Thus, for example, a billet can revolve in the extrusion container and this makes it possible to use a spirally grooved die to provide spirally finned rod at a low extrusion ratio, and a coiled billet can easily be extruded into wire (figure 4).

In addition, frictional forces at the die may be substantially decreased because pressurized fluid acts as a lubricant immediately accessible at the die entry. Because of this absence of friction, extrusion pressures are considerably lower in the hydrostatic method than in a comparable conventional operation. The pressure level in hydrostatic extrusion is practically constant throughout extrusion—a fact which contributes greatly to accuracy of product dimensions. There is a general tendency for the extrusion pressure at any extrusion ratio to fall, with decreasing die angle, to a minimum value, which is determined by a combination of increasing friction and decreasing 'redundant work' (the work involved in changing the direction of metal flow at the die entrance and die throat). As the die angle is further decreased past the point of minimum extrusion pressure, lubrication becomes less effective and frictional forces continue to increase, requiring an increased extrusion pressure for a given extrusion ratio.

The ability to use small angle dies in hydrostatic extrusion, and therefore to eliminate much of the redundant work, and the virtual elimination of frictional forces have already been mentioned. Thus, for a given extrusion ratio, less force on the ram is required than in conventional cold extrusion, or for a given available force, higher extrusion ratios can be attained in a single operation.

Because of the lower degree of deformation in small angle dies, the amount of cold work put into the material and emerging in the product is less than in conventional extrusion and there is a more uniform pattern of metal flow into the product. For this reason and also because of the complete absence of friction between billet and container the characteristic 'extrusion defect'—a pipe appearing at the base of a billet towards the end of conventional extrusion—is eliminated, making it possible to extrude a higher proportion of the billet, although to achieve this certain additional features must be incorporated in the tooling design.

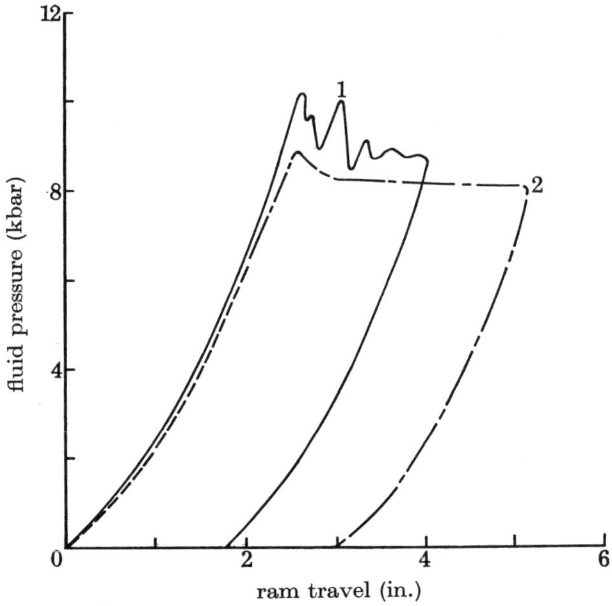

FIGURE 5. Variation of pressure during extrusion of aluminium alloy. Curve 1: 'stick–slip' motion, poor lubrication; curve 2: uniform motion, good lubrication. Extrusion ratio, 20 : 1; die angle 45°.

The mechanical properties are generally more uniform across the diameter of a hydrostatically extruded rod, and the tendency to grain growth in some aluminium alloys as the result of cold-work and recrystallization is suppressed. Subsequent heat treatment, without grain growth, may also be possible. There may also be the advantages of 'retained ductility' as the result of the material having been subjected to hydrostatic stress.

The accuracy of extruded sections is outstanding because the pressure, the strain on the tools and the rate of heat generation remain constant during extrusion. The surface finish obtained on simple sections in hydrostatic extrusion is generally better than that in conventional extrusion and with appropriate precautions can be excellent.

The die in hydrostatic extrusion may be supported by the pressure in the extrusion liquid and so may be much lighter and cheaper than a conventional die, and because of this the hand mounting of dies in large machines presents no difficulty.

There are, of course, disadvantages and difficulties. One of the most important is illustrated by the pressure time traces in figure 5, plotted by Fiorentino and his colleagues of the Battelle Memorial Institute. The pressure in the fluid rises to a maximum, and falls slightly when metal begins to move through the die; the coefficient of friction falls, allowing the billet to accelerate and the fluid to expand until the driving force falls below that needed for extrusion; and the cycle is repeated. Not only can this 'stick–slip' motion give rise to undesirable markings on the surface of the extruded product, but it makes control of the extrusion, especially towards the end of the billet, very difficult, and it may be impossible to avoid complete extrusion, with loss of fluid. 'Stick–slip' may be minimized by a proper combination of fluid and lubricant and virtually eliminated by the recent development by Green of the U.K.A.E.A. of augmented hydrostatic extrusion (*The Engineer* 1966), where an additional controlling force is applied to the billet.

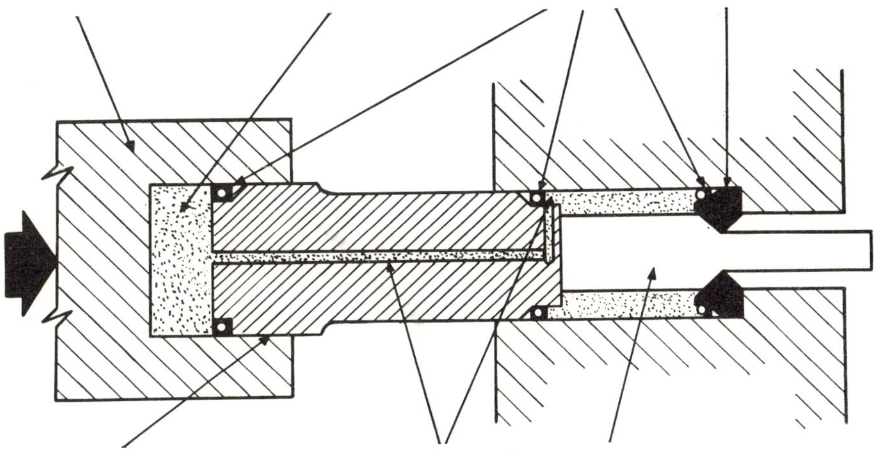

FIGURE 6. Augmented hydrostatic extrusion.

In hydrostatic drawing, where a tensile stress is applied to the product, the usual limit to reduction of area by a single die, imposed by the yield stress in tension of the product, is increased by the extra deformation allowed by the hydrostatic stress. Alternatively, the hydrostatic pressure may be regarded as supplemented by the tensile stress exerted by the drawbench. Current work by Thompson of the U.K.A.E.A. on hydrostatic drawing of stainless steel tubes promises, with further development, to give products that compare favourably with those obtained by the conventional methods of cold reducing and drawing, in a smaller number of operations and with fewer interstage anneals.

In augmented hydrostatic extrusion an additional pushing force is applied to the billet producing an axial compressive stress higher than that which would be caused by hydrostatic pressure by itself. This allows extrusion to be carried out at lower liquid pressures than by hydrostatic pressure alone, for a given extrusion ratio.

As shown in figure 6, a ram generates fluid pressure in the cylinder on the left, which is linked by the augmenting punch to the extrusion container on the right, of smaller diameter. A channel through the augmenting punch ensures that pressures in the two cylinders are equal and a constant out-of-balance force arising from the difference in cross-sections is transmitted directly to the billet.

Fatigue of pressure vessels

The main pressure vessels and ancillary high-pressure equipment in a hydrostatic extrusion press must be designed to withstand repeated applications of the operating pressure. At relatively low extrusion pressures, conventional monobloc vessels are in the low endurance range, e.g. a low alloy steel monobloc vessel with diameter ratio of 4 operating at 7 to 8 kbar can withstand only 10^4 pressure repetitions before failure. It is essential to know the fatigue behaviour of all proposed high-pressure vessel designs, to ensure a reliable and economic choice of container. Until recently, little was known of the fatigue behaviour of simple vessels with high diameter ratios operating at high pressures, and virtually no fatigue work has been attempted on more complex designs.

Much of the early fatigue work on pressurized thick cylinders was carried out at Bristol by Morrison, Crossland & Parry (1960) who examined a wide range of vessels with low diameter ratios (external diam./internal diam. less than 3) at pressures up to 5 kbar; because of this relatively low pressure the failures were all in the high endurance régime, i.e. exceeding 10^5 cycles to failure. The work has since been extended (Austin & Crossland 1965) into the lower endurance region on cylinders with diameter ratios of 1·4 and 1·8. Their results indicated that a maximum elastic shear stress criterion applied, although some deviation was apparent at the lower endurances. This is consistent with the occurrence of plastic deformation in the bore of a cylinder in this region.

In considering the fatigue behaviour of monobloc vessels with high diameter ratios at higher pressures, it became essential to take plastic deformation into account. Jones & Tomkins (1967) therefore considered the fatigue behaviour of very thick tubes using a plastic strain criterion of failure, which is well established in low endurance fatigue work and is consistent with the basic fatigue mechanisms. On the basis of this criterion, the endurance of a cylinder is determined by the magnitude of the maximum plastic shear strain range experienced by the cylinder during each pressure cycle and pressure-endurance curves can be predicted for high diameter ratio vessels using basic materials fatigue data. The predictions were confirmed by tests on cylinders of mild steel with diameter ratios of up to 7.

It is likely that greatly increased endurances in vessels operating at very high pressures may be achieved if very high precompressions can be set up in the bore material. Multibloc segmented and fluid interface vessels offer the possibility of doing this, but difficulties in the construction of such vessels offset the benefits gained in fatigue endurance which must be considerable for them to be worthwhile.

The inner cylinder of a duobloc system is sliced radially to produce individual segments or sectors, which act as load spreaders, their only function being to transmit radial compressive forces to an outer cylinder. Consequently the segments may be made in material of high compressive strength but perhaps weak in tension as is common in some of the hardest materials. The origin of the segmented vessel is not clear, but perhaps the first published work is attributable to Poulter (1951). Another variation was due to Ballhausen (1963) who supported the outer surface of his segments with a low liquid pressure. This aspect is being studied at Imperial College (Lengyel, Burns & Prasad 1963), with particular reference to the importance of achieving high fatigue life for industrial applications, where a sudden failure could disrupt production schedules. Designs incorporating static and fluctuating supporting fluid pressures have been constructed and are being evaluated.

Semi-continuous operation

Although the length of a billet that can be extruded hydrostatically is, in theory, limited only by the length of the billet container, it is inconvenient to construct very long high-pressure containers; and there may be instances where it is desirable to feed into the apparatus very long billets that are not entirely contained by a high-pressure cylinder.

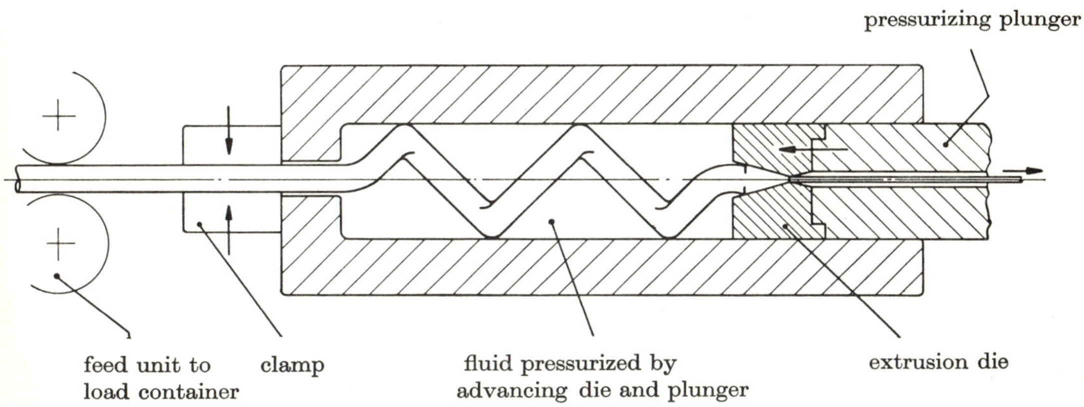

Figure 7. Semi-continuous hydrostatic extrusion of wire.

Truly continuous operation has not yet been obtained, but there are at least two methods for semi-continuous operation, where new material is fed into the pressure vessel between successive extrusion strokes. The problem is to introduce the material into the high-pressure liquid chamber without leakage of fluid, without danger of expelling the workpiece backwards, and without undue interruption of the process.

In the semi-continuous hydrostatic extrusion of wire ('SCHEW' process) invented at Imperial College, London (Alexander & Lengyel 1965), advantage is taken of the ability of coiled billets to be extruded by hydrostatic pressure. The three stages of operation can be followed from figure 7. With the liquid at atmospheric pressure, the clamp at the right-hand end of the cylinder is released and wire is forced into the apparatus, with a tapered end protruding through the die, and allowed to bend into a convenient shape, with the aim of accommodating as much wire as possible in the container. The clamp is tightened, sealing the system, and pressure is applied—in the original design, by forcing the die which is mounted on the ram, into the cylinder. The fluid pressure extrudes the wire through the die and into the atmosphere.

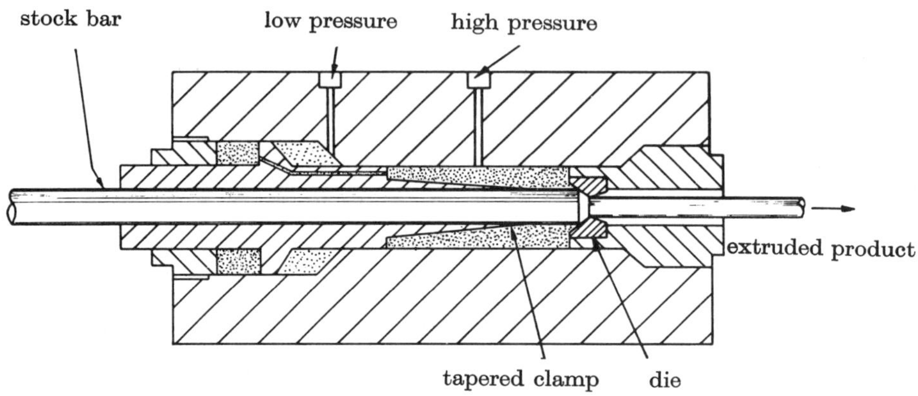

FIGURE 8. R.F.L. apparatus for semi-continuous extrusion of rigid bar.

Movement of the ram is stopped and the pressure released before the wire in the cylinder has completely straightened, and the clamp is released. Finally the ram and die are withdrawn to their starting positions, more wire is fed through the clamp and the cycle is repeated.

A design for semi-continuous extrusion of rigid bar developed at the Reactor Fuel Laboratories of the U.K.A.E.A. (Slater & Green 1967) is based on the augmenting punch previously described in that the feedstock passes longitudinally through the bore of the punch, and through a 'tail tube' into the atmosphere (figure 8). A tapered clamp operated by the hydrostatic pressure holds the feedstock securely without causing deformation, and transmits the extrusion force from the augmenting ram to the feedstock. As the ram moves, liquid flows through the channel to maintain the augmenting load.

THE 1600/80 'HYDROSTAT'

The principle of augmentation has been incorporated in a 1600-ton press (figure 9) built and designed by Fielding and Platt Ltd. to the specification of the U.K.A.E.A. and installed at the Reactor Fuel Laboratories of the U.K.A.E.A. at Springfields, near Preston, and financed jointly by the Ministry of Technology and by the manufacturers.

Billets up to $4\frac{1}{2}$ in. diameter and 20 in. long can be extruded under hydrostatic pressures up to 12 kbar, with augmenting loads up to 150 tons force. Dies are mounted externally on a movable carriage and are easily changed as required.

Operation of the press, after commissioning troubles had been overcome, has presented few problems. Most of the work so far carried out has been on extrusion of aluminium and its alloys and of copper, both in simple rod and tube shapes and in complicated sections.

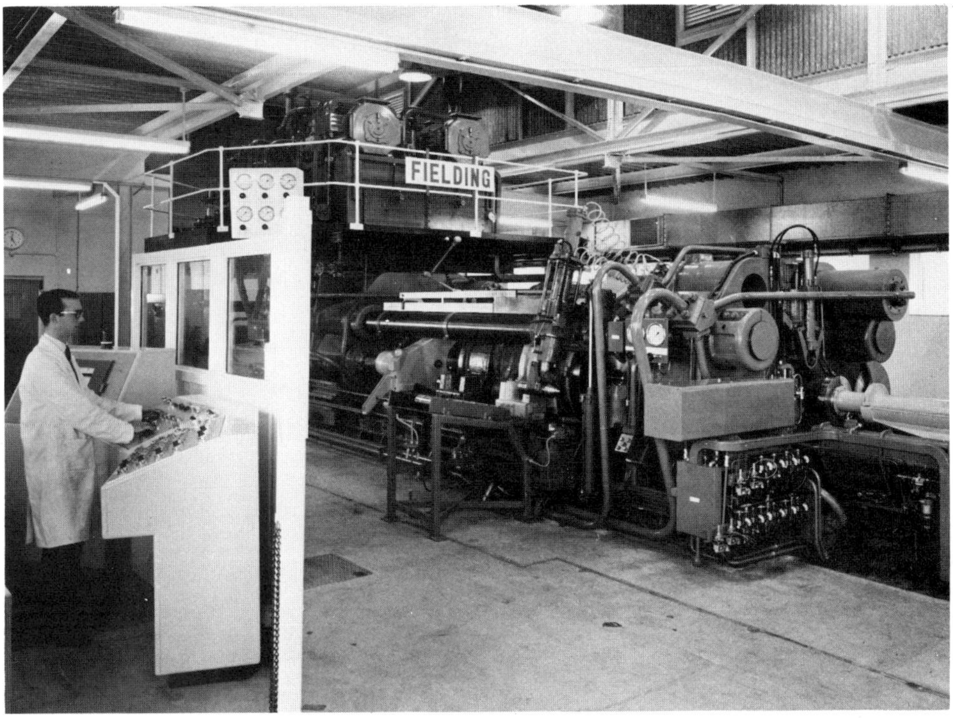

(Photograph by courtesy of Fielding & Platt Ltd.)

FIGURE 9. 'Hydrostat' at Reactor Fuel Laboratories, U.K.A.E.A.

The dimensional accuracy has been unprecedented—for example, on a 1 in. diameter bar of aluminium, extruded from a $4\frac{1}{2}$ in. diameter billet with excellent surface finish, the variation in diameter over a length of 20 ft. was less than 5 μm.

Very high extrusion ratios—greater than 1000 to 1—have been achieved with aluminium at a pressure of only 10 kbar, assisted by augmentation; and substantial lengths of product have been obtained with no measurable wear on the dies and consequently no changes in dimensions.

The present state of the art

To sum up: although hydrostatic extrusion was first proposed at about the same time as conventional hot extrusion of metals was introduced on a commercial scale, it has only recently become of real interest outside the laboratory.

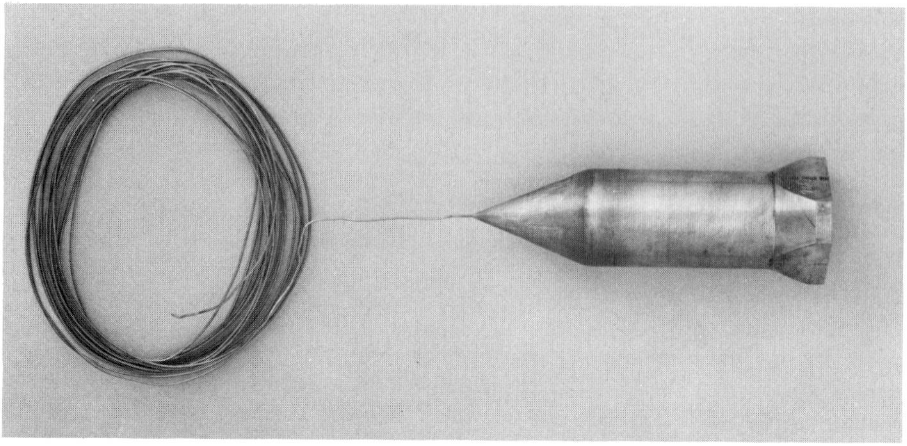

FIGURE 10a. Large reduction (99·5 % Al).

FIGURE 10b. Cladding: lead sheath on piano wire.

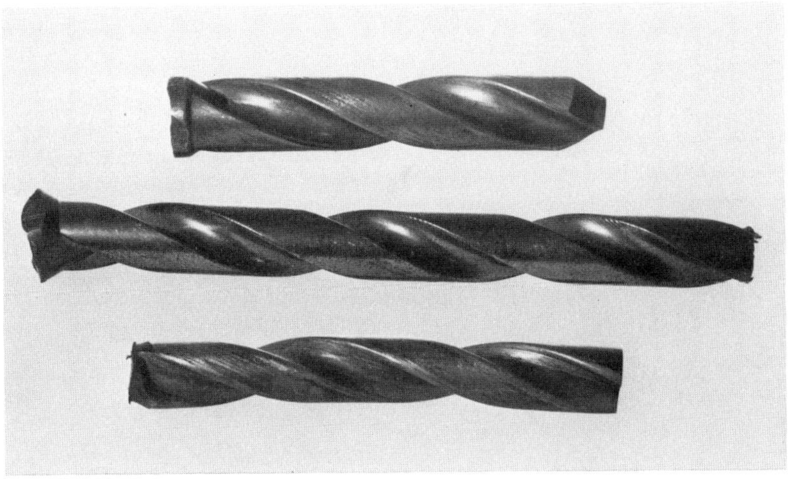

FIGURE 10*c*. Twist drill section products.

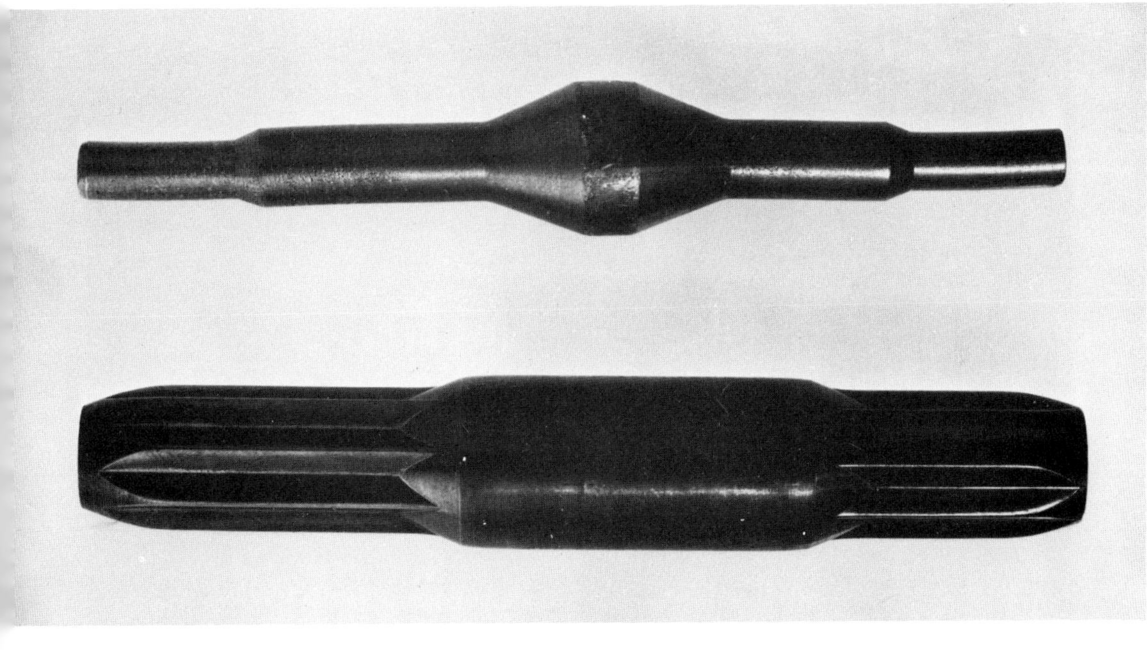

FIGURE 10*d*. Double-ended products.

FIGURE 11. Extrusions and discards produced on U.K.A.E.A 1600/80 'Hydrostat' in aluminium, copper and mild steel.

The pressures involved are high; it was not possible to contain them satisfactorily until steels strong enough for pressure vessels and dies had become available, and seals and other equipment had been developed.

The main advantages over conventional extrusion are:

(1) There is no friction between billet and container, so that the extrusion force at a given temperature and extrusion ratio is reduced. Billets of any length or shape can be extruded, and it is possible to develop semi-continuous methods for manufacturing long single rods, tubes or wires.

(2) The friction between billet and die is greatly reduced by the lubricating action of the pressurized fluid, so that a die with small included angle, although longer than a conventional die, can be used without increasing the frictional force; and the work of deformation is therefore reduced.

(3) The environment of compressive stress increases the ductility of the material, by extending the stress-strain curve, and is of value not only in hydrostatic extrusion, but also in extrusion of flanges and in the forming of tubes, cups, etc. under hydrostatic pressure.

The first phase of investigating the effects of hydrostatic pressure on metals and devising techniques to harness these effects, is virtually complete; during the last few years several workers, notably in the United Kingdom, have progressed far in the second phase of development—to convert the former laboratory curiosity into practical industrial processes, by constructing a variety of production scale machines capable of operation by industrial staff. There is still work to be done along these lines.

We are now entering the third phase of development—the widespread adoption of the technique by industry. To provide opportunities for exploitation, the Ministry of Technology has provided funds for equipment in Government establishments, and the Science Research Council is supporting research in universities.

The large-scale equipment is now available for commercial firms to use in tests of their own materials. Several firms have already taken advantage of this opportunity at the Reactor Fuel Laboratories of the U.K.A.E.A., for a gratifyingly wide range of materials and applications.

It is not our responsibility in the U.K.A.E.A. to assess the commercial value of these proposed applications, but it is a fact that several firms are becoming increasingly reluctant to allow the results of their tests to be disclosed, from which we may conclude that they expect to gain advantages over their competitors by making use of hydrostatic extrusion for particular purposes.

Figures 10 and 11 show a number of examples of extrusions made by this technique.

Future developments

It seems likely that there may be a demand for hydrostatic extrusion to produce certain sections in common materials where surface finish, dimensional accuracy or enhanced mechanical properties are of prime importance, while for the newer, stronger metals and alloys it should be possible to replace some established processes by making the same items more cheaply and to extrude more complicated sections than can currently be manufactured.

It seems unlikely that there will be an ever-upward tendency in pressure, but with improved container materials, and perhaps the use of segmented and pressure supported vessels, pressures of 20 kbar may become common.

Should these prove inadequate for the most arduous duties, there is the possibility of warm hydrostatic extrusion, with billets preheated and rapidly extruded before temperatures fall, as in conventional extrusion. Provision for heating would however add to manufacturing costs and remove one of the main attractions of hydrostatic extrusion

As industry gains confidence in the technique and identifies areas where its advantages are clearly defined, it seems more probable that machines will be installed for specific metal forming operations than that general purpose equipment like the 1600-ton press at the U.K.A.E.A.'s laboratory will be built in substantial numbers.

The complaint is often heard that British industry is more reluctant than its overseas competitors to exploit new techniques. It is to be hoped that hydrostatic extrusion, which owes much to pioneering work by British scientists and engineers, will not provide an example of this alleged lack of commercial enterprise.

I wish to record my appreciation of the courteous help of Professor J. M. Alexander, Imperial College of Science and Technology, in providing the illustration used in figure 7, and of Dr H. Ll. D. Pugh, National Engineering Laboratory, for the photographs, reproduced in figure 10, of extruded products made in his laboratory.

References

Alexander, J. M. & Lengyel, B. 1965 *J. Inst. Metals* **93**, 137–144.

Alexander, J. M. & Lengyel, B. 1966 *Proc. Instn Mech. Engrs, Lond.* **180**, pt 31, 317.

Austin, B. A. & Crossland, B. 1965 *Proc. Instn Mech. Engrs, Lond.* **180**, 134–145.

Ballhausen, C. 1963 U.S. Patent No. 3084388, April 1963.

Beresnev, B. I., Vereschagin, L. F. & Ryabinin, Yu. N. 1957 *Izv. Akad Nauk SSSR, Otdel. Tekh Nauk* **5**, 48–55.

Beresnev, B. I., Vereschagin, L. F. & Ryabinin, Yu. N. 1958 *Fiz. Metal Metalloved.* **6**, 144–146.

Beresnev, B. I., Vereschagin, L. F., Ryabinin, Yu. N. & Livshits, L. D. 1963 *Some problems of large plastic deformation of metals at high pressures.* New York: Macmillan and Co.

Bobrowsky, A. & Stack, E. A. 1965 *Final Report on NASA contract NASW*-742. February 1965.

Bridgman, P. W. 1949 *The physics of high pressure.* London: G. Bell and Sons.

Bridgman, P. W. 1950 *Proc. Roy. Soc.* A **203**, 1–17.

Bridgman, P. W. 1952 *Large plastic flow and fracture.* New York and London: McGraw-Hill.

Fiorentino, R. J., Richardson, B. D., Sabroff, A. M. & Boulger, F. W. 1967 *Proceedings of International Conference on Manufacturing Technology* 1967, pp. 941–954. Dearborn, Mich.: American Society of Tool and Manufacturing Eingineers.

Fuchs, F. J. 1965 *Am. Soc. Mech. Engrs Publ.* no. 65-PROD-17.

Green, D. 1965 *J. Inst. Metals* **93**, 65–70.
Institute of Metals 1962 Discussion on five papers on extrusion, London, 23 March 1961. *J. Inst. Metals* **90**, 200–215.
Johnson, W. 1965 *J. Inst. Metals* **85**, 403–408.
Jones, P. M. & Tomkins, B. 1967 *Proc. Instn Mech. Engrs, Lond.* **182**, pt. 3C, 81–90.
Lengyel, B., Burns, D. J. & Prasad, L. V. 1967 *Advances in machine tool design and research.* Oxford and New York: Pergamon Press Ltd.
Morrison, J. L. M., Crossland, B. & Parry, J. S. 1960 *Proc. Instn Mech. Engrs, Lond.* **174**, 95–117.
Poulter, T. C. 1951 U.S. Patent no. 2554499, May 1951.
Pugh, H. Ll. D. 1964 *A.S.T.M. Spec. Tech. Publ.* no. 374, pp. 68–137.
Pugh, H. Ll. D. 1965 *Recent developments in cold farming.* Bulleid Memorial Lectures, 1965, 111B, p. (3) 8. University of Nottingham.
Pugh, H. Ll. D. & Ashcroft, K. 1962 *Symposium on the physics and chemistry of high pressures,* pp. 163–176. London: The Society of Chemical Industry.
Pugh, H. Ll. D. & Green, D. 1958 *MERL Plast. Rep.* no. 147.
Pugh, H. Ll. D. & Gunn, D. 1962 *Symposium on the physics and chemistry of high pressures,* pp. 157–162. London: The Society of Chemical Industry.
Robertson, J. 1893 British Patent no. 19356.
Sauvè, C. 1965 *Rev. Metall.* **62**, 587–602.
Slater, H. K. & Green, D. 1967 *Proc. Instn Mech. Engrs, Lond.* **182**, pt. 3C, 217–222.
The Engineer 1966 Development in hydrostatic extrusion. *The Engineer*, **221**, 59–61.
Thompson, P. J. 1968 M.Sc. Thesis, University of Aston.

The Croonian Lecture, 1968

The haemoglobin molecule

By M. F. Perutz, F.R.S.

Medical Research Council Laboratory of Molecular Biology, Cambridge

(*Lecture delivered* 30 *May* 1968—*MS received* 30 *September* 1968)

Proceedings of the Royal Society, B, *volume* 173, *pp.* 113–140, 1969

Introduction

Haemoglobin is the respiratory protein of the red blood cells which carries oxygen from the lungs to the tissues and facilitates, both directly and indirectly, the return transport of carbon dioxide. Mammalian haemoglobin has a molecular weight of 64500 and contains two pairs of polypeptide chains: the α-chains with 141 amino acid residues each and the β-chains with 146. Each chain is combined with one haem. Myoglobin, the oxygen carrier of muscle, is closely related to haemoglobin, but has a simpler constitution: it consists of only one polypeptide chain of 153 residues and a single haem. The amino acid sequences of the myoglobins and haemoglobins of man and of several animals have been determined (Dayhoff & Eck 1968).

By means of the method of isomorphous replacement with heavy atoms, X-ray analysis of sperm whale myoglobin at 2·0 Å resolution provided the first solution of the structure of a protein (Kendrew *et al.* 1960; Watson 1969). All but 21 of its 153 residues form part of helices; over most of their length these helices have conformations closely resembling the right-handed α-helix of Pauling & Corey (1951). The chain is divided into 8 helical segments, separated by corners or non-helical regions. Together these form a kind of basket into which the haem group fits neatly, so that only its propionic acid side-chains protrude into the surrounding liquid (figures 1, 2). X-ray analysis at 5·5 Å resolution showed each chain of horse haemoglobin to be folded in much the same way as the single chain of sperm whale myoglobin. The 4 chains are arranged tetrahedrally, each carrying one haem in a pocket near the protein surface. The chemically identical halves of the molecule are related by a twofold symmetry axis (figure 3, plate 6; Cullis *et al.* 1962).

Each haem group has one atom of iron capable of combining reversibly with molecular oxygen while remaining in the ferrous state. The reaction of myoglobin with oxygen follows bimolecular kinetics, so that its oxygen equilibrium curve is a simple hyperbola; its oxygen affinity is independent of pH. On the other hand, the oxygen affinity of a haemoglobin solution rises with the degree of its saturation with oxygen, giving the oxygen equilibrium curve a sigmoidal shape which is attributed to interaction between the 4 haems in one molecule. The oxygen affinity varies

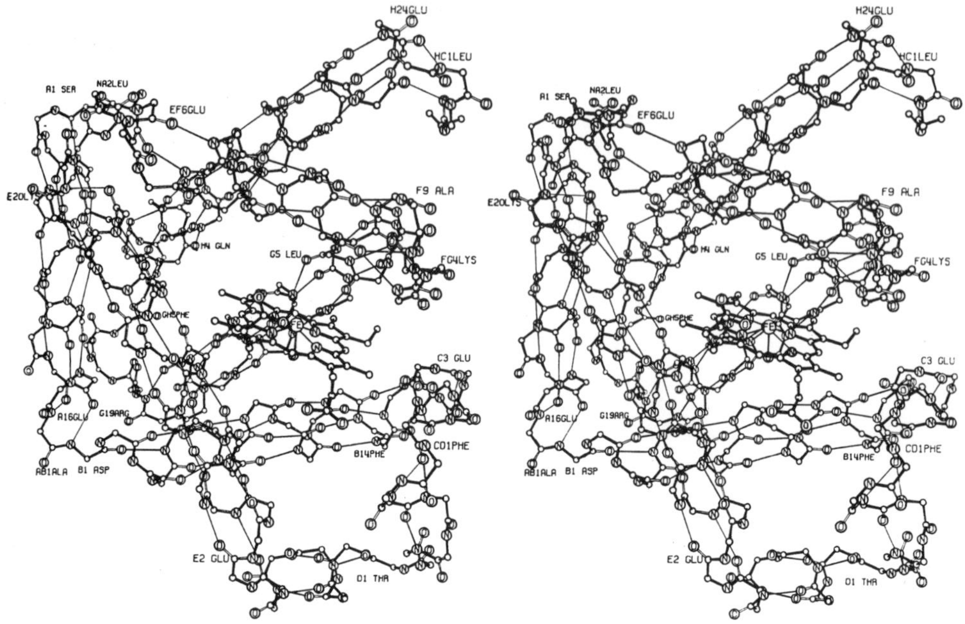

FIGURE 1. Stereoscopic drawing of the main polypeptide chain, without the side chains, and of the haem groups of myoglobin. Covalent bonds of the main chain are shown as double lines, hydrogen bonds as single lines and the bonds of the haem group as bold lines. The displacement of one of the pyrollic nitrogens from the plane of the porphyrin ring is due to error. (This drawing was prepared by Dr C. K. Johnson, using a modified version of his computer program OR.TEP and is reproduced, by permission, from Watson (1969)).

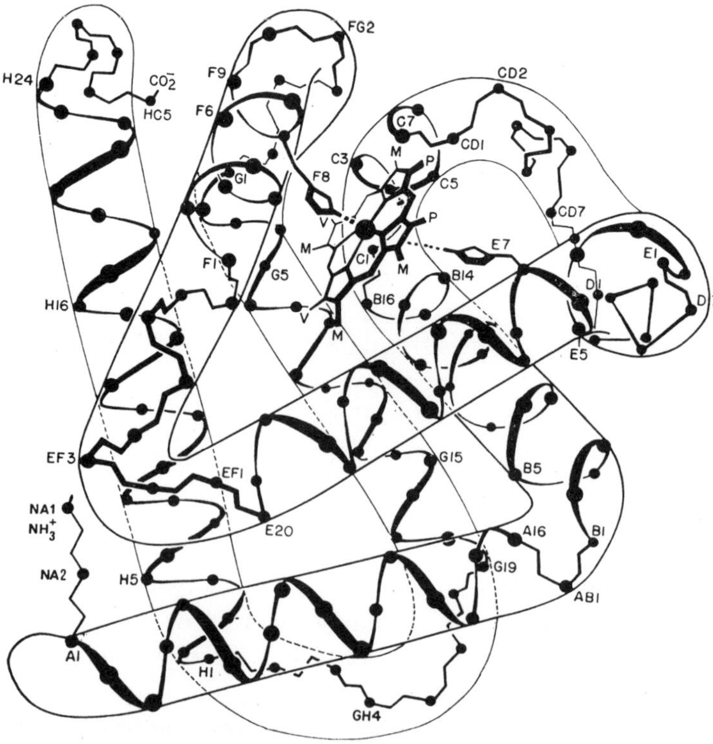

FIGURE 2. Diagrammatic drawing showing the folding of the polypeptide chain in myoglobin and the notation of amino acid residues in helical and non-helical regions (Kendrew *et al.* 1961; reproduced, by permission, from Dickerson (1964).

with pH, being lowest near pH 6 and high near pH 9; this is known as the Bohr effect and means that low concentrations of hydrogen ion facilitate the uptake of oxygen. Conversely, low concentrations of oxygen facilitate the uptake of hydrogen ion by haemoglobin. By raising the pH of the venous blood and thereby allowing it to take up bicarbonate, this provides the main mechanism for the return transport of carbon dioxide from the tissues back to the lungs. A subsidiary mechanism depends on the direct combination of deoxyhaemoglobin with carbon dioxide (Roughton 1964). Deoxyhaemoglobin also has a high affinity for certain phosphates which are present in red blood cells, especially for 2,3-diphosphoglycerate; these appear to have the function of lowering its oxygen affinity in a physiologically advantageous manner (Chanutin & Curnish 1967; Benesch, Benesch & Yu 1968; Benesch & Benesch 1969). The affinities of haemoglobin for oxygen, carbon ioxide, hydrogen ion and 2,3-diphosphoglycerate are interdependent, so that the concentration of any one of these ligands affects its combination with all the others (Wyman 1968).

The interaction between the ligands depends on a change of structure which the haemoglobin molecule undergoes on combination with oxygen. The molecule exists in two alternative forms: the deoxy-form, which is taken up when the oxygen-combining sites at the iron atoms are empty, and the oxy-form, which is taken up when they are filled. However, they need not be filled with oxygen, nor need the iron atoms remain in the ferrous state. Carbonmonoxyhaemoglobin and all the ferric derivatives of haemoglobin have the same structure as oxyhaemoglobin, which implies that the valency and spin state of the iron atom have no influence on the conformation of the haemoglobin molecule (Perutz 1953). Oxy- and carbonmonoxyhaemoglobin have been reported to differ in the free energy changes accompanying certain reactions (Guidotti 1967*a*, *b*), but these amount to no more than a few hundred calories and may be based on conformational changes too small to be detected by X-ray analysis.

Physiologically, the two states of haemoglobin are distinguished by their affinity for the different ligands. The deoxy-form has a low affinity for oxygen and a high affinity for hydrogen ion, carbon dioxide and 2,3-diphosphoglycerate. The oxy-form has a 300 times greater affinity for oxygen and a lower affinity for the other three ligands. The moment of transition from the deoxy to the oxy-form is not yet clear, but it seems that it needs combination with oxygen by more than two of the 4 iron atoms (Gibson & Parkhurst 1968).

So far, X-ray analysis has furnished three-dimensional electron density maps of deoxyhaemoglobin of man and horse at 5·5 Å resolution (Muirhead *et al.* 1967; Bolton, Cox & Perutz 1968), and of oxyhaemoglobin of horse at 5·5 Å and at 2·8 Å resolution (Cullis *et al.* 1962; Perutz, Muirhead, Cox, Goaman, Mathews, McGandy & Webb 1968; Perutz, Muirhead, Cox & Goaman 1968). The twofold symmetry axis relating the chemically identical halves of the molecule persists in the deoxy-form. At the limited resolution of 5·5 Å the structures of the deoxyhaemoglobins of man and horse were the same within the experimental error. Oxy- and deoxyhaemoglobin of horse, on the other hand, were found to exhibit marked differences in quaternary structure (figure 4, plate 7, table 1), but differences in secondary and tertiary structure, though present, were no larger than the experimental error. This showed that the stereochemical changes produced within the individual subunits by the combination with oxygen must be small and require a resolution higher than 5·5 Å for their detection. It was with this aim in view that the

Table 1. Change in quaternary structure in going from oxy- to deoxyhaemoglobin

Mathematically, any movement of a body can be most simply represented by a rotation about, combined with a shift along, a suitably placed screw axis. In order to describe the change from oxy- to deoxyhaemoglobin in this way, Muirhead *et al.* (1967) chose a Cartesian coordinate system with its origin at the centre of mass of the four haem groups in the oxyhaemoglobin molecule. One of its axes (X_2) was chosen parallel to the molecular dyad, the second (X_1) as the bisector of the larger of the two angles joining corresponding haem groups, and the third (X_3) so as to complete a right-handed system.

(*a*) Defines the positions and orientations of the four screw axes about which each of the subunits has to be moved. Each axis is defined by the Cartesian coordinates O_1, O_2 and O_3 of a point on the axis and by the angles RX_1, RX_2 and RX_3 which it encloses with the Cartesian coordinate axes. δ^0 gives the angles of rotation of the subunits about each of the screw axes and S the shifts along them. The latter turn out to be negligible. The screw axes are illustrated in figure 4, plate 7.

	point on screw axis (Å)			direction of screw axis			angle of rotation about screw axis	shift along screw axis
	O_1	O_2	O_3	RX_1^0	RX_2^0	RX_3^0	(δ^0)	S (Å)
α_1	10·5	7·4	−3·1	124·4	57·2	128·5	+9·38	−0·19
	±0·5	±0·5	±0·5	±1·5	±1·5	±1·5	±0·25	±0·05
α_2	−10·5	7·4	3·1	55·6	57·2	51·5	+9·38	−0·19
	±0·5	±0·5	±0·5	±1·5	±1·5	±1·5	±0·25	±0·05
β_1	12·1	12·8	−1·6	135·4	50·7	107·7	+7·43	−0·18
	±0·8	±0·8	±0·8	±2·0	±2·0	±2·0	±0·25	±0·08
β_2	−12·1	12·8	1·6	44·6	50·7	72·3	+7·43	−0·18
	±0·8	±0·8	±0·8	±2·0	±2·0	±2·0	±0·25	±0·08

(*b*) Comparison of distances between iron atoms in ångströms in deoxy- and oxyhaemoglobin

		deoxyhaemoglobin	oxyhaemoglobin
Fe_1–Fe_2	β_1–β_2	39·9	33·4
Fe_3–Fe_4	α_2–α_1	34·9	36·0
Fe_1–Fe_3	α_2–β_1	24·7	25·0
Fe_2–Fe_4	α_1–β_2	24·6	25·0
Fe_1–Fe_4	α_1–β_1	36·9	35·0
Fe_2–Fe_3	α_2–β_2	36·9	35·0

Description of plate 6

Figure 3. Model of the horse haemoglobin molecule derived from the three-dimensional Fourier synthesis at 5·5 Å resolution (Cullis *et al.* 1962). (*a*) View along the *a* axis, showing segments involved in the contact $\alpha_1\beta_1$. (*b*) View perpendicular to the *a* axis, showing segments involved in the contact $\alpha_1\beta_2$.

For legend see facing page.

PLATE 6

(*Facing p.* 176)

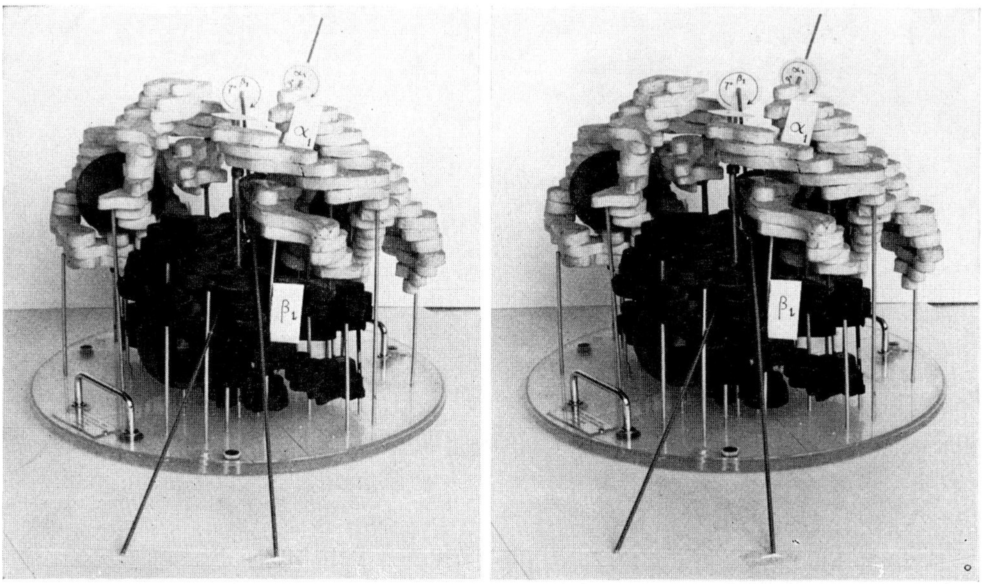

FIGURE 4. Stereophotograph showing the change in quaternary structure from oxy- to deoxyhaemoglobin. The model illustrates horse oxyhaemoglobin. On deoxygenation the subunit marked α_1 turns about the axis marked α_1 by 9·4°; the subunit marked β_1 turns about the axis marked β_1 by 7·4°. Both rotations take place in a clockwise direction relative to the twofold axis of symmetry at the centre of the molecule. The two subunits at the back of the model turn about axes which are related to the ones at the front by the twofold axis of symmetry.

PLATE 7

present Fourier synthesis of horse oxyhaemoglobin at 2·8 Å resolution was determined. It has given a picture of the electron density distribution sufficiently detailed for the construction of an atomic model of the molecule. Its description forms the main topic of this lecture. The model rules out certain mechanisms of haem–haem interaction which have been proposed, but it still does not reveal what the actual mechanism is.

The electron density map

To obtain an electron density map at 2·8 Å resolution, the phase angles of 8000 X-ray reflexions were determined by the method of isomorphous replacement, using crystals of native haemoglobin and of three heavy atom derivatives. The map shows the positions of most of the amino acids residues, the orientation of their side chains and some details of the haem groups.

In its overall shape the oxyhaemoglobin molecule resembles a spheroid with a length of 64 Å, a width of 55 Å and a height of 50 Å. The packing of the molecules in the crystal is pseudo-face-centred-cubic (Cullis *et al.* 1962). Neighbouring molecules are in contact at only a few points, and salt solution fills the wide interstices between them; it also fills an internal cavity surrounding the dyad symmetry axis. In the contour maps these liquid regions form a low undulating landscape from which the protein peaks stand out sharply (figures 5, 6). The polypeptide backbone is visible as a continuous chain of high density from which side chains branch off in different directions (figure 6).

Some of the side chains are sufficiently characteristic in size and shape to be identified uniquely. These include tryptophan where the indole ring is unmistakable (figure 5), cysteine and methionine, recognizable by their high sulphur peaks (figure 8), and some of the lysines (figures 5, 7), arginines, tyrosines and prolines. Nearly all phenylalanines and histidines are recognizable as rings (figure 8) but cannot be distinguished from each other. Leucines, valines, threonines, the acidic amino acids and amides form bulges of various shapes, but not often of a recognizable kind. Serine is hard to tell from alanine, and alanine from glycine.

Despite these ambiguities, nearly every amino acid residue can be located by making use of the known chemical sequence of the 141 residues in the α-chains (Matsuda, Gehring-Müller & Braunitzer 1963) and the 146 residues in the β-chain (Smith 1963, 1968; table 2). Starting from a clearly recognizable side chain such as tryptophan 14α or 15β, one can proceed along the polypeptide chain in steps of 3·6 Å and identify each side chain in turn, checking one's advance by the recognizable signposts which are passed on the way. The terminal residues are visible only faintly or not at all, presumably because they are able to take up two or more alternative conformations. For the same reason many of the polar side chains are either wholly or partly invisible.

The haem groups stand out as disks surrounding a central peak, linked to the polypeptide chain on one side by a continuous region of high density representing the proximal histidine (figures 6, 7). The distal histidines are also distinct. The four

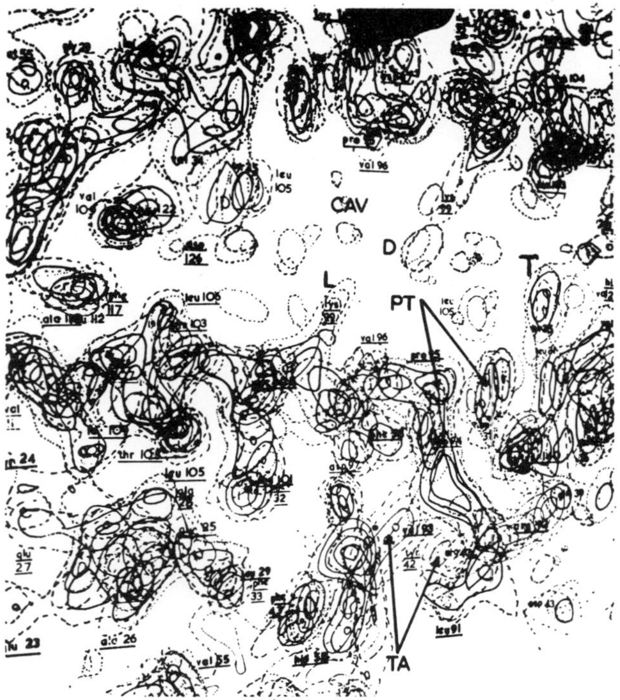

5

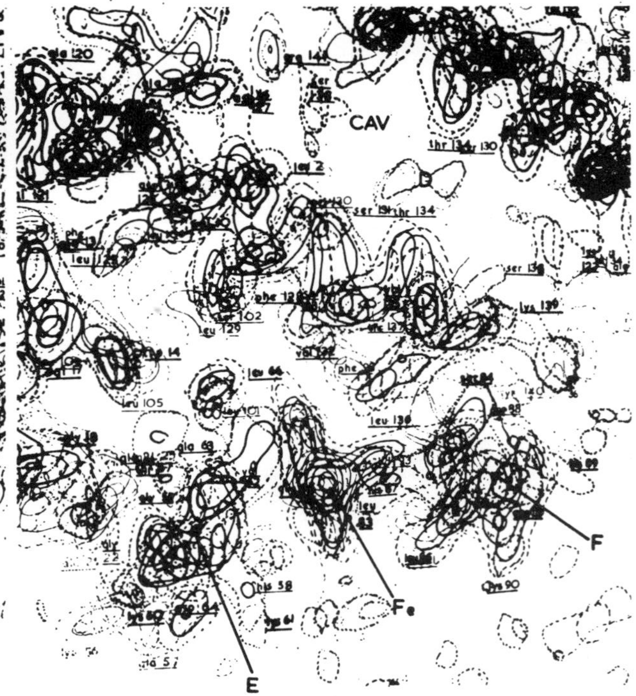

6

FIGURES 5 to 8. Contour maps showing the distribution of electron density in the molecule of horse oxyhaemoglobin. The first contour, shown as a broken line, was drawn at 0·30 electron/Å^3 above the average background; higher contours are drawn as solid lines at intervals of approximately 0·30 electron/Å^3 above the first contour. Each figure shows a view through several sheets of Perspex stacked on top of each other, representing a different part of the molecule. Underlined residues form part of the α-chains, plain ones part of the β-chains.

FIGURE 5. The central cavity (CAV) with parts of the four subunits. The α-chains are on the lower left and upper right; D marks the molecular dyad. PT refers to proline 95α (left) and tryptophan 37β (right) at the contact $\alpha_1\beta_2$. TA refers to tyrosine 42α (left) and the guanidinium group of arginine 40β (right) which form another part of the same contact (see figure 15). Note lysine 99α (L) above and below the molecular dyad and tyrosine 35β (T) slightly above and to the right of tryptophan 37β. Helix G runs horizontally from right to left just below the central cavity. The interstitial liquid appears at the bottom right.

FIGURE 6. The haem pocket of the α-chain. The haem group is seen edge-on, with helices E and F to the left and right.

7

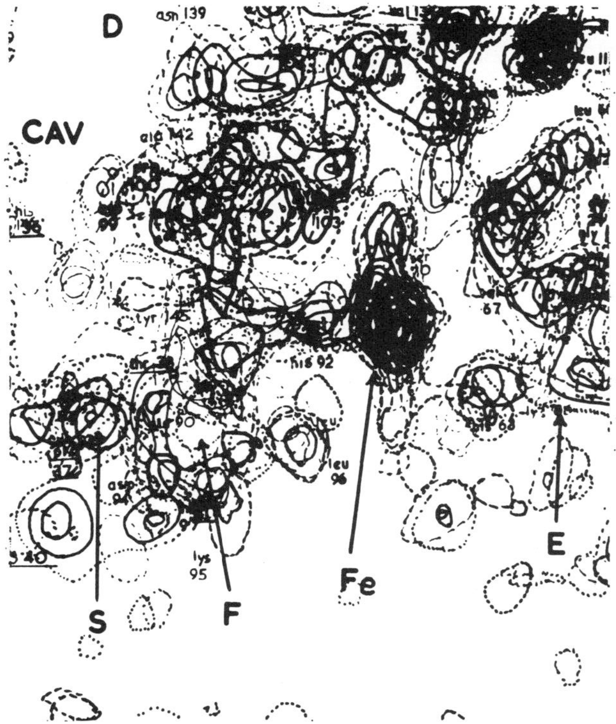

8

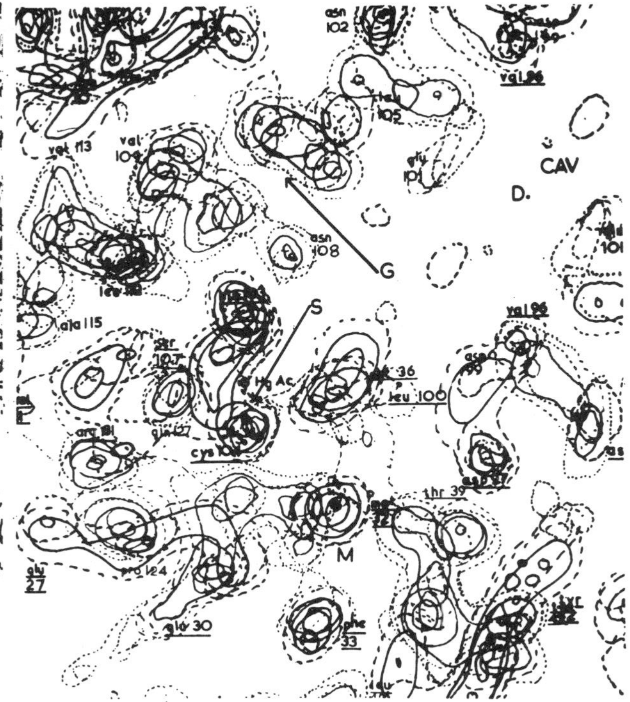

FIGURE 6. (*cont.*)

The haem-linked histidine is visible as a bridge connecting the iron atom to helix F. Helix H traverses the picture from the top left to the middle right. The interstitial liquid appears at the bottom. (See also figures 17 and 19.)

FIGURE 7. The haem pocket of the β-chain. The haem group is seen edge-on, with helices F and E to the left and right. The haem-linked histidine (92) and the distal one (68) are both visible. On the left of helix F lies the reactive cysteine (93, marked S) with the attached mercury atom shown as a black dot. The interstitial liquid is at the bottom of the figure. (See also figures 17 and 18.)

FIGURE 8. The unreactive cysteine 104α, marked S, forming part of helix Gα is shown embedded between a variety of other residues. Note helix Gβ, which runs from the top right to the middle left. M marks methionine 32α which forms part of helix B.

Table 2. Amino acid sequence of horse haemoglobin

	α	residue number in segments		β			α	residue number in segments		β	
		NA	1	Val	1	49	Ser		8	Ser	49
1	Val		2	Gln	2			D	1	Asx	50
2	Leu		3	Leu	3				2	Pro	51
3	Ser	A	1	Ser	4				3	Gly	52
4	Ala		2	Gly	5				4	Ala	53
5	Ala		3	Glu	6				5	Val	54
6	Asp		4	Glu	7	50	His		6	Met	55
7	Lys		5	Lys	8	51	Gly		7	Gly	56
8	Thr		6	Ala	9	52	Ser	E	1	Asn	57
9	Asn		7	Ala	10	53	Ala		2	Pro	58
10	Val		8	Val	11	54	Gln		3	Lys	59
11	Lys		9	Leu	12	55	Val		4	Val	60
12	Ala		10	Ala	13	56	Lys		5	Lys	61
13	Ala		11	Leu	14	57	Ala		6	Ala	62
14	Trp		12	Trp	15	58	His		7	His	63
15	Ser		13	Asp	16	59	Gly		8	Gly	64
16	Lys		14	Lys	17	60	Lys Gln		9	Lys	65
17	Val		15	Val	18	61	Lys		10	Lys	66
18	Gly		16			62	Val		11	Val	67
19	Gly	AB	1			63	Ala		12	Leu	68
20	His	B	1	Asn	19	64	Asp		13	His	69
21	Ala		2	Glx	20	65	Gly		14	Ser	70
22	Gly		3	Glx	21	66	Leu		15	Phe	71
23	Glu		4	Glu	22	67	Thr		16	Gly	72
24	Tyr Phe		5	Val	23	68	Leu		17	Glx	73
25	Gly		6	Gly	24	69	Ala		18	Gly	74
26	Ala		7	Gly	25	70	Val		19	Val	75
27	Glu		8	Glu	26	71	Gly		20	His	76
28	Ala		9	Ala	27	72	His	EF	1	His	77
29	Leu		10	Leu	28	73	Leu		2	Leu	78
30	Glu		11	Gly	29	74	Asp		3	Asp	79
31	Arg		12	Arg	30	75	Asp		4	Asn	80
32	Met		13	Leu	31	76	Leu		5	Leu	81
33	Phe		14	Leu	32	77	Pro		6	Lys	82
34	Leu		15	Val	33	78	Gly		7	Gly	83
35	Gly		16	Val	34	79	Ala		8	Thr	84
36	Phe	C	1	Tyr	35	80	Leu	F	1	Phe	85
37	Pro		2	Pro	36	81	Ser		2	Ala	86
38	Thr		3	Trp	37	82	Asp		3	Ala	87
39	Thr		4	Thr	38	83	Leu		4	Leu	88
40	Lys		5	Gln	39	84	Ser		5	Ser	89
41	Thr		6	Arg	40	85	Asp		6	Glu	90
42	Tyr		7	Phe	41	86	Leu		7	Leu	91
43	Phe	CD	1	Phe	42	87	His		8	His	92
44	Pro		2	Asp	43	88	Ala		9	Cys H	93
45	His		3	Ser	44	89	His	FG	1	Asp	94
46	Phe		4	Phe	45	90	Lys		2	Lys	95
47	Asp		5	Gly	46	91	Leu		3	Leu	96
			6	Asp	47	92	Arg		4	His	97
48	Leu		7	Leu	48	93	Val		5	Val	98

TABLE 2 (*cont.*)

	α	residue number in segments		β			α	residue number in segments		β	
94	Asp	G	1	Asp	99	118	Thr	H	1	Thr	123
95	Pro		2	Pro	100	119	Pro		2	Pro	124
96	Val		3	Glu	101	120	Ala		3	Glx	125
97	Asn		4	Asn	102	121	Val		4	Leu	126
98	Phe		5	Phe	103	122	His		5	Gln	127
99	Lys		6	Arg	104	123	Ala		6	Ala	128
100	Leu		7	Leu	105	124	Ser		7	Ser	129
101	Leu		8	Leu	106	125	Leu		8	Tyr	130
102	Ser		9	Gly	107	126	Asp		9	Gln	131
103	His		10	Asn	108	127	Lys		10	Lys	132
104	Cys H		11	Val	109	128	Phe		11	Val	133
105	Leu		12	Leu	110	129	Leu		12	Val	134
106	Leu		13	Ala	111	130	Ser		13	Ala	135
107	Ser		14	Leu	112	131	Ser		14	Gly	136
108	Thr		15	Val	113	132	Val		15	Val	137
109	Leu		16	Val	114	133	Ser		16	Ala	138
110	Ala		17	Ala	115	134	Thr		17	Asn	139
111	Val		18	Arg	116	135	Val		18	Ala	140
112	His		19	His	117	136	Leu		19	Leu	141
113	Leu	GH	1	Phe	118	137	Thr		20	Ala	142
114	Pro		2	Gly	119	138	Ser		21	His	143
115	Asn		3	Lys	120	139	Lys	HC	1	Lys	144
116	Asp		4	Asx	121	140	Tyr		2	Tyr	145
117	Phe		5	Phe	122	141	Arg		3	His	146

(Matsuda *et al.* 1963; Smith 1963, 1968).

pyrrolic rings form a cross of high density within each disk. At first their side groups were thought to be indistinct, but careful correlation with a model later allowed some of them to be traced. Note that access to the haem groups and to cysteine 93β from the interstitial liquid is not obstructed by neighbouring molecules. The unreactive cysteine 104α, on the other hand, is enclosed by other residues on all sides and is accessible only from the central cavity (figure 8).

The horse haemoglobin crystals used here contain up to four components which arise by substitution of glutamine for lysine at 60α and of tyrosine for phenylalanine at 24α (Kilmartin & Clegg 1967). The two electrophoretically distinct components have been crystallized separately and were found to be isomorphous (Perutz *et al.* 1959). There is no reason to believe that the substitution at 24α, being internal, would affect the crystal structure. In the electron density maps the side chains at 24 and 60α are both distinct, but the lysine is cut short and the tyrosine hydroxyl is less dense than that of the other tyrosines.

The results which emerge from the Fourier synthesis are remarkably clear. Although the electron density maps by themselves would not suffice to solve the structure, when combined with the chemical sequence and the known stereochemistry of the amino acids they provide enough information for the construction of an atomic model.

Method of construction and determination of atomic coordinates

At the present resolution, the amide groups of the polypeptide chains appear as straight segments of density hinged at the α-carbon atoms which are also the points where the side chains can be seen to branch off. To start with, the approximate coordinates of the α-carbon atoms and the centres of the side chains were read off the electron density maps. A skeletal model of the molecule was constructed on a scale of 2 cm = 1 Å. The coordinates of all its atoms were read with a plumb-line and marked on the maps. Shifts indicated by atomic positions being 'off' the electron density peaks were then used to adjust the model. The process was repeated until the atomic positions coincided exactly with the peaks. At the same time, care was taken to construct only stereochemically allowed conformations and to keep non-bonded atoms at appropriate minimum distances.

Because of the finite size of the brass rods and other errors inherent in a mechanical model, the final coordinates were not sufficiently accurate to conform with accepted bond distances and angles. Our rough coordinates were therefore subjected to a mathematical model building and refinement procedure programmed in Fortran for a digital computer (Diamond 1966). This builds up a stereochemically correct polypeptide chain on the basis of images of amino acids as determined in small structures; it enables both the main chain and the side chains to rotate about single bonds, and uses the method of least squares to bring the atoms of the stereochemically correct model as close as possible to the coordinates determined directly from the brass model. At the end of the refinement the root mean square difference between initial and final coordinates was 0·4 Å, which is as good as can be expected. Uncertainties in the coordinates caused by the limited resolution of the Fourier synthesis are probably greater than this. It was difficult to judge the orientation of some of the amide groups, both in the main chain and the amide side chains, which made the coordinates especially of their oxygens subject to large errors. The coordinates of the C-terminal residues of the α-chains (140 and 141) and of some side chains which are not visible on the electron density maps were omitted from the refinement.

Notation

We shall use the notation which assigns to each residue letters and numbers corresponding to its structural rather than its sequential position (figure 2) (Kendrew *et al.* 1961). Starting from the amino end of the myoglobin chain, the letters A to H are assigned to the eight helical segments, and AB, BC, etc., to the corners or non-helical segments. NA and HC denote the non-helical segments at the amino and carboxyl ends of the chain. Residues within each segment are numbered from the amino end. They are taken to be part of a helix if either their amino or their carbonyl group is hydrogen-bonded within the helix. Table 2 correlates the sequential and structural notations. In the description which follows, each residue will be given its structural notation followed by its sequential number in parentheses. The positions of many of the helical and non-helical segments of haemoglobin are shown in figure 3.

In the α-chains the positions of amides are known (Matsuda *et al.* 1963). In the β-chains amides are known to occur in positions 2 and 131 (Smith 1968). The positions of the remaining amides in the β-chain are still to be determined. In table 2 and in the text I have assigned amides and acids to correspond with homologies in the β-chain of human haemoglobin. Where there are no homologies I have referred to the residues as asx and glx. In the figures I labelled all residues which might be acids or amides as asx and glx.

Description of the model

The α-chains

The α-chains contain 141 residues each, compared with 153 in myoglobin. Of the twelve missing residues, six lie in helix D, which is absent in the α-chain, and six at the carboxyl end. Over most of its length the conformation of the chain is similar to that in myoglobin, but certain differences appear in both helical and non-helical segments (table 3). In myoglobin the short segment C is a distorted α-helix, while its

Table 3. Differences between the α and β-chains and myoglobin (M)

	α	β	M
no. of residues ...	141	146	153
segment			
NA	= M	contains 3 residues	contains 2 residues
A	= M	= M	—
AB	2 amides A16–B1 differ	−2 residues	—
B	= M	= M	—
BC	= M	= M	—
C	$3{\cdot}0_{10}$ helix	$3{\cdot}0_{10}$ helix	α-helix
CD	all different	2 amides CD5–7 differ	—
D	absent	= M	—
E	irregular	E18–20 irregular	1 kink
EF	3 amides EF2–5 differ, but $\alpha = \beta$	—	—
F	= M	= M	—
FG	= M	= M	—
G		G1–3 in $3{\cdot}0_{10}$ helix differ	—
GH	= M	amide GH1–2	—
H	contains 21 residues		contains 26 residues
HC	contains 3 residues		contains 4 residues

conformation in the α-chain approaches more closely that of a 3_{10} helix. In myoglobin the long helix E which carries the distal histidine contains one kink of 7°, but is an otherwise regular α-helix. In the α-chain it is irregular throughout, so that linear hydrogen-bonding from NH in one turn to CO in the next is the exception rather than the rule. In myoglobin the C-terminal helix H extends over twenty-six residues (from gly 124 to leu 149), but in the α-chain it extends over only twenty-one (from thr 118 to ser 138), the last three residues being non-helical. In myoglobin the last turn of each of the helices A, E and H tends towards the 3_{10} conformation; the same happens in the helices A, E, G and H of the α-chain. The carboxyl end of the helix F, on the other hand, is opened out towards the $4{\cdot}4_{16}$ (π) helix (Low & Grenville-Wells 1953) in both myoglobin and the α-chain. This consists of just one turn and is made up of residues FG1 – 4 (89 to 92 in the α-chain).

Of the non-helical segments only NA, BC and FG are similar to myoglobin. It is interesting to inquire why the others should be different. In myoglobin AB is a right-angled corner in which all residues are part of either helix A or B. Alanine AB1(19) has its imino group hydrogen-bonded to helix A and aspartic acid B1(20) has its carbonyl group hydrogen-bonded to helix B. In the α-chain the equivalent residues are glycines, which allows the amide group connecting them to adopt a conformation which would otherwise be forbidden (Ramachandran, Ramakrishnan & Sasisekharan 1963; Ramakrishnan & Ramachandran 1965). In consequence, glycine AB1(19) is not part of either helix. The absence of helix D and the occurrence of a proline at CD2(44) give the CD region of the α-chain a unique conformation. The EF region differs in the conformation of the three amide groups between the α-carbons of FG2 – 5. There are two reasons for this: glycine EF3(90) in myoglobin allows the preceding amide group to adopt a conformation which is forbidden by an aspartic acid at the equivalent position of the α-chain; moreover, EF2 in myoglobin is occupied by a lysine whose polar side chain extends into the surrounding liquid; while in the α-chain the equivalent residue is a leucine whose non-polar side chain is tucked away in a surface crevice.

The β-chains

The β-chains contain 146 residues each. Compared with myoglobin they contain one extra residue in the NA segment, two fewer at the AB corner and six fewer at the C-terminus. Helices C and H have conformations similar to those in the α-chain, HC1 – 3 being again non-helical. The last three residues of helix E (gly E18(74), val E19(75) and his E20(76)) form an irregular helix. FG1 – 4 form one turn of a π-helix as in the other two chains (figure 10). G1 – 3 (asp 99, pro 100 and glu 101) tend towards the 3_{10} conformation, which is usually found at the carboxyl rather than at the amino end of α-helices.

It is noteworthy that each of the five prolines of the β-chain occurs in position 2 of a helix. Generally, position 1 is occupied by a residue whose side chain forms a hydrogen bond with the main chain imino group of residue 4, thus 'closing off' the helix and forcing the main chain to turn away from it. Position 1 is occupied by asx in D, asx in E, asp in G and thr in H (figures 11 to 13). Similar sequences also occur in the helices G and H of the horse α-chains and in the helices A of the human α and β chains. In helix C a slightly different arrangement occurs; a proline at C2 is followed at C4 by a threonine whose OH group makes a hydrogen bond with the main chain CO of C1 (Watson 1969).

Of the non-helical segments, the AB corner has a unique conformation on account of the deletion of residues A16 and AB1 (figure 9). Valine A15(18) forms the last residue of helix of helix A and asn B1(19) the first of helix B; the carbonyl of its side chain makes a hydrogen bond with the main chain imino group of B4. Segments BC, FG and GH are similar to myoglobin and EF is similar to the α-chain. CD probably differs from myoglobin in the conformation of the amide groups between CD5–7. This may be due to glycine CD5(46) in the β-chain replacing a lysine in myoglobin, and allowing a conformation in the former which would be forbidden in the latter (Ramachandran *et al.* 1963; Ramakrishnan & Ramachandran 1965).

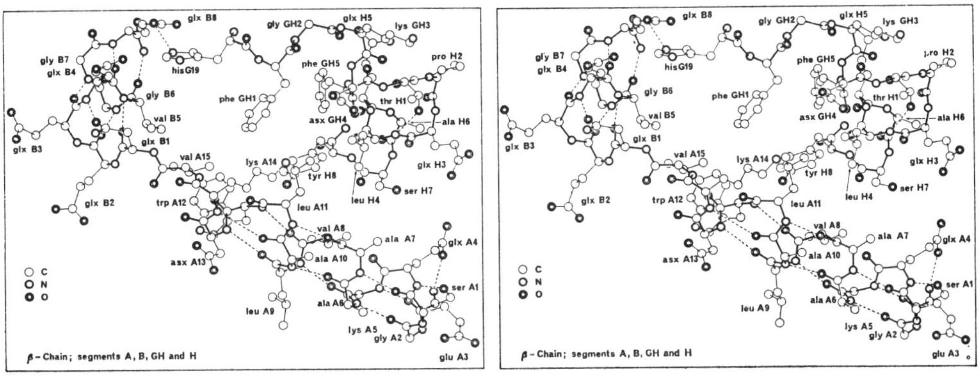

FIGURES 9 to 18. Stereoscopic drawings of the atomic model of horse oxyhaemoglobin. These should be viewed with a stereoviewer. (An inexpensive one is available from The Taylor Merchant Corporation 25 West 45th Street, New York, N.Y. 10036. More elaborate ones are made by C. F. Casella, Ltd., Regent House, Britannia Walk, London, N.1, and by the Lansing Instrument Corporation, Lansing, Michigan, U.S.A. A pair of enlarging lenses of focal length 85 mm, mounted 70 mm apart between cardboard is also adequate.) The main chains are drawn in bold lines and the side chains in thin lines. Broken lines indicate hydrogen bonds. Hydrogen bonds within helices have been omitted from some of the figures. Residues belonging to the α-chains are underlined; those belonging to the β-chains are not. The atomic positions in figures 17 and 18 are derived from photographs of the model, those of the others from Dr R. Diamond's model building program and Mr T. H. Gossling's model drawing program, as described by Perutz, Muirhead, Cox & Goaman (1968). Of the polar side chains on the surface of the β-chain, the only ones whose conformation emerges clearly from the electron density maps are lys E3, glx B2, B3, glu B8 and asx GH4. Lys E10 and GH3 and glu B4 are clear in parts. The conformation of the remaining ones has been drawn arbitrarily.

FIGURE 9. Segments A, B, GH and H of the β-chain. The side chain of lys A5 has been omitted. The amino ends of helices A, B, and H are stabilized by hydrogen bonds between the α-NH of residue 4 in the helix with the oxygen of serine A1, asn B1 and thr H1 respectively. Also note that the AB corner contains only one amide group which does not form part of either helix A or B.

General features

Polar residues are excluded from the interior of the α and β subunits, except for the occasional serine or threonine whose hydroxyl group is hydrogen bonded to a carbonyl group within the same α-helix. Glycines and alanines occur anywhere.

Larger non-polar side chains generally lie either in the interior or in surface crevices of the subunits or at the boundaries between unlike subunits. Non-polar side chains protruding into the surrounding water are rare, but occur occasionally; for example, cysteine F9(93) (figure 10) or leucine E17(68).

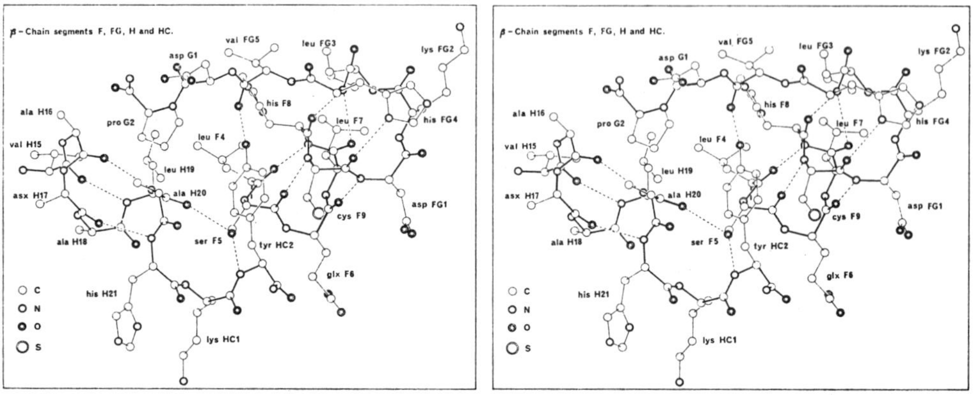

FIGURE 10. Segments F, FG, H and HC of the β-chain. Note the single turn of π-helix between his F7 and his FG4.

Contacts between α- and β-chains

The contacts are of two different kinds (figure 3); the residues involved in each of them are listed in table 4. $\alpha_1\beta_1$ is the more extensive of the two contacts made up of thirty-four residues and about 110 atoms coming within a distance of 4 Å of each other. The great majority of the interactions are non-polar. The five probable hydrogen bonds are all in contact with water, so that their contributions to the binding energy should be small. This expectation is not, however, borne out by the properties of certain abnormal human haemoglobins (Perutz & Lehmann 1968).

The greater part of the contact is formed by those residues of the two chains which lie between G10 and H9. It is flanked on the left by helices B and Dβ and on the right by helix Bα (figure 11). Figures 12 to 14 show some of the stereochemical details, such as the contact between proline H2(119) in α, and proline D2(51) and methionine D6(55) in β; the contact between the C-termini of the two helices G; and that between proline H2(124) in β, and arginine B12(31) and leucine B15(34) in α. The guanidinium groups of the three arginines in the contact all lie in surface crevices. There is a pair of reciprocal contacts between arginine B12 of one chain and the carbonyl of phenylalanine GH5 of its partner. In addition, arginine G18(116) β, which is linked strongly to glu B8(26)β, appears to make a secondary link with the main chain carbonyl of proline GH2(114)α. Two other probable hydrogen bonds are between aspartic acid H9(126)α and tyrosine C1(35)β (COO^- to HO); and between histidine G10(103)α and the main chain carbonyl of asn G10(108)β. Both these bonds are in the internal cavity.

The contact $\alpha_1\beta_2$ is made up of nineteen residues contributing about eighty atoms within a distance of 4 Å of each other (table 4*b*; figure 15). It contains one clearly marked hydrogen bond between the side chains of aspartic acid G1(94)α and asn G4(102)β, and another possible one between the side chains of threonine C6(41)α and histidine FG4(97)β. All other interactions are non-polar. The contact is made up chiefly of residues in helices C and G and in the non-helical segments FG. Figure 15 shows the contact between the indole ring of tryptophan C3(37) in β and aspartic acid G1(94) and proline G2(95) in α; this faces the internal cavity. Figure 16

TABLE 4. THE CONTACT α, β

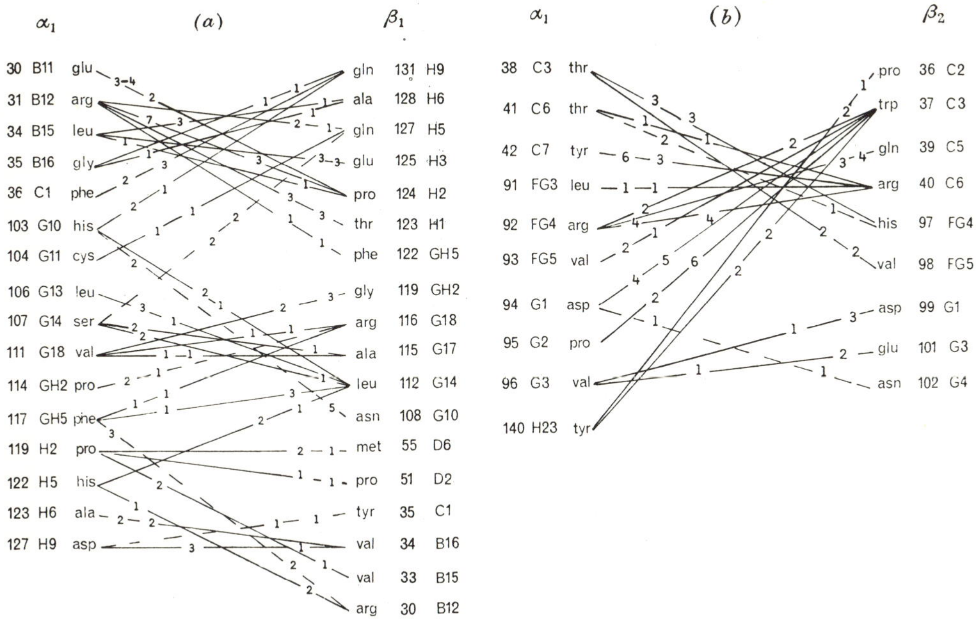

(*a*) 34 residues, including about 110 atoms in contact.

(*b*) 19 residues, including about 80 atoms in contact.

illustrates the contact at the molecular surface between the phenolic group of tyrosine C7(42)α and the guanidinium group of arginine C6(40)β. The two groups are parallel, which suggests a π-π type of interaction. The aliphatic chain of arginine C6(40)β, which crosses from β to α, touches that of arginine FG4(92)α, which crosses from α to β and lies directly above it. The complex (phe CD1-tyr C7)α-(argC6-phe C7-phe CD1)β, shown in figure 16, is the shortest bridge between any pair of haem groups in the molecule. Figure 16 also shows the contact between threonine C6(41)α and histidine FG4(97)β which may or may not contain a hydrogen bond.

In the transition from oxy- to deoxyhaemoglobin, movement in the contact $\alpha_1\beta_1$ is slight; relative to the α_1 subunit, the β_1 subunit rotates by 3·7° about a screw axis and moves by 0·3 Å along it (Muirhead *et al.* 1967; Bolton, Cox & Perutz 1968). The relative displacement of atoms at the contact is only about 1 Å. This contact is now seen to be extensive and to contain many interlocked side chains. Because of

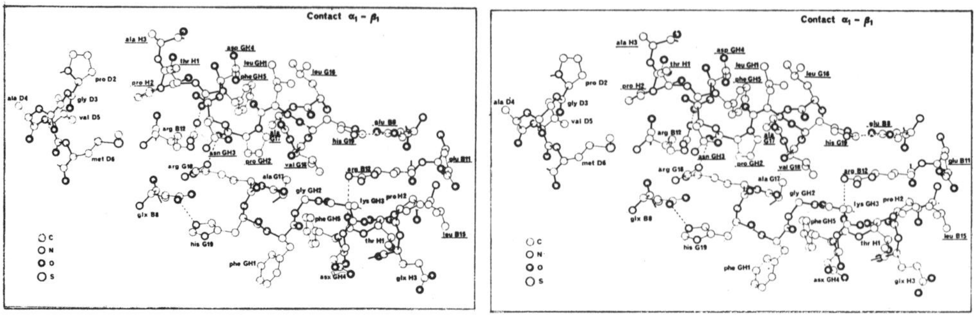

FIGURE 11. Contact $\alpha_1\beta_1$, viewed approximately parallel to a; b is vertical as in figure 3. Only those residues which lie near the surface of the molecule are shown. Note the three arginines in surface crevices.

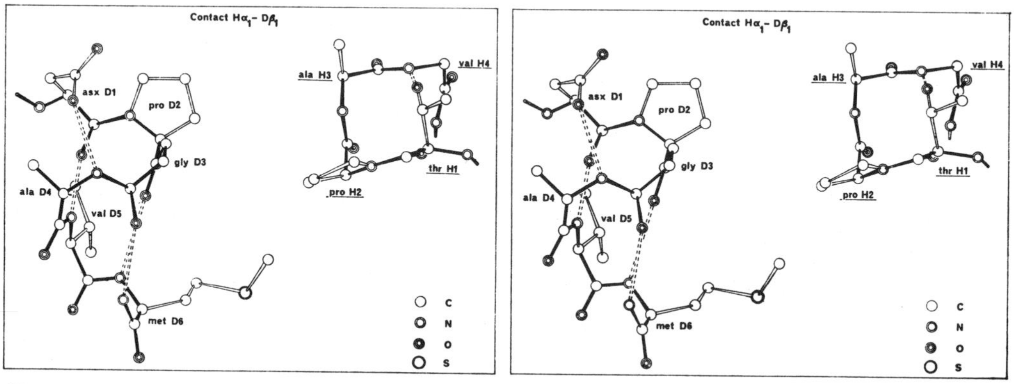

FIGURE 12. Enlarged view of the top left-hand corner of figuro 11, showing the van der Waals contact of proline H2α with proline D2β and methionine D6β. Note how the amino end of helix Hα is closed by the hydroxyl oxygen of threonine H1, making a hydrogen bond with NH of valine H4. The amino end of helix Dβ is closed by a hydrogen bond between the carbonyl of asx D1 with NH of alanine D4.

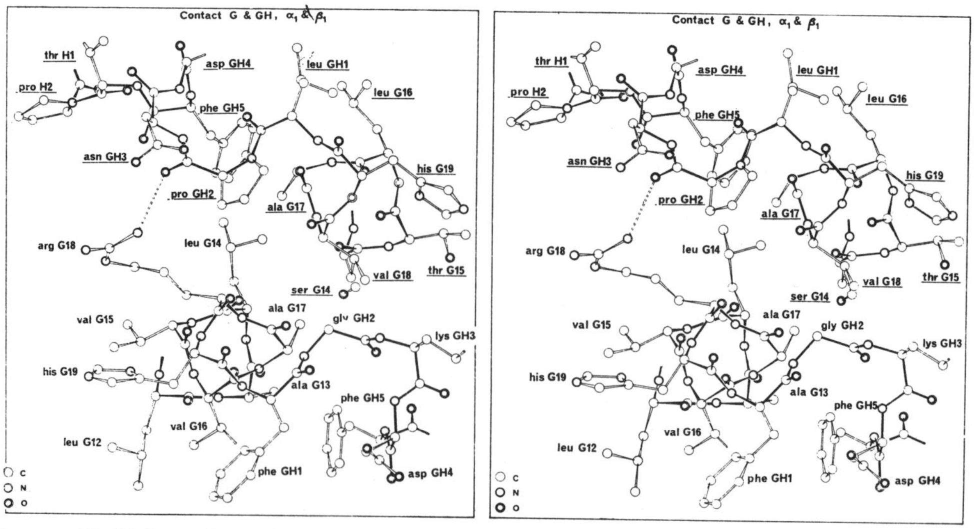

FIGURE 13. Enlarged view of the centre of figure 11, showing mainly the contacts between residues belonging to helices Gα and β. The entire depth of the contact is not shown because the overlap would make the diagram too confusing.

this interlocking a small change in secondary and tertiary structure might be translated into a larger change in quaternary structure.

Movement in the contact $\alpha_1\beta_2$, on the other hand, is large. Relative to the α_1 subunit, the β_2 subunit rotates by 13·5° about a screw axis and moves by 1·9 Å

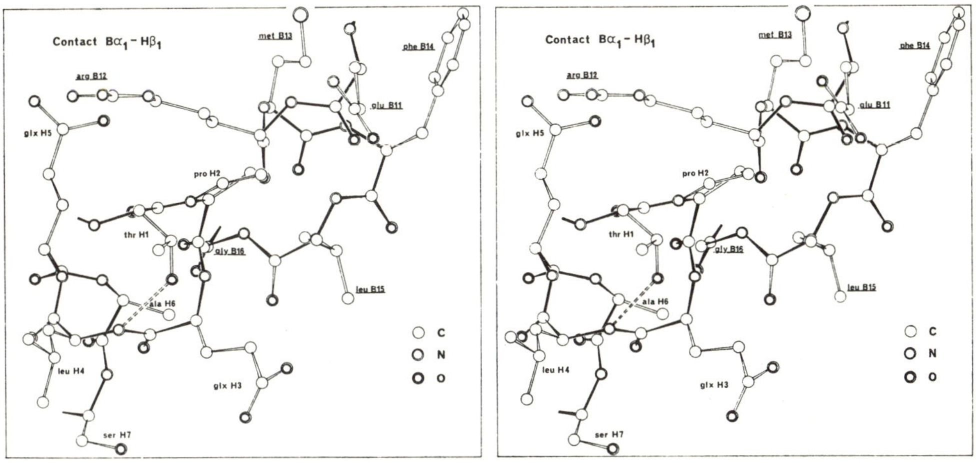

FIGURE 14. Enlarged view of the bottom right of figure 11, showing contacts between helices Bα and Hβ. Pro H2β makes contact with glu B11, arg B12 and leu B15α. Gly B16α makes contact with ala H6β and gln H9β (the latter not shown). Note that the amino end of helix Hβ is closed off by a threonine as in Hα.

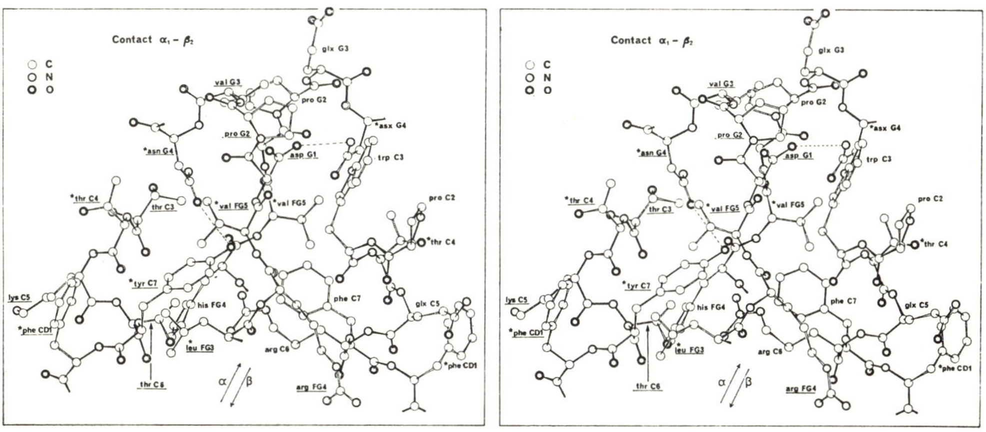

FIGURE 15. Contact $\alpha_1\beta_2$ viewed approximately along *b*; that is, looking vertically down parallel to the dyad axis in figure 3*b*. Residues in Cα are in contact with those of FGβ, and vice versa. The diagram includes all residues in this contact with the exception of tyrosine H23α which lies vertically above, and in van der Waals contact with, tryptophan C3β. The haem groups are not shown, but would be seen approximately end-on, one of them on top of the lower left and the other underneath the lower right of the diagram. Residues which are in contact with the haem groups are starred. The arrows at the bottom indicate the relative movements of the subunits on deoxygenation. The internal cavity is at the top, the molecular surface at the bottom. See also figure 5, which shows the wider surroundings of this contact, viewed from the same direction.

along it. This leads to relative displacements of atoms at the contact by as much as 5·7 Å, in the approximate directions indicated by the arrows in figure 15. The contact $\alpha_1\beta_2$ is seen to be both smaller and smoother than $\alpha_1\beta_1$, and is constructed so as to allow the two subunits to slide past one another. It is closely connected to the haem groups, so that any changes in the area of contact would be expected to affect their environment.

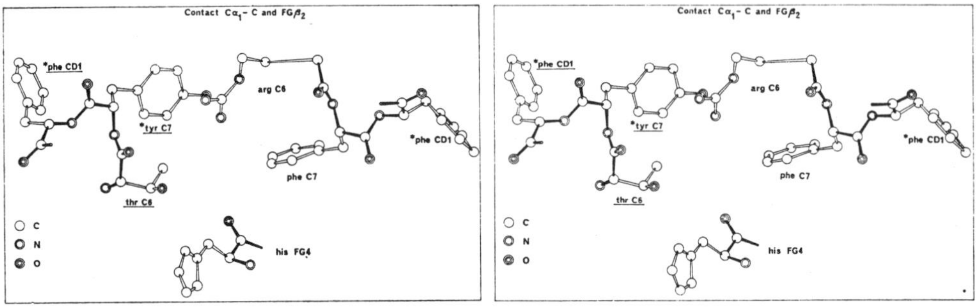

FIGURE 16. Detail of figure 15 viewed parallel to c; the b-axis is now vertical. It shows the van der Waals contact between tyrosine C7α and arginine C6β; also what may be either a van der Waals contact or a hydrogen bond between threonine C6α and histidine FG4β. For lack of knowledge the atoms of the imidazole ring are all shown as carbons. Residues in contact with the haem groups are starred. Leucine FG3β and valine FG5β, which immediately precede and follow histidine FG4β, are both haem contacts, so that any movement by them would affect the histidine.

Contacts between like chains

No contacts between like chains are visible on the maps, but they probably exist, at least at low salt concentrations, and may consist of salt bridges involving terminal residues. The only terminal residue which is clearly visible is valine 1α, and its amino group appears to be free. The others are not visible on the map. If one investigates their possible conformations, one finds that they could form salt bridges or hydrogen bonds of the kind listed here.

$$\alpha\text{–}\alpha\left\{\begin{array}{llll}\text{arg H24(141) COO}^- & \text{—} & \text{NH}_3^+ \text{ lys H10(127)} & \\ \text{arg H24(141) gua}^+ & \text{—} & \left[\begin{array}{c}\text{internal}\\ \text{anion}\end{array}\right] & \begin{array}{l}\text{— gua}^+ \text{ arg H24(141)}\\ \text{— NH}_3^+ \text{ lys G6(99)}\end{array} \\ \text{lys G6(99) NH}_3^+ & \text{—} & & \end{array}\right.$$

$$\beta\text{–}\beta\left\{\begin{array}{lll}\text{his H24(146) COO}^- & & \text{NH}_3^+ \text{ lys H10(132)} \\ \text{his H24(146) im} & \text{—} & \text{NH}_3^+ \text{ val NA1(1)}\end{array}\right.$$

Such salt bridges probably exist in dilute buffer solution but are weakened by the high concentration of ammonium sulphate in the crystal lattice. On deoxygenation the contact between the two β-chains is broken, and that between the two α-chains is altered (Muirhead *et al.* 1967; Bolton, Cox & Perutz 1968).

The internal cavity

This extends all the way along the molecular dyad axis, for a depth of 50 Å. Its shape is most easily represented by two boxes, each about 20 Å long, 8 to 10 Å wide and 25 Å deep; one separates the α-chains and the other the β-chains from each other. The two boxes are set one above the other with their 20 Å axes at right angles to the dyad and to each other, each box being open at the top and bottom (figures 5 to 7).

The internal cavity is linked with polar residues of all kinds. Especially noteworthy is the large number of serines and threonines facing the cavity along helix Hα. J. C. Kendrew has pointed out that the hydroxyl groups of these residues usually form a hydrogen bond with the main chain carbonyl four residues back along the helix. A. D. McLachlan suggested that these may actually protect the helix from attack by water.

The haem groups

As in myoglobin, the haems lie in non-polar pockets of the globin chains. Figure 19 shows an attempt at mapping the contacts between the α and β-chains and their haem group from measurements made on the model. Figures 17 and 18 show stereo-drawings of the haem groups with those parts of their amino acid side chains which lie near the surface of the molecule. Not counting the covalent bond between iron and histidine F8, there are about sixty interactions between atoms of the globin chains coming to within 4 Å of atoms of the haems; all but one of those in the α-chain and two of those in the β-chain are non-polar. In the α-chain the polar contact extends from one of the two propionic acid carboxyls to histidine CD3(45). In the β-chain, one propionic acid carboxyl makes contact with serine CD3(44) and the other probably with lysine E10(66). All polar contacts are in water so that their contribution to the binding energy should be small. Comparison of the amino acid sequences of different mammalian species shows all but one of the contacts in figure 19*a* to be common to all normal α-chains, and all but two of those in figure 19*b* to all normal β-chains. The exceptions are leucine H12(129)α and phenylalanine E15(71)β, which are both doubtful contacts, and serine CD3(44)β; this can be replaced by a histidine which would make an even better hydrogen bond with the propionic acide side chain of the haem group. The invariance of the residues surrounding the haem group implies that they are nearly all essential for the function of the haemoglobin molecule.

The ability of the ferrous iron in haemoglobin to combine reversibly with molecular oxygen is clearly due to its non-polar surroundings. This was predicted by Nakahara & Wang (1958) and first established in myoglobin (Kendrew 1962).

Tyrosines and tryptophans

Each of the chains contains three tyrosines. A tyrosine at H23 has been found in all vertebrate haemoglobins and myoglobins analysed so far. In myoglobin its phenolic hydroxyl is hydrogen-bonded to the main chain carbonyl of residue FG5

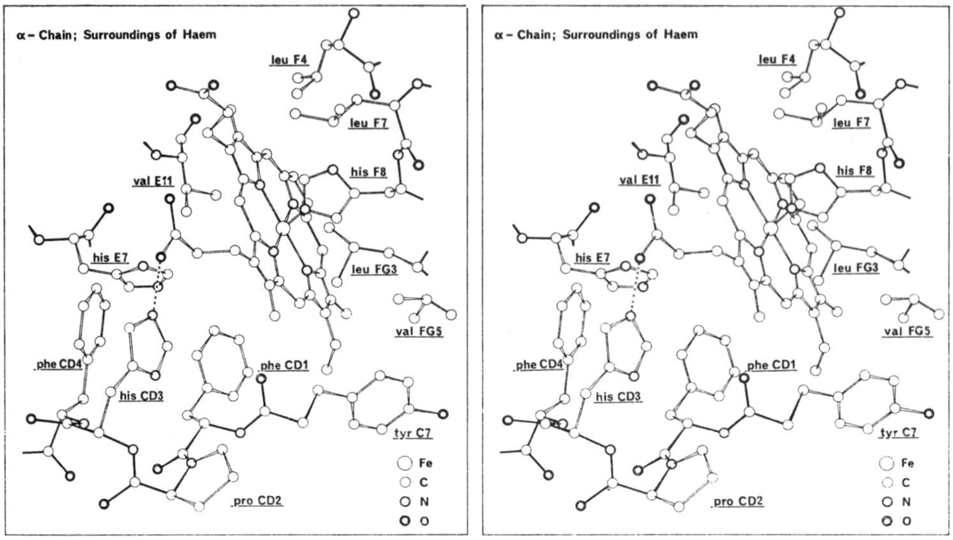

FIGURE 17

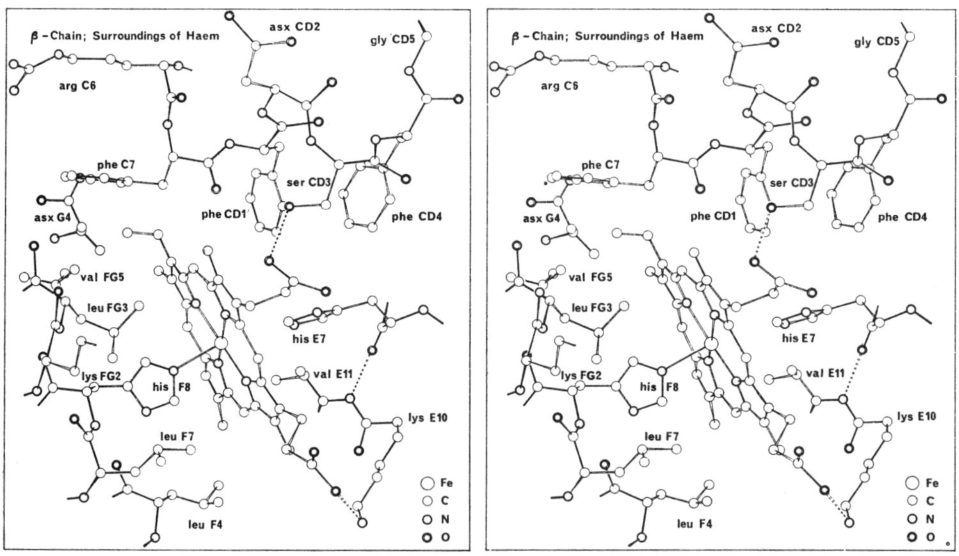

FIGURE 18

FIGURES 17, 18. Haem group with surrounding residues of the α and β-chains. Only the residues belonging to segments CD, E and F, which lie near the surface, are shown. The interior contacts belonging to segments G and H are omitted. Distortions of the porphyrin ring in figure 18 have resulted because some of the atoms were obscured in the photographs used for making the drawing, so that their positions had to be guessed. The orientation of the vinyl groups is arbitrary.

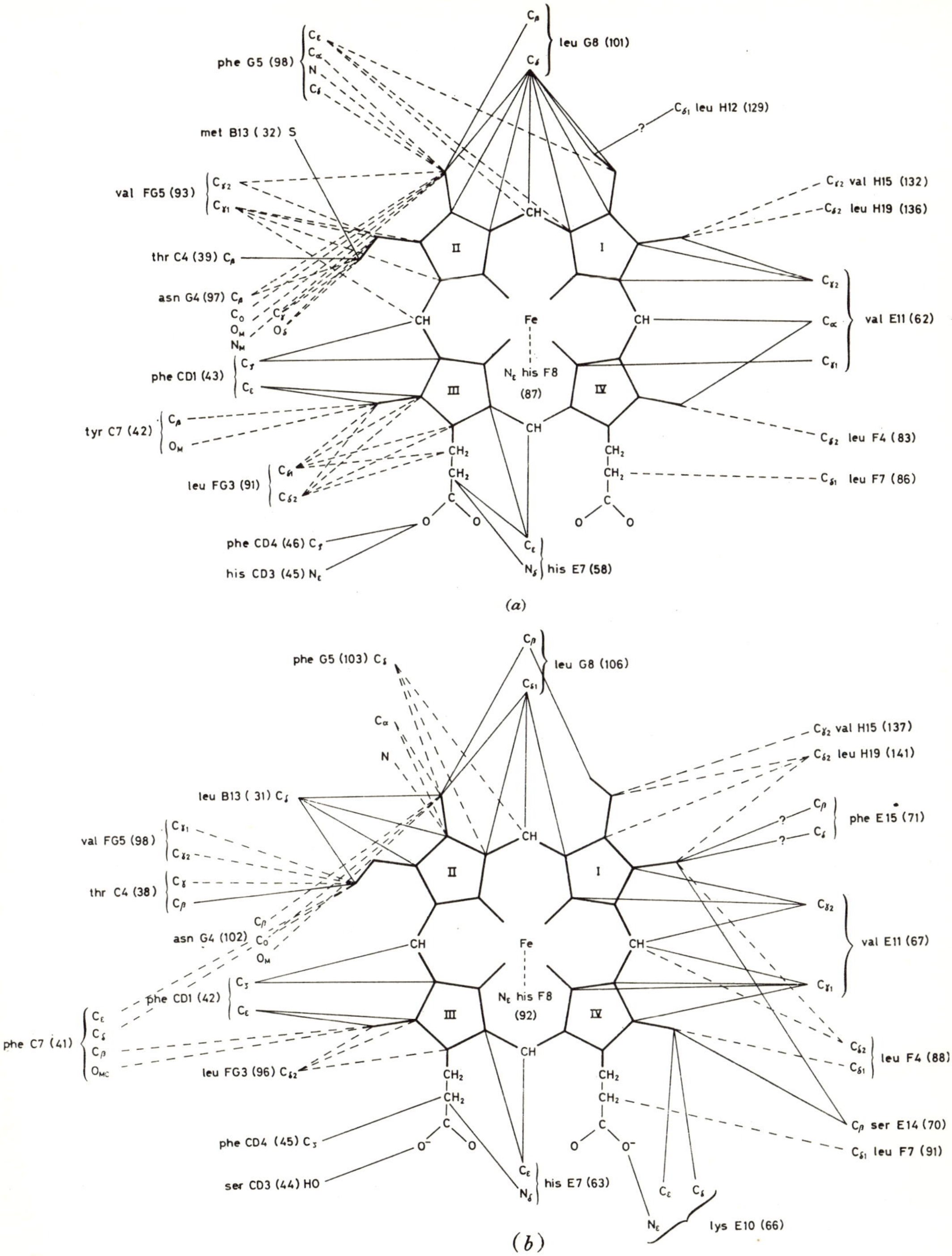

FIGURE 19. Contacts between the haem group and residues of the globin in (*a*) the α-chain and (*b*) the β-chain. All contacts of approximately 4 Å or less are listed. Plain lines indicate contacts on the side of helix E (distal histidine); broken lines, contacts on the side of helix F (proximal histidine). The suffix M means 'main chain'.

(figure 10). In our electron density maps, the corresponding phenolic groups show up as weak and ill-defined peaks, but their positions are consistent with the hydroxyls being hydrogen-bonded as in myoglobin. Probably the free rotation of the C-terminal residues, which follow the tyrosines, makes them wobble, thus blurring their images on the maps.*

Tyrosine C7(42)α is hydrogen-bonded internally to the side chain of asparagine G4(97)α (figure 15). The hydroxyl group of B5(24)α is external, while the ring itself is internal. C1(35)β is in the internal cavity, probably hydrogen-bonded to asp H9(126)α (table 4). H8(130)β is internal and its OH group is probably bonded to the carbonyl of valine A8(11)β (figure 9).

The α-chain contains one tryptophan and the β-chain two. The tryptophans in position A12 seem to serve as spacers between helices A and E (figure 9). The planes of the indole rings are parallel to the haem groups and lie at a distance of 14 Å from the iron atoms. These tryptophans must be responsible for the well-known photodissociation of carbonmonoxyhaemoglobin in the ultraviolet. Tryptophan C3(37)β has already been described (table 4 and figure 15). The indole imino group of tryptophan A12α is probably hydrogen-bonded to the side chain of threonine E16(67)α; the imino groups of the other indole rings are free.

Discussion

Polar and non-polar residues

The fold of the polypeptide chain in haemoglobin and myoglobin appears to be one of the fundamental patterns of Nature designed to hold a haem group so that its iron atom can combine reversibly with molecular oxygen. The pattern has now been found in four mammalian species (man, horse, sperm whale and seal) and also in two invertebrates (the fly *Chironomus thummi* (Huber, Formanek & Epp 1968) and the annelid worm *Glycera dibranchiata* (Love & Padlan 1968)). This suggests that the haemoglobin gene may have come into existence at an early stage of evolution.

What determines the fold of the chain? Comparison of the amino acid sequences of different haemoglobins and myoglobins shows that only seven out of more than 140 sites are occupied by the same residue throughout. The most striking feature common to all globin chains is the almost complete exclusion of polar residues from interior sites (Perutz, Kendrew & Watson 1965), suggesting that the pattern of invariantly non-polar side chains, together with the non-polar parts of the porphyrin ring, may be decisive in determining the conformation of the globin chain. This expectation is borne out by the properties of the abnormal human haemoglobins, which show that the haemoglobin molecule is insensitive to most replacements on its surface, except those which affect the solubility of either the oxy-or the

* *Added in proof.* The tyrosine H 23 in the β-chains has since emerged clearly on difference electron density maps of haemoglobin containing the inhibitor bis(N-maleidomethyl)-ether. These maps confirm that this tyrosine occupies a position equivalent to that in myoglobin (J. K. Moffatt & M. F. Perutz, unpublished), as shown in figure 10.

deoxy-form, and those which disrupt some essential contact between neighbouring segments of polypeptide chain. On the other hand, the molecule is extremely sensitive to even quite small alterations in the pattern of non-polar contacts, especially of those near the haem group. Introduction of an internal polar group has a disastrous effect on the stability of the molecule (Perutz & Lehmann 1968).

The reasons for the interior location of the non-polar side chains are similar to those underlying the insolubility of hydrocarbons in water or the high surface energies at interfaces between non-polar substances and water. These were first explained by Frank & Evans (1945). The presence of a hydrocarbon immobilizes the water molecules in its immediate vicinity, with a resulting reduction in the entropy of the system. Hence the exposure of any large and rigid non-polar side chain would tend to make the structure of a protein molecule unstable by reducing the entropy of the surrounding water. The increase in unitary free energy on exposing one mole of non-polar side chain to water has been estimated as: tryptophan 3, tyrosine 2·9, phenylalanine 2·65, leucine 2·4 and valine 1·7 kilocalories (Tanford 1962). Conversely, the packing of the non-polar side chains away from contact with water results in a gain of entropy of the native, folded structure of the protein and its surrounding water relative to its unfolded, denatured form. This gain is believed to make up for the loss of entropy due to the rigid configuration imposed by the native form on the polypetide chain itself (Tanford 1962; Némethy & Scheraga 1962). In view of their long hydrocarbon side chains, it may seem paradoxical that many lysines protrude from the surfaces of haemoglobin and myoglobin. However, their flexibility may reduce the loss of entropy resulting from their exposure to water.

To sum up, the general principles of construction appear to be such as to lead to the least free energy and the greatest possible entropy of the protein molecule and its surrounding water.

Dissociation

In solution, the haemoglobin tetramer is in dynamic equilibrium with dimers and monomers. The balance of the equilibrium depends upon several factors, including the nature of the ligand at the iron atom, and the concentration of haemoglobin, of neutral electrolyte, and of hydrogen ion (for reviews see Rossi-Fanelli, Antonini & Caputo 1964; Wyman 1964). Symmetrical dissociation into dimers of the type $\alpha_2\beta_2 \leftrightharpoons 2\alpha\beta$ takes place at high concentrations of neutral electrolyte, or of urea, and under the influence of acid or alkaline pH. Unsymmetrical dissociation into subunits of the type $\alpha_2\beta_2 \leftrightharpoons \alpha_2 + \beta_2$ is not observed. Dissociation into monomers is detectable, by suitably sensitive methods, even under physiological conditions of pH and electrolyte concentration (Guidotti, Konigsberg & Craig 1963), but appreciable concentrations of monomers appear only at pH below 4·9 or above 11·0.

Human haemoglobin can, however, be split into free α and β-chains by treatment with paramercuribenzoate at pH 6 and ionic strength of 0·1. Under these conditions the reactive SH-groups at cysteine 93β as well as the normally unreactive SH-groups of cysteines 104α and 112β which are close to the contact $\alpha_1\beta_1$ react with paramercuribenzoate. The attachment of the strongly polar benzoate groups appears to prevent reassociation of the α and β-chains by inhibiting the closing of $\alpha_1\beta_1$ contact.

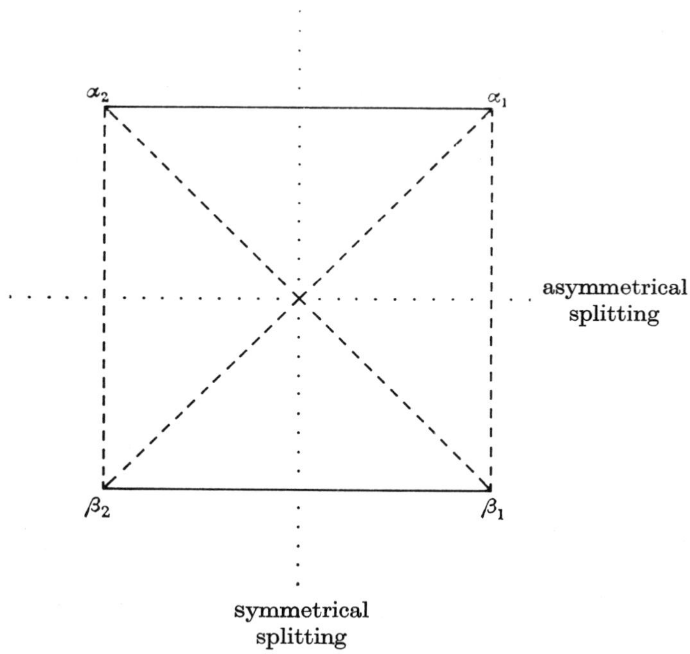

FIGURE 20. Diagrammatic sketch explaining the splitting of the haemoglobin tetramer into $\alpha\beta$ dimers under the influence of strong salt solutions. ——, Polar contacts; - - - -, non-polar contacts.

The dissociation of the tetramer into $\alpha\beta$ dimers at high concentration of neutral electrolyte is readily explained by the model. High concentrations of neutral electrolyte are known to weaken polar interactions and to strengthen non-polar ones. They would therefore be expected to weaken the polar contacts between like subunits and to strengthen the predominantly non-polar ones between unlike subunits. This explains why only symmetrical dissociation occurs; it requires splitting of only one of the pairs of non-polar contacts and of two polar contacts, whereas asymmetrical dissociation would require both pairs of non-polar but no polar contacts to be split (figure 20). Of the two types of non-polar contacts, $\alpha_1\beta_2$ has the smaller area and would therefore be the first to break (table 4). The break would bring the side chains of tryptophan C3β and tyrosine C7β into contact with water and might be accompanied by changes in their ultraviolet absorption spectrum.

The properties of certain abnormal human haemoglobins suggest that at physiological electrolyte concentrations hydrogen bonds make appreciable contributions to the binding energy between unlike subunits. In haemoglobin Kansas asparagine G4β is replaced by threonine (Bonaventura & Riggs 1968). This replacement increases the dissociation constant of its oxy form from tetramers to $\alpha\beta$ dimers by two orders of magnitude. In haemoglobin A, the side chain of asparagine G4β_2 makes a hydrogen bond with aspartic acid G1α_1. In the mutant, the hydroxyl group of threonine, which replaces the asparagine, cannot make this hydrogen bond, but instead it could make one with the main chain carbonyl of aspartic acid G1β, so that the hydrogen bond stabilizing the contact between subunits in the wild type would be replaced in the mutant by another which stabilizes the β-chain internally. In addition the γ-carbon atom of the threonine may introduce conformational changes in helix G, because it probably makes a short contact with the vinyl and methyl side chains of pyrrol II of the haem group.

In the recently discovered haemoglobin Philly, replacement of tyrosine C1(35)β by phenylalanine causes all six cysteines to be reactive, so that after exposure for 30 min to a low concentration of paramercuribenzoate about half the haemoglobin obtained from a heterozygote was present in the form of free α and β-chains (personal communication from R. F. Rieder, F. A. Oski and J. B. Clegg).

The tyrosine lies in the internal cavity of the tetramer and its hydroxyl group is probably hydrogen-bonded to the carboxyl group of aspartic acid H9(126)α at the contact $\alpha_1\beta_1$. The side chain of the phenylalanine cannot make the hydrogen bond because it lacks the hydroxyl group, but is otherwise so similar to that of tyrosine that conformational changes are unlikely to accompany the replacement. The tipping of the equilibrium from tetramers to monomers must therefore be a result of the loss of one pair of hydrogen bonds which loosens the contact $\alpha_1\beta_1$ and $\alpha_2\beta_2$. It is remarkable that this should cause haemolytic anaemia by precipitation of the haemoglobin in the red cell.

In view of the large number of non-polar interactions at the contacts between unlike subunits, the dramatic effect caused by the loss of a single pair of hydrogen bonds is surprising. Normally, the contribution of hydrogen bonds to an aqueous system would be expected to be weak, because the free energy of the bond between two polar groups in the protein would only exceed by a few hundred calories the sum of the free energies of the bonds which each of the polar groups could make with water. What special features do the hydrogen bonds at the $\alpha\beta$ contacts have to invalidate this argument?

The hydrogen bond between asparagine G4β_2 and aspartic acid G1α_1 lies in an environment from which water is largely excluded, which would raise the bond energy by lowering the local dielectric constant. On dissociation into $\alpha\beta$ dimers the polar groups of both asparagine and aspartic acid would be exposed to water where the dielectric constant is high, so that each of the two polar groups would form only weak hydrogen bonds with water. The hydrogen bond between tyrosine C1β_1 and aspartic acid H9α_1 lies in a narrow part of the internal cavity where the few surrounding water molecules may be in fixed, rather than random, positions and form a network of hydrogen bonds surrounding the two polar groups. This network may be disorganized by the removal of the phenolic hydroxyl group, leading to a large loss in bond energy.

The increased dissociation of haemoglobin Kansas into $\alpha\beta$ dimers and of haemoglobin Philly into free α and β-chains confirms the prediction that dissociation into dimers occurs by the splitting of the contacts $\alpha_1\beta_2$ and $\alpha_2\beta_1$, while splitting into free α and β-chains occurs at the contacts $\alpha_1\beta_1$ and $\alpha_2\beta_2$ (Rosemeyer & Huehns 1967). This point becomes important when different mechanisms of haem-haem interaction are considered.

Interaction between ligands

There have been indications that dimers consisting of one α and one β-chain exhibit all the cooperative effects of which haemoglobin is capable and that interaction between dimers is of only secondary importance (Rossi-Fanelli, Antonini & Caputo 1964; Guidotti 1967*c*, *d*). Because the dimers have the structure $\alpha_1\beta_1$, the contact $\alpha_1\beta_2$ would be left open and inactive, except in the event of a suitable collision between dimers.

The crystallographic results suggest that, on the contrary, the contact $\alpha_1\beta_2$ is of primary importance. It contains several side chains forming bridges between residues which touch the haems on either side, thus offering a direct pathway for interactions between the two pairs of haems which are closest together (25 Å between the iron atoms; figures 15, 16). Interactions between the haem groups α_1 and β_1, on the other hand, would have to be transmitted through an intricate network of many residues acting over a longer distance (37 Å between the iron atoms; see figures 11 to 14). On oxygenation and deoxygenation the greater relative movement takes place in $\alpha_1\beta_2$ (Muirhead *et al.* 1967; Bolton, Cox & Perutz 1968). All but one of the residues making up this contact are invariant in all α and all β-chains of mammalian species analysed so far, whereas more than one-third of the residues in the contact $\alpha_1\beta_1$ are variable. All six mutant human haemoglobins containing amino acid substitutions in $\alpha_1\beta_2$ show diminished haem-haem interaction (Perutz & Lehmann 1968). On the other hand, haemoglobin E, which has a substitution at the contact $\alpha_1\beta_1$, exhibits normal haem-haem interaction (Bellingham & Huehns 1968). Finally, two reagents which affect the conformation of the β-chain in the neighbourhood of the contact $\alpha_1\beta_2$ inhibit all cooperative effects (Simon & Konigsberg 1966; Simon *et al.* 1967; and unpublished work of S. R. Simon, D. Arndt, W. H. Konigsberg, L. E. Webb, J. K. Moffat, W. Bolton & M. F. Perutz). Taken as a whole, this evidence suggests that both contacts are essential for the expression of the cooperative effects and that the tetramer, rather than the dimer, is the functional unit of haemoglobin. Similar conclusions have been reached by Gibson & Parkhurst (1968) on the basis of kinetic evidence.

Ferguson & Roughton (1934*a*, *b*) and Ferguson (1936) predicted that carbon dioxide probably combines with the amino groups of the globin chains by forming carbamino compounds. Later work indicated that only the α-amino groups have a pK sufficiently low to allow them to form such compounds (Roughton & Rossi-Bernardi 1966). At constant pH and $p(CO_2)$ deoxyhaemoglobin has a higher affinity for carbon dioxide than oxyhaemoglobin. This difference is responsible for a significant fraction of the total carbon dioxide exchanged during respiration. Kilmartin & Rossi-Bernardi (1969) have now shown that this difference vanishes when all four α-amino groups are blocked by cyanate. However, it is not yet clear why the affinity of the α-amino groups for carbon dioxide should be higher in deoxy- than in oxyhaemoglobin.

Long before the two histidines on either side of each haem group were demonstrated directly, their presence was inferred from the Bohr effect, and it was suggested that the combination with oxygen lowered the pK of one of them and raised that of the other (Wyman 1948). This explanation is now difficult to maintain because the two histidines occur in both haemoglobin and myoglobin, and the latter exhibits no Bohr effect. It seems more likely, therefore, that histidines at the boundaries between subunits are involved in the Bohr effect, since their environment would be changed by the structural transition between the two forms. So far, however, they have not been identified. A possible contribution by α-amino groups to the Bohr effect has also been suggested (Tanford & Nozaki 1966; Rossi-Bernardi & Roughton 1967).

I should like to thank all my colleagues who have contributed to the solution of the structure of haemoglobin, and especially Dr Hilary Muirhead and Miss Joyce M. Cox, who are responsible for the mathematical part of the work.

References

Bellingham, A. J. & Huehns, E. R. 1968 *Nature, Lond.* **218**, 924.
Benesch, R. & Benesch, R. E. 1969 *Nature, Lond.* **221**, 618.
Benesch, R., Benesch, R. E. & Yu, C. I. 1968 *Proc. natn. Acad. Sci., U.S.A.* **59**, 526.
Bolton, W., Cox, J. M. & Perutz, M. F. 1968 *J. mol. Biol.* **33**, 283.
Bonaventura, J. & Riggs, A. 1968 *J. biol. chem.* **243**, 980.
Chanutin, A. & Curnish, R. R. 1967 *Arch. Biochim. Biophys.* **121**, 96.
Cullis, A. F., Muirhead, H., Perutz, M. F., Rossmann, M. G. & North, A. C. T. 1962 *Proc. Roy. Soc.* A **265**, 161.
Dayhoff, M. O. & Eck, R. V. 1968 *Atlas of protein sequence and structure* 1967–68. National Biomedical Research Foundation, Silver Spring, Md. 20901, U.S.A.
Diamond, R. 1966 *Acta Cryst.* **21**, 253.
Dickerson, R. E. 1964 *The Proteins* (ed. Neurath, H.), 2nd. ed. **2**, 603. New York: Academic Press.
Ferguson, J. K. W. 1936 *J. Physiol.* **88**, 40.
Ferguson, J. K. W. & Roughton, F. J. W. 1934*a* *J. Physiol.* **83**, 68.
Ferguson, J. K. W. & Roughton, F. J. W. 1934*b* *J. Physiol.* **83**, 87.
Frank, H. W. & Evans, M. W. 1945 *J. Chem. Phys.* **13**, 507.
Gibson, Q. H. & Parkhurst, L. J. 1968 *J. biol. Chem.* **243**, 5521.
Guidotti, G. 1967*a* *J. biol. Chem.* **242**, 3673.
Guidotti, G. 1967*b* *J. biol. Chem.* **242**, 3685.
Guidotti, G. 1967*c* *J. biol. Chem.* **242**, 3694.
Guidotti, G. 1967*d* *J. biol. Chem.* **242**, 3704.
Guidotti, G., Konigsberg, W. & Craig, L. C. 1963 *Proc. natn. Acad. Sci., U.S.A.* **50**, 774.

Huber, R., Formanek, H. & Epp, O. 1968 *Naturwissenschaften* **55**, 75.
Kendrew, J. C. 1962 *Brookhaven symposia in biology*, **15**, 216.
Kendrew, J. C., Dickerson, R. E., Strandberg, B. D., Hart, R. G., Davies, D. R., Phillips, D. C. & Shore, V. C. 1960 *Nature, Lond.* **185**, 422.
Kendrew, J. C., Watson, H. C., Strandberg, B. E., Dickerson, R. E., Phillips, D. C. & Shore, V. C. 1961 *Nature, Lond.* **190**, 663.
Kilmartin, J. V. & Clegg, J. B. 1967 *Nature, Lond.* **213**, 269.
Kilmartin, J. V. & Rossi-Bernardi, L. 1969 *Nature, Lond.* (In the Press.)
Low, B. W. & Grenville-Wells, H. J. 1953 *Proc. natn. Acad. Sci., U.S.A.* **39**, 785.
Matsuda, G., Gehring-Müller, R. & Braunitzer, G. 1963 *Biochem. Z.* **338**, 669.
Muirhead, H., Cox, J. M., Mazzarella, L. & Perutz, M. F. 1967 *J. mol. Biol.* **28**, 117.
Nakahara, A. & Wang, J. H. 1958 *J. Am. chem. Soc.* **80**, 6526.
Némethy, G. & Scheraga, H. A. 1962 *J. Phys. Chem.* **66**, 1773.
Padlan, E. A. & Love, W. E. 1968 *Nature, Lond.* **220**, 376.
Pauling, L. & Corey, R. B. 1951 *Proc. natn. Acad. Sci., U.S.A.* **37**, 205.
Perutz, M. F. 1953 *Proc. Roy. Soc.* B **141**, 69.
Perutz, M. F., Steinrauf, L. K., Stockell, A. & Bangham, A. D. 1959 *J. mol. Biol.* **1**, 402.
Perutz, M. F., Kendrew, J. C. & Watson, H. C. 1965 *J. mol. biol.* **13**, 669.
Perutz, M. F., Muirhead, H., Cox, J. M., Goaman, L. C. G., Mathews, F. S., McGandy, E. L. & Webb, L. E. 1968 *Nature, Lond.* **219**, 29.
Perutz, M. F., Muirhead, H., Cox, J. M. & Goaman, L. C. G. 1968 *Nature, Lond.* **219**, 131.
Perutz, M. F. & Lehmann, H. 1968 *Nature, Lond.* **219**, 902.
Ramachandran, G. N., Ramakrishnan, C. & Sasisekharan, V. 1963 *J. mol. biol.* **7**, 95.
Ramakrishnan, C. & Ramachandran, G. N. 1965 *Biophys. J.* **5**, 909.
Rosemeyer, M. A. & Huehns, E. R. 1967 *J. mol. Biol.* **25**, 253.
Rossi-Bernardi, L. & Roughton, F. J. W. 1967 *J. biol. Chem.* **242**, 785.
Rossi-Fanelli, A., Antonini, E. & Caputo, A. 1964 *Advanc. Protein Chem.* **19**, 74.
Roughton, F. J. W. 1964 *Handbook of physiology; respiration I.* pp. 767–825. Washington D.C.: American Physiological Society.
Roughton, F. J. W. & Rossi-Bernardi, L. 1966 *Proc. Roy. Soc.* B **164**, 381.
Simon, S. R. & Konigsberg, W. H. 1966 *Proc. natn. Acad. Sci., U.S.A.* **56**, 749.
Simon, S. R., Konigsberg, W. H., Bolton, W. & Perutz, M. F. 1967 *J. mol. Biol.* **28**, 451.
Smith, D. B. 1963 *Canad. J. Biochem. Physiol*, **42**, 755.
Smith, D. B. 1968 *Canad. J. Biochem.* **46**, 825.
Tanford, C. 1962 *J. Am. chem. Soc.* **84**, 4240.
Tanford, C. & Nozaki, Y. 1966 *J. biol. Chem.* **241**, 2832.
Watson, H. C. 1969 In *Progress in stereochemistry IV* (ed. Harris, M. M. and Aylett, B. J.) London: Butterworths (in the Press).
Wyman, J. 1948 *Advanc. Protein Chem.* **4**, 407.
Wyman, J. 1964 *Advanc. Protein Chem.* **19**, 224.
Wyman, J. 1968 *Quart. Rev. Biophys.* **1**, 35.

The Ferrier Lecture, 1968

Motor apparatus of the baboon's hand

By C. G. Phillips, F.R.S.
University Laboratory of Physiology, Oxford

(*Lecture delivered* 9 *May* 1968—*MS received* 30 *September* 1968)

Proceedings of the Royal Society, B, *volume* 173, *pp*. 141–174, 1969

Introduction

It is exactly 40 years since Ferrier died, and 39 years since Sherrington gave the first Ferrier Lecture in honour of his work. Sherrington (1906) had dedicated *The integrative action of the nervous system* to Ferrier 'in token of recognition of his many services to the experimental physiology of the central nervous system'. At Aberdeen Ferrier had been taught by Bain, the Professor of Logic, who kept a model of the human brain upon his lecture table. Ferrier moved to London in 1870, and was excited by Hughlings Jackson's revolutionary discoveries and ideas about the nature and localization of the sensory and motor functions of the human brain. Early in 1873 he discussed Fritsch & Hitzig's (1870) pioneer galvanic stimulations of the dog's cortex with his friend James Crichton-Browne. In the spring and summer of that year he performed his pioneer faradic stimulations of the brains of monkeys in the laboratory of the West Riding Lunatic Asylum, of which Crichton-Browne was Medical Director (Sherrington 1928). He sought to 'put to experimental proof the views entertained by Dr Hughlings Jackson on the pathology of Epilepsy, Chorea and Hemiplegia, by imitating artificially the "destroying" and "discharging lesions" of disease which his writings have defined and differentiated' (Ferrier 1873). 'The phenomena of localized and unilateral convulsive movements, depending, as Hughlings Jackson shows, on vital irritation of certain regions of the cortex, are essentially of the same nature as those caused by electrisation of the same regions' (1876, p. 133). But Ferrier believed that many of the movements he mapped by electrical stimulation had 'evidently a purposive or volitional character' (1876, p. 163). In 1876 he transferred his monkey map to the brain of man, and described of the relation between brain fissures and bony landmarks; and in 1883 he suggested that the time had come when localized cerebral lesions should be excised by the surgeon. For this, as he later recalled with amusement, he was criticized in an editorial in the *Lancet* (Ferrier, 1888); but in 1884 Godlee removed a localized tumour 'of the size of a walnut' (Sherrington 1928) from the Rolandic cortex through an opening exactly overlying it and but little larger than itself. Ferrier and Hughlings Jackson both witnessed this memorable operation (Trotter 1934).

As time went on, experimentalists interested themselves more in defining details of localization than in developing concepts of function (Phillips 1966). 'More penetrative modes and aims of analysis came to be little pursued,' wrote Sherrington (1928). 'A localization vogue reigned for nearly a quarter of a century, and became in due course tedious and relatively infertile.'

To Jackson's 'destroying lesions' the functions of the hand are pre-eminently vulnerable, and Ferrier (1886, p. 365) agreed that after cortical injury 'the varied and delicate movements of the hand are most of all impaired, and the last to be re-established'. Focal convulsions due to 'discharging lesions' are specially apt to begin in the thumb and index finger (Jackson 1873*a*), which Jackson (1873*b*) regarded as the 'most voluntary or specialized parts of the body'. No explanation of this selective effect of coarse, diffuse lesions seems to have been forthcoming until Walshe (1943) proposed that 'Jacksonian fits have their characteristic form of onset because the movements concerned are those that have the widest fields of low-threshold excitability' in the cortex. Applying single 5 ms rectangular current pulses to the baboon's cortex in very light anaesthesia, Liddell & Phillips (1950)

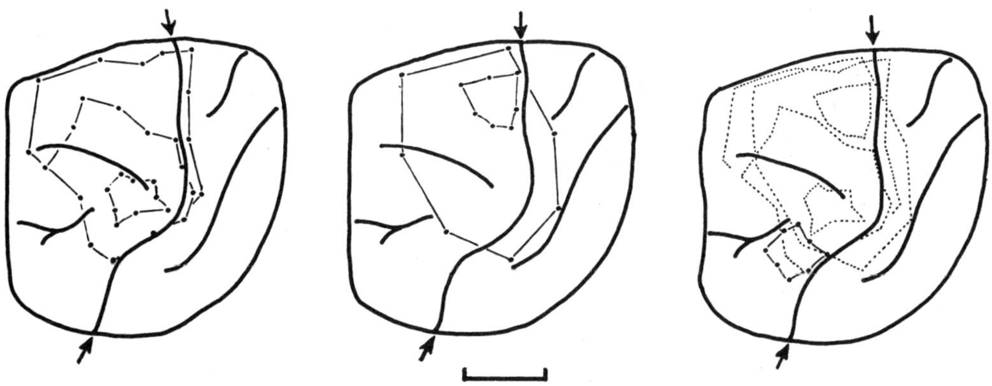

a *b* *c*

FIGURE 1. Left Rolandic region of baboon, with arrows to mark the Rolandic fissure. Precentral gyrus to left of fissure. Scale at *b* = 10 mm. Stimulation by single surface-cathodal rectangular pulses, duration 4·5 ms. (*a*) Extent of area for flick movements of thumb, index and little finger at strengths 1·2, 1·6, and 2·7 mA. (*b*) Area for hallux and middle toe, strengths 2·05 and 4·7 mA. (*c*) Area for angle of mouth, 2·35 mA. Areas of (*a*) and (*b*) shown by dotted lines (Liddell & Phillips 1950).

were able to map these wide fields (figure 1), and to show that the threshold was lower for thumb and index than for foot or face. Hand, foot and face were the only peripheries to respond to these single pulses. Ferrier himself (1876, p. 132) must have come very near to getting this result. When comparing the merits of faradism and galvanism, he obtained by galvanizing a particular region of the precentral gyrus, not the smooth elbow flexion which led him to prefer faradism, but 'spasmodic and sudden jerks of the hand and forearm'. If he had weakened the galvanic stimulus to threshold, he could hardly have failed to observe that the 'spasmodic and sudden jerks' had been limited to the monkey's thumb and index.

Wide cortical fields of low-threshold excitability suggest that the densest and most direct corticofugal projections are sent to those segmental structures that are preferentially used by the primates in their 'most voluntary' activities. Of these the hand is the leading exponent. The experiments to be reviewed have set out to test this hypothesis. Electrical stimulation does not evoke natural functioning of an area of cortex (Phillips 1966). But it is an invaluable tool for exciting the neurons that project from the area. The actions of these projections on the motoneurons of the hand, and on the fusimotor neurons which control the feedback from its muscle spindles, can now be investigated by physiological methods far surpassing in sensitivity any that were available in Ferrier's time. Such investigations, admittedly, do no more than specify the properties of a motor *apparatus*, but this is an essential preliminary to future, more sophisticated investigations of motor *function*.

Actions of cortical stimulation on corticospinal pyramidal neurons in the baboon

Direct electrical excitation of corticospinal pyramidal neurons depends on the polarity of the cortical stimulus and the orientation of the long axes of these neurons in relation to the lines of current flow. The experiments to be described have all been made on the baboon's brain, but essentially similar results have been obtained from the sensorimotor cortex of the cat (Phillips 1956; Hern, Phillips & Porter 1962; Gorman 1966).

The maps shown in figure 1 were made with surface-cathodal pulses of 4·5 ms duration, the stimulating current flowing between a ball-pointed focal cathode touching the pia-covered cortical surface and a remote, diffuse anode beneath the scalp. Fritsch & Hitzig (1870), however, had found in the dog's cortex that the anodal threshold was lower than the cathodal. In the baboon the surface-anodal (S+) and surface-cathodal (S−) thresholds for minimal movement of thumb and index are of the same order, but the cortical fields are different: the S+ field is centred on the Rolandic fissure, spreading on to the post-central gyrus, and the S− field, which overlaps it, is centred on the convexity of the precentral gyrus (Hern, Landgren, Phillips & Porter 1962, figure 2). Further analysis of these differences requires still weaker currents. The resulting corticofugal discharges, which are too small to evoke movement, can be detected by recording with microelectrodes from axons in the contralateral corticospinal tract (in which case the destination of the impulses is unknown); or by intracellular recording of monosynaptic excitatory post-synaptic potentials (EPSP) from α motoneurons in the contralateral cervical enlargement, proved to belong to particular muscles or muscle groups by antidromic stimulation of motor nerves in the forearm.

Corticospinal cells on the convexity of the precentral gyrus are directly stimulated by an overlying focal anode. Figure 2*a* (Phillips & Porter 1962) shows superimposed responses of the axon of such a cell to about fifty brief (0·2 ms) anodal pulses of near-threshold strength (1·15 mA). The unvarying latency (2·2 ms at the recording site in the lateral corticospinal tract at C_5 level) attests direct stimulation of a specific region of the cell, probably by a cathodal action on the initial segment or first node of the axon (cf. Landau, Bishop & Clare, 1965). Further evidence of direct stimulation comes from the cell's ability to follow stimulating pulses delivered at frequencies of 250/s or higher. The explanation of surface-anodal stimulation in terms of the 'virtual cathode established in the deeper and more excitable regions' was proposed by Ferrier himself (1886, p. 232) when discussing Hitzig's results.

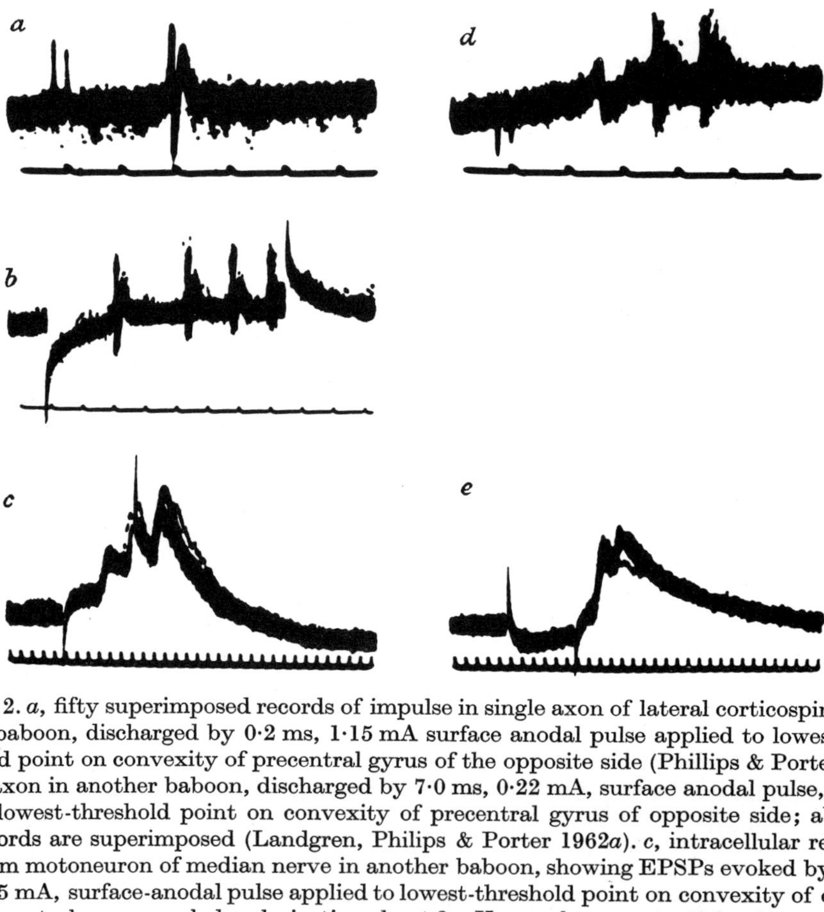

FIGURE 2. *a*, fifty superimposed records of impulse in single axon of lateral corticospinal tract of baboon, discharged by 0·2 ms, 1·15 mA surface anodal pulse applied to lowest-threshold point on convexity of precentral gyrus of the opposite side (Phillips & Porter 1962). *b*, axon in another baboon, discharged by 7·0 ms, 0·22 mA, surface anodal pulse, applied to lowest-threshold point on convexity of precentral gyrus of opposite side; about 30 records are superimposed (Landgren, Philips & Porter 1962*a*). *c*, intracellular recording from motoneuron of median nerve in another baboon, showing EPSPs evoked by 5·0 ms, 0·55 mA, surface-anodal pulse applied to lowest-threshold point on convexity of opposite precentral gyrus; peak depolarization about 2 mV; membrane potential −60 to −62 mV; about 20 records superimposed (Hern, Landgren, Phillips & Porter 1962). *d*, same axon as in *a*, discharged by 0·2 ms, 1·85 mA, surface-cathodal pulse; axon discharged at one or other of two preferred times; small tract wave (D wave) precedes the axonal impulses; about 50 records superimposed (Phillips & Porter 1962). *e*, same motoneuron as in *c*; EPSP evoked by 5·0 ms, 0·50 mA, surface-cathodal pulse; peak depolarization about 1·5 mV; membrane potential −60 to −62 mV; about 20 records superimposed (Hern, Landgren, Phillips & Porter 1962). Time scale 1000/s below each tracing.

The actions of S+ pulses of the longer duration used for drawing the single-pulse maps are of special interest. Figure 2*b* (Landgren, Phillips & Porter 1962*a*) shows superimposed responses of another corticospinal neuron on the convexity of the precentral gyrus to twenty S+ pulses of 7·0 ms duration, delivered at 1/s. The current was about one-eighth of the threshold for minimal flick movement of the thumb. Yet the neuron fired four impulses within 5 ms, and at precisely the same times, in response to every stimulating pulse. Such constancy of response suggests direct stimulation: the repetitive firing is the rhythmic response of the pacemaker region of the cell membrane, presumably the unmyelinated initial segment of the axon (Phillips 1961), to the steady outward (cathodal) fraction of the current entering the cortex at the surface anode, and comparable to the repetitive response of sensory axons to anodal polarization of receptor organs (cf. Hern, Landgren, Phillips & Porter 1962).

Evidence of repetitive response of corticospinal cells located on the convexity of the precentral gyrus is also found in intracellular records from cervical α motoneurons. In figure 2*c* (Hern, Landgren, Phillips & Porter 1962) the lowest-threshold area for evoking an EPSP in a motoneuron of the median nerve was stimulated with S+ pulses of about 6·0 ms duration and well below the threshold for movement. The twenty superimposed EPSPs all show the same three accurately timed steps. Each step is the monosynaptic response to a small, well-synchronized corticospinal volley, arising from a small colony of cells each of which discharged three impulses in response to the cathodal current crossing the membrane of its initial axonal segment. The increasing amplitude of the successive steps is explained by an increase in the transmitting potency of the corticomotoneuronal synapses (see figure 16). The smooth repolarization of the α motoneuron shows that there was no after-discharge from the corticospinal colony. Figure 3 (top two lines) (Hern, Landgren, Phillips & Porter 1962) shows responses of an ulnar motoneuron to S+ pulses of 5·0 ms duration, below the threshold for movement. The superimposed EPSP records (top left) show four distinct monosynaptic impacts. When these pulses were repeated at 44/s there was no cumulative action and no after-discharge. Not only do S+ pulses directly stimulate the pacemaker membrane of corticospinal cells by cathodal action in the depth of the cortex: they presumably also depress, by anodal action, the afferent axons and interneurons of the outer cortical layers and hence do not give rise to synaptic actions on the corticospinal cells (Hern, Landgren, Phillips & Porter 1962).

The contrasting effects of S− pulses on corticospinal cells of the precentral convexity must now be considered. Figure 2*d* (Phillips & Porter 1962) shows the response of the axon of figure 2*a* to brief (0·2 ms) S− cortical pulses. The threshold for discharge was higher, the latency longer; since the axonal conduction-time would have been the same as in figure 2*a*, the increased latency can be explained only in terms of intracortical synaptic delay. Synaptic excitation occurred at one or other of two preferred times: if an impulse was fired at the first of these, no impulse was fired at the second, and vice versa. The slight temporal scatter of the impulses at each preferred time is characteristic of synaptic transmission: the triphasic spikes are not exactly superimposed as in the case of the directly stimulated responses of figure 2*a*. Figure 2*c* (Hern, Landgren, Phillips & Porter 1962) shows intracellularly recorded EPSPs from the same α motoneuron as in figure 2*c*.

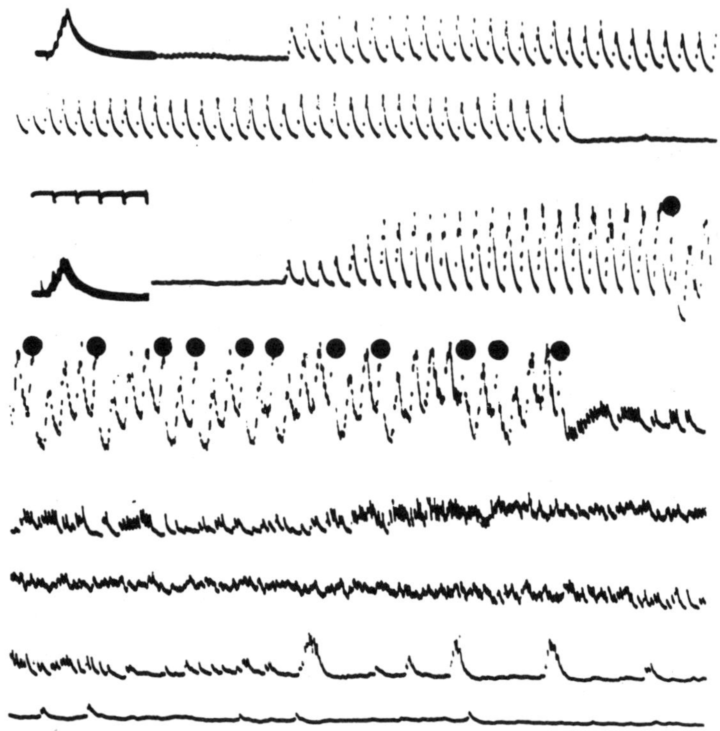

FIGURE 3. Intracellular recording from ulnar motoneuron in a baboon, membrane potential −69 to −74 mV.

Top line begins with superimposed records of EPSPs (peak depolarization 7·5 mV) evoked by single 5·0 ms, 0·7 mA surface-anodal cortical pulse, followed by repetition of same stimuli at 44/s for 1·4 s.

Third line begins with superimposed records of EPSPs (peak depolarization 6·0 mV) evoked by single 5·0 ms, 1·0 mA surface-cathodal cortical pulse, followed by repetition at 44/s for 1·4 s. Note growth of EPSPs to about 15 mV. The motoneuron discharged an impulse (marked by a dot) whenever the depolarization reached firing level (−58 mV). Last impulse coincides with end of period of stimulation. Note persistence of small EPSPs due to after-discharges of cells of corticomotoneuronal colony.

Time for superimposed records 100/s (Hern, Landgren, Phillips & Porter 1962).

Its responses to long S− pulses have a longer latency than the responses to S+ pulses (figure 2*c*), but still show no after-discharge. Finally, corticospinal discharges set up by repetitive S− stimulation with 5·0 ms pulses at 44/s may be compared with the series of stereotyped discharges evoked by S+ stimulation, and detected as excitatory synaptic impacts on the membrane of an ulnar motoneuron (figure 3). After the first three pulses the EPSPs increase in size and complexity, as the corticospinal discharge becomes complicated by increasing intracortical synaptic and interneuronal activity. Stimulation was for 1·4 s, and the ensuing 4 s of record show that the corticospinal cells gave a prolonged after-discharge, at first continuous and irregular, later grouped into clonic bursts. Hern, Landgren, Phillips & Porter (1962) argued that these complex and prolonged effects of repetitive S− pulses were due to prolonged intracortical synaptic bombardment of corticospinal cells and not to bombardment of the α motoneuron by spinal interneurons. If spinal interneuron pools had been stirred up by corticospinal impulses, the complex and prolonged effects should have been seen also in response to repetitive S+ pulses.

The sharp distinction between direct (S+) and synaptic (S−) stimulation of corticospinal cells obtains only in the case of near-threshold stimulation of cells on the convexity of a gyrus; that is, of neurons whose long apical dendrites are near the point of contact of the focal electrode and at right angles to the pia-covered surface of the brain. Stronger stimulation not only reaches cells distant horizontally along the precentral gyrus, but also cells in the anterior wall of the Rolandic fissure (Landgren, Phillips & Porter 1962*b*; Phillips & Porter 1964). Such distant cells may be directly stimulated by stronger S− as well as by S+ pulses. Thus, the synaptically generated corticospinal impulses evoked by S− pulses in figure 2*d* are preceded by a small triphasic wave. This is a multi-unit wave, and it shows that a population of corticospinal cells was directly stimulated by the S− pulses. It is probable that such a population is made up of neurons lying in the anterior wall of the Rolandic fissure. The long axes of such cells, which are at right angles to the plane of the fissure, are parallel to the surface of the brain. The thresholds for exciting these cells from the surface are higher than in the case of cells on the convexity of the precentral gyrus, the difference between S+ and S− thresholds is smaller, and the cells are directly excited by either polarity (Phillips & Porter 1962). Further, the S+ threshold tends to be lower when the focal anode is at the edge of the Rolandic fissure, that is, when it is nearest to the dendritic pole of the cell: the S− threshold is lower when the focal cathode is on the convexity of the precentral gyrus, that is, nearest to the axonal pole (Phillips & Porter 1962). The simplest explanation is that the necessary outward flow of current through the initial segment of the axon would be provided by a focal anode nearest the dendritic pole or by a focal cathode nearest the axonal pole (Phillips 1967).

In sum, the foregoing experiments make a strong case for choosing brief (0·2 ms) focal anodal pulses in preference to brief focal cathodal pulses if one's object is to send small well-synchronized corticospinal volleys from circumscribed areas of the precentral gyrus into the spinal cord in order to analyse their actions at the segmental level. Conventional bifocal stimulation with ball-pointed electrodes about 2 mm apart has mixed anodal and cathodal effects, which are not confined to the interpolar region, and offers no advantages (Phillips & Porter 1962). But as focal anodal or focal cathodal stimuli are further increased in strength, corticospinal populations of similar size are eventually directly stimulated; the characteristic differences between S+ and S− disappear (Hern, Landgren, Phillips & Porter 1962) as cells further and further from the point of contact of the focal electrode— both horizontally along the gyrus and in the depth of the fissure—are involved in the response (Landgren *et al.* 1962*b*; Phillips & Porter 1964).

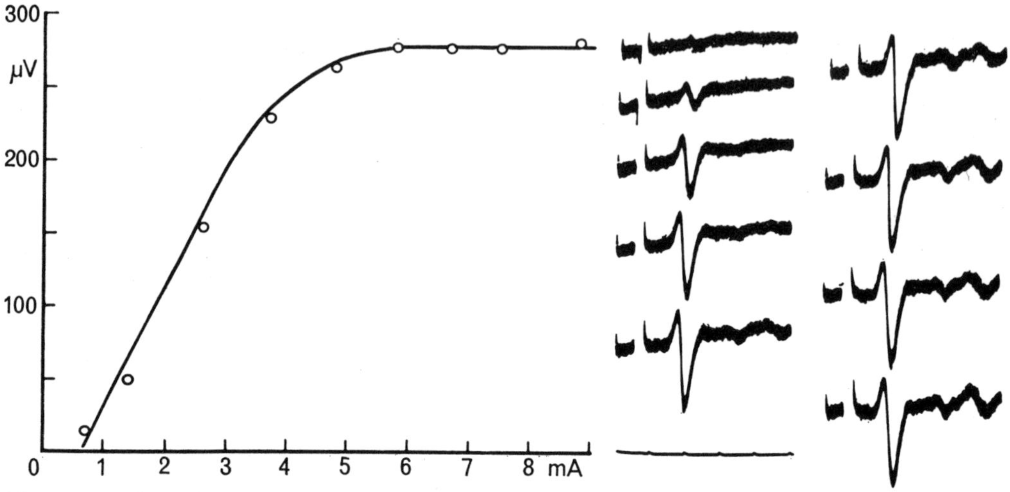

FIGURE 4. Superimposed records of tract wave (D wave) recorded from uninjured lateral corticospinal tract of baboon in response to stimulation of opposite cortical hand area by 0·2 ms surface-anodal pulses. Note growth of D wave to maximum at strength 6·0 mA (plotted on left of figure), and appearance of I waves at strengths exceeding 3·0 mA. (Phillips & Porter (1964), in J. C. Eccles & J. P. Schadé (eds.) *Physiology of spinal neurons*, *Prog. Brain Research*, **12**, 222–242. Reproduced by kind permission of the publishers, Elsevier, Amsterdam.)

Figure 4 (Phillips & Porter 1964) shows how the population of directly excited corticospinal cells increases to the maximum that can be reached from a point in the middle of the precentral 'arm area' as 0·2 ms S+ pulses are strengthened from threshold to 5·0 mA. The response is recorded as the corticospinal tract wave ('D wave' of Patton & Amassian 1954) from the dorsolateral surface of the cord at C_2 level. This is the response of the cell population which supplies the largest axons to the corticospinal tract, conducting at 60 to 70 m/s (Phillips & Porter 1964; Kernell & Wu Chien-ping 1967*a*). These experiments give no information about the large majority of small-calibre axons which make up the pyramidal tract. In figure 4, pulses stronger than 3·0 mA evoke later waves ('I waves' of Patton & Amassian 1954). These have been proved by Kernell & Wu Chien-ping (1967*a*) to be due to repetitive firing of the fast axons that fired first during the D wave. The remarkably stable latencies of the successive I waves, and of the repetitive unitary impulses of which they are made up (figure 6), suggests that these very strong stimuli have not only stimulated the corticospinal cells directly (causing the D wave) but have also started an intracortical interneuronal 'clockwork' which causes a 'periodic bombardment of Betz cells through chains of neurons with fixed temporal characteristics' (Patton & Amassian 1960).

The cortico-motoneuronal projection

The existence of a monosynaptic connexion between corticospinal axons and α motoneurons in the primates has been established by anatomical methods (cf. Kuypers 1964), and experimental investigations have shown that this connexion is excitatory (Cooper & Denny-Brown 1927; Bernhard, Bohm & Petersén 1953; Preston & Whitlock 1960, 1961; Landgren *et al.* 1962*a*; Phillips & Porter 1964). The time relations (figure 5) show that the monosynaptic cortico-motoneuronal (CM) excitation is mediated by the fastest corticospinal axons, i.e. those responsible for the D wave. Monosynaptic excitation is sometimes followed by disynaptic inhibition (Landgren *et al.* 1962*a*; Kernell & Wu Chien-ping 1967*b*) which differs from the monospynaptic excitation in being greatly depressed by barbiturates (Preston & Whitlock 1960, 1961; Clough, Kernell & Phillips 1968). The onset of inhibitory synaptic action is always after the peak of the CM EPSP. Thus the peak amplitude of the maximal CM EPSP can always be measured for accurate

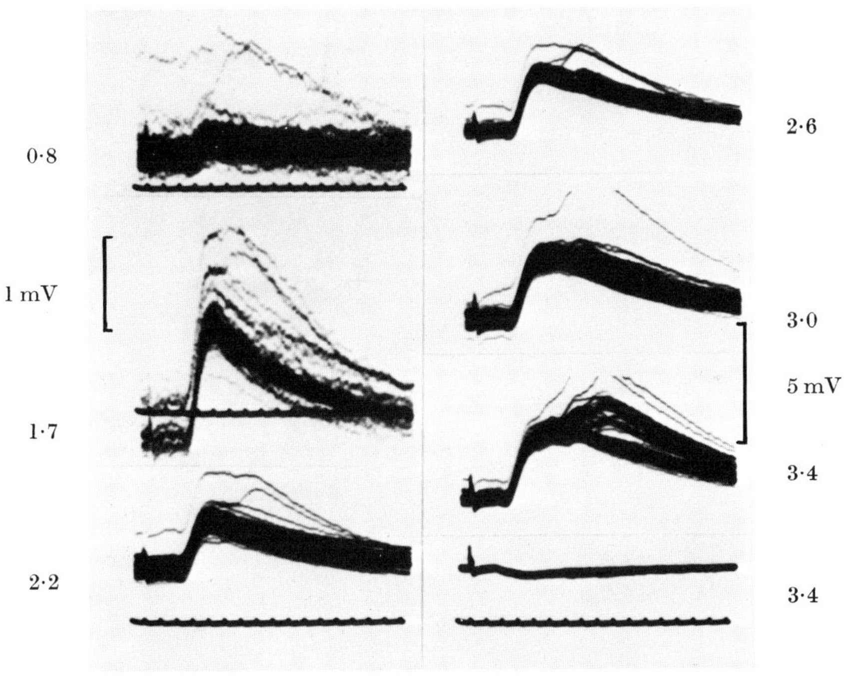

Figure 5. Superimposed intracellular records of monosynaptic EPSPs evoked in motoneuron of median nerve (membrane potential −74 mV) by 0·2 ms, surface-anodal pulses to lowest-threshold area of opposite precentral gyrus. Strength (mA) noted beside each record. Final trace is control of extracellular potential field recorded after withdrawal of microelectrode from cell. Delay between arrival of earliest corticospinal impulses and start of EPSP = 0·66 ms; between arrival of majority of impulses and start of EPSP = 0·32 ms.

Calibrations. Amplitude: 1·0 mV scale applies to responses to 0·8 and 1·7 mA stimuli; 5·0 mV scale to remaining responses. Time: 1000/s (Phillips & Porter (1964), in J. C. Eccles and J. P. Schadé (eds.), *Physiology of spinal neurons*, *Prog. Brain Research*, **12**, pp. 222–242. Reproduced by kind permission of the publishers, Elsevier, Amsterdam).

comparison of the maximum quantities of monosynaptic action that can be generated by corticospinal cells in α motoneurons of different muscles and muscle groups of the baboon's arm, forearm & hand (Landgren *et al.* 1962*b*; Phillips & Porter 1964; Clough *et al.* 1968). Figure 5 (Phillips & Porter 1964) shows the growth to a maximum of the CM EPSP evoked in an α motoneuron of the median nerve by brief S+ pulses of increasing strength, applied to the 'best' cortical point for this motoneuron. This growth of the EPSP shows that the motoneuron receives synapses from more than one corticospinal cell: the attainment of the maximal EPSP signifies that all the cells connected monosynaptically to it are being stimulated. Such a collection of corticospinal cells is defined as a colony (Landgren *et al.* 1962*b*). Saturation of the colony of figure 5 required S+ pulses of about 3 mA intensity. At this strength, corticospinal cells as far as 10 mm distant horizontally along the precentral gyrus to either side of the 'best point', and cells in the depth

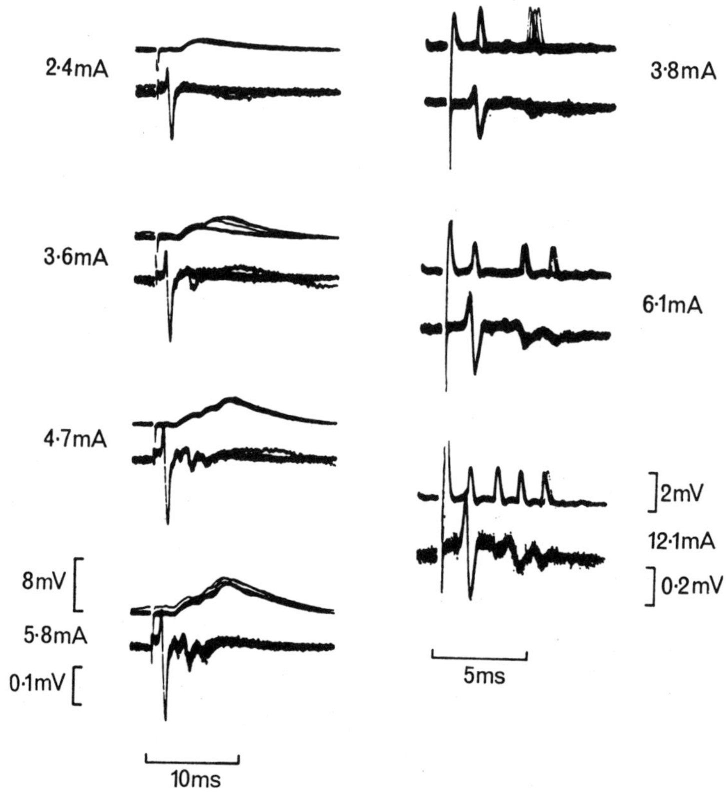

FIGURE 6. *Left.* Upper traces: superimposed intracellular records from ulnar motoneuron in a baboon, showing EPSPs evoked by 0·2 ms, surface-anodal pulses to opposite arm area at strengths noted beside each trace. Lower traces: simultaneous superimposed records of tract waves from lateral corticospinal tract (Kernell & Wu Chien-ping 1967*b*).

Right. Superimposed records of discharges of single corticospinal axon (upper traces) and of simultaneously recorded tract waves (lower records). Strengths of 0·2 ms surface-anodal pulses noted beside each trace (Kernell & Wu Chien-ping 1967*a*).

of the Rolandic fissure, would have been stimulated: this gives an approximate estimate of the spatial extent of the colony (for scale, cf. figure 1).

With pulses stronger than 2·2 mA, a secondary peak or peaks appear on the CM EPSP of figure 5. Such secondary peaks have been shown by Kernell & Wu Chien-ping (1967*b*) to be due to repetitive corticospinal firing (I waves: figure 6).

Preferential accessibility of motoneurons belonging to the baboon's hand

The corticospinal colonies connected to motoneurons of different muscles and muscle groups of the baboon's forelimb vary in their spatial extent, and command different quantities of monosynaptic action in their target α motoneurons (Landgren *et al.* 1962*b*; Phillips & Porter 1964; Clough *et al.* 1968).

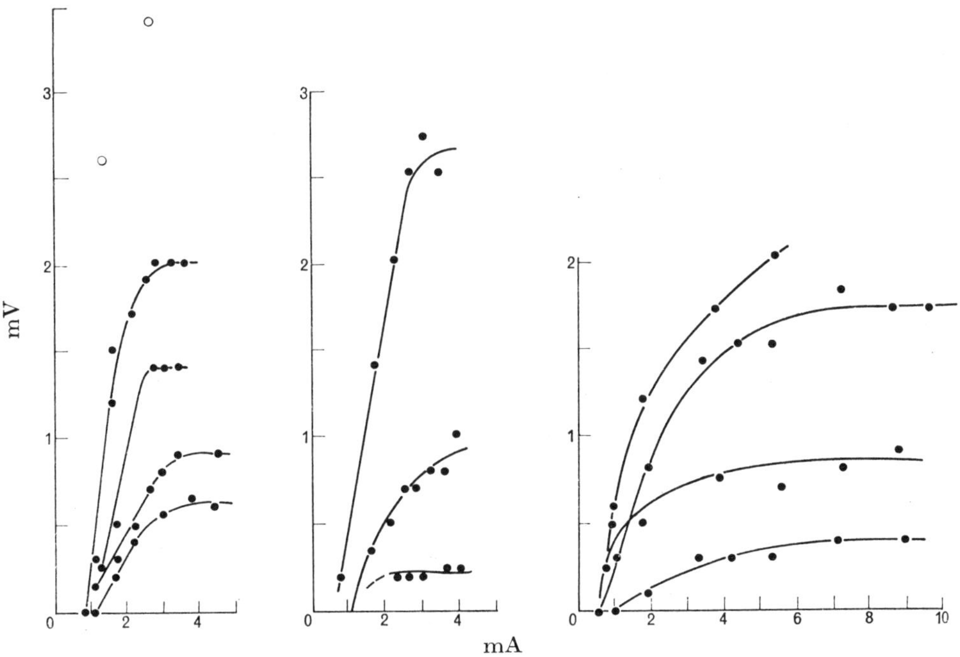

Figure 7. Curves showing growth of monosynaptic EPSPs to maxima for some motoneurons of baboons' ulnar (left), median (middle) and musculocutaneous (right) nerves. Ordinates: amplitude of monosynaptic EPSP in mV. Abscissae: strength of 0·2 ms, surface-anodal cortical pulses in mA (Phillips 1967).

Figure 7 illustrates these points in a few curves of the growth to their maxima of CM EPSPs evoked in motoneurons of the ulnar (left), median (centre) and musculocutaneous (right) nerves. The tallest curve in the centre plot is for the median motoneuron of figure 5. The broader curves of the musculocutaneous motoneurons (elbow flexors) indicate more extensive colonies, the narrower curves of the ulnar and median motoneurons (forearm and hand) indicate less extensive colonies. The spatial extent of colonies of these different muscle groups is such that they must overlap one another in the cortex (Landgren *et al.* 1962*b*). Triceps motoneurons (not illustrated) show smaller, broad curves; about 50% of those sampled by Phillips & Porter (1964) showed no CM EPSPs (but showed early IPSPs and polysynaptic EPSPs). The narrower curves for forearm and hand tended to be taller (more monosynaptic action) than the broader curves for elbow (less monosynaptic action).

These impressions, based on 266 motoneurons of proximal and distal muscle groups, have now been strengthened by rigorous statistical study of a new sample of 157 α motoneurons of the baboon's forearm and hand (Clough *et al.* 1968). The

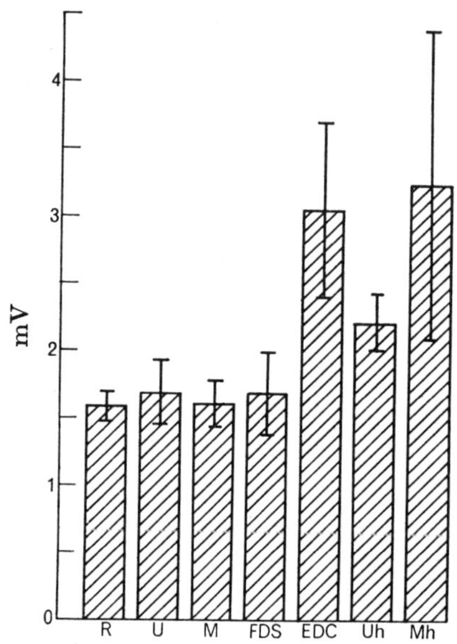

FIGURE 8. Mean amplitudes of maximal monosynaptic EPSPs evoked by corticospinal volleys in motoneurons identified by antidromic stimulation of whole interosseous nerve ('radial', R), whole ulnar (U), whole median (M) (including four palmaris longus motoneurons) and nerves to flexor digitorum sublimis (FDS), extensor digitorum communis (EDC), ulnar nerve to intrinsic muscles of hand (Uh) and median nerve to intrinsic muscles of hand (Mh) (Clough *et al.* 1968).

sample was better subdivided than in the earlier experiments: the motoneurons were classified not only by antidromic stimulation of the whole median, ulnar and posterior interosseous ('radial') nerves, but also by antidromic stimulation of the ulnar and median nerves at the wrist (intrinsic muscles of hand) and of the muscular branches to flexor digitorum sublimis, palmaris longus and extensor digitorum communis (EDC). The mean amplitudes of the maximal CM EPSPs in these groups are shown in figure 8. (Four palmaris longus motoneurons are here grouped with the median motoneurons.) The histogram shows that the mean amplitudes for motoneurons of EDC and intrinsic muscles of the hand (Uh and Mh) are larger than those for flexor digitorum sublimis and for the whole radial, median and ulnar groups, whose means would be kept down by the admixture of motoneurons of more proximal forearm muscles.

The conclusion to be drawn from the experiments described so far is that the preferential accessibility of baboon's thumb and index to single 5·0 ms pulses at any point within a wide cortical field (Liddell & Phillips 1950) depends on the distribution to their α motoneurons of the densest monosynaptic projections that issue from that wide field. The long pulse is effective because it causes a brief, high-frequency burst of corticospinal impulses, which increases the transmitting potency of cortico-motoneuronal synapses and depolarizes the α motoneurons steeply (Landgren *et al.* 1962*a*). Since the threshold for 'flick movements' is higher than the threshold for corticospinal cells the discharge of large parts or wholes of CM colonies is proved to be necessary, and is effective by spatial summation at the motoneuron membrane. It is probable that the early involvement of thumb and index in human Jacksonian epilepsy has a similar explanation.

Liddell & Phillips (1951) found that the thresholds and spatial extension of the wide fields for flick movements were very sensitive to the level of anaesthesia, the fields shrinking and disappearing when small doses of barbiturate were injected, to return again when anaesthesia became lighter. Since the amplitude of directly excited corticospinal volley (D waves) discharged from the area is not diminished even when anaesthesia is deepened to the level of respiratory arrest, such failure of motor response is presumably due to a withdrawal of background facilitation from the spinal α motoneurons.

It is an everyday clinical observation that in 'destroying lesions' of the cortical motor projections, the movements most severely impaired are those of isolated fingers; next in severity come *extension* of the fingers (cf. EDC, figure 8) and wrist (Walshe 1963). It is probable that damage to the CM projection is responsible for such selective impairment.

The wider variety of motor responses to stimulation by trains of pulses whose duration is measured in seconds (the 'faradic stimulation' preferred by Ferrier 1876) are presumably less dependent on the CM projection. There is ample time for temporal summation at interneurons at subcortical and spinal level (Phillips 1966). Similar motor maps can be drawn before and after cutting the baboon's pyramidal tracts, though the surviving responses are weaker (Lewis & Brindley 1965). For these responses the cortico-rubro-spinal projection (Kuypers & Lawrence 1967) is probably important.

Motor unit organization of a sample muscle of the baboon's hand

It is important to know the mechanical properties of the motor units innervated by the motoneurons that receive the densest monosynaptic corticospinal projections. These properties would be expected to provide for fine gradation, in strength and velocity, of the delicate movements of the hand. The intrinsic muscles are difficult to isolate, and extensor digitorum communis (EDC) was therefore preferred (R. M. Eccles, Phillips & Wu Chien-ping 1968), since its α motoneurons stand nearest to those of the intrinsic muscles on the scale of preferential cortical accessibility.

We made records from about 100 motor units. We began in each case by stimulating antidromically the muscle nerve of EDC and recording from subdivided filaments of the ventral roots until we found one which contained a single fibre; that is to say, transmitted an all-or-nothing antidromic impulse (figure 9, upper traces). We could thus measure the conduction velocity of the motor axon. We

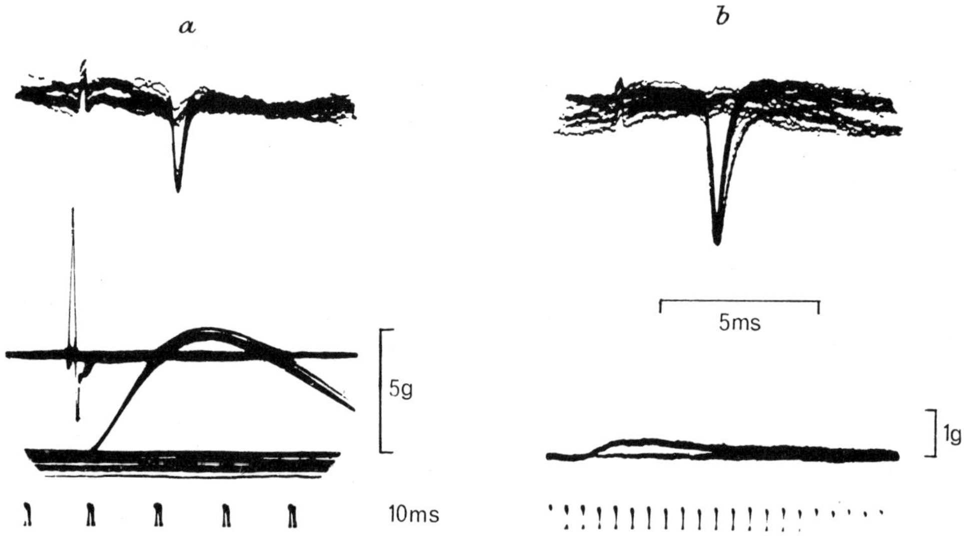

FIGURE 9. Upper records: all-or-nothing antidromic impulses in two motor axons, *a* and *b*, recorded from ventral root filaments while stimulating the motor nerve of EDC. Lower records: *a*, all-or-nothing muscle impulses and all-or-nothing twitch recorded from EDC when ventral root filament *a* was stimulated; *b*, all-or-nothing twitch recorded from EDC when ventral root filament *b* was stimulated (R. M. Eccles *et al.* 1968).

then stimulated the filament orthodromically and recorded the mechanical responses of the motor unit with a very sensitive isometric myograph attached to the EDC tendons (figure 9, lower traces). We observed the surface of the muscle with a microscope and in those cases in which we could see localized twitchings of muscle bundles we made electrical records of the motor unit action potentials with fine surface electrodes (figure 9*a*, middle trace). The twitch tensions ranged from $< 0{\cdot}5$ to $9{\cdot}0$ g wt. Figure 9*a* illustrates one of the larger units and figure 9*b* one of the smaller. From our sample the general structure of EDC can be reconstructed. About 66 % of the units were small, with twitch tensions $< 2{\cdot}0$ g wt; these would account for about 40 % of the force developed by the whole muscle. About 9 % of the units had twitch tension $> 5{\cdot}0$ g wt. Units with twitch tensions > 2 g wt would account for about 60 % of the total force. EDC is a 'fast' muscle (Cooper & Eccles 1930): the contraction times of 75 % of the motor units were $<$ 30 ms and 97 % were $<$ 35 ms. The conduction velocities of the motor axons ranged from 49 to 84 m/s, and there was no relation between conduction velocity and contractile properties of the units (Eccles *et al.* 1968).

In these particulars EDC differs strikingly from those muscles of the cat's hind limb with which the neurophysiologist is familiar. Some comparisons are given in table 1. The motor units are small. They do not fall into two populations, 'slow' and 'rapid' (Denny-Brown 1929), innervated by 'small' and 'large' motoneurons respectively (cf. Clough *et al.* 1968). Thus, in recruiting the α motoneurons of EDC by the CM system, the cortex would be making use of a segmental motor apparatus that would allow a very fine-stepped control of force and velocity, appropriate to the delicate movements required of a primate's hand.

Table 1. Comparison of motor units of EDC with those of some muscles of cat's hind limb

muscle	range of twitch tensions (g wt.)	authors
EDC (baboon)	0·5–9·0	Eccles *et al.* (1968)
plantaris (cat)	6·7–39·0	Devanandan, Eccles & Westerman (1965)
flexor digitorum longus (cat)	11·0–26·0	
flexor hallucis longus (cat)	3·4–9·0	
medial gastrocnemius (cat)	13·6–59·0	Wuerker, McPhedran & Henneman (1965)
	2·5–97·0	Burke (1967)

Precentral and segmental apparatus for the possible servo-control of movement

The hypothesis of the servo-control of movement (Merton 1953; Eldred, Granit & Merton 1953) has been, and continues to be, very fertile in suggesting experiments on motor control and in interpreting their results. In its original or 'follow-up length servo' version, it proposes that the length of a stationary muscle is reflexly maintained by excitatory feedback from the length-measuring muscle spindles to the α motoneurons; that the brain can initiate movement by exciting the fusimotor ('γ') neurons, thereby creating misalinement between muscle length and spindle length; that the error signal is the increased feedback, which reflexly shortens the muscle, removes the misalinement, and leaves the muscle to maintain the new length by defending itself reflexly against any misalinements imposed by external forces (Hammond, Merton & Sutton 1956).

Two important conditions must be satisfied and have been studied experimentally in the cat's hind limb. The first concerns the central distribution of the spindle feedback. At the time of Eldred *et al.*'s (1953) pioneer work it was believed that the monosynaptic excitatory feedback from the spindles of a given muscle was directed only to the motoneurons of the same muscle and to those of its closely related synergists. J. C. Eccles, R. M. Eccles & Lundberg (1957) and R. M. Eccles & Lundberg (1958) then discovered that the feedback from some muscles was more widespread, reaching the motoneurons of muscles not immediately related in function. In discussing this unexpected result Lundberg (1959) concluded that the fusimotor system could hardly initiate precise movements in the cases in which such widespread connexions had been shown to exist. The second condition for fusimotor control is that the feedback circuit shall have sufficient gain (Matthews 1966, 1967; Granit 1968).

If the follow-up length servo theory is to be applicable to the special case of the cortex and hand of the baboon, it is necessary to show: (1) that the hand muscles contain length receptors coupled in parallel with their motor units; (2) that signalling from the length receptors can be increased by projections from the cortex to a system of fusimotor (γ) neurons and, if precise movements are to be possible (3) that the feedback from the spindles of any muscle is directed only to motoneurons of the same muscle.

In his original statement of the servo-theory Merton (1953) considered the probability of direct (α) control for 'urgent' movements, which would save about 50 ms in initiating a movement of the human hand. Such α initiation would, of course, unload the spindles and cut off the feedback. Matthews (1964) has proposed that voluntary movement could be initiated by the α route without sacrificing the advantages of a servo, if 'employed in conjunction with sufficient fusimotor activity to prevent any decrease in spindle discharge occurring during the contraction; this would be achieved if the relative amounts of α and γ activity were adjusted to be appropriate for the velocity of shortening "expected"...then if shortening proceeded faster than "intended" by the higher centres it would be slowed by servo action, and if shortening were hindered by some unexpected load it would be speeded up by servo action.' Granit (1955, 1968) introduced the concept of 'α γ linkage' to describe the co-ordinated distribution of suprasegmental motor output between the α and γ motoneurons, and their coactivation in reflexes.

Figure 10 presents an oversimplified and partly speculative diagram of the possible working of such a system, drawn with movements of the fingers in mind, and designed to concentrate discussion on the force and velocity of movements rather than on the maintenance of stationary postures. The 'intended' movement is from *a* to *c* (figure 10) at uniform velocity, and is supposed to be brought about by a graded recruitment of α motoneurons. The movement would progressively reduce the total feedback from the spindles of the prime movers (figure 10, unloading effect), but the diagram supposes that fusimotor neurons are also recruited (figure 10, fusimotor effect), and, for the sake of simplicity, that the fusimotor effect and the unloading effect are exactly balanced, so that there would be a null effect on the spindle feedback as long as the movement continued at the 'prescribed' velocity towards its target *c*.

Figure 10*A* shows the effect on the spindle feedback of slowing of the movement by an 'unexpected' resistance encountered at *b*. This reduces the unloading effect (figure 10*A*, full line), but the fusimotor effect is unchanged, so the total spindle feedback must increase. The primary endings (figure 10*A*, *p*) which are supposed to be under the influence of both dynamic and static fusimotor neurons (Matthews 1962), will increase their discharge abruptly at time *b*, and, because they are sensitive to velocity as well as to length, the further increase in their discharge will be the greater, the faster the actual unloading effect (figure 10*A*, full line) deviates

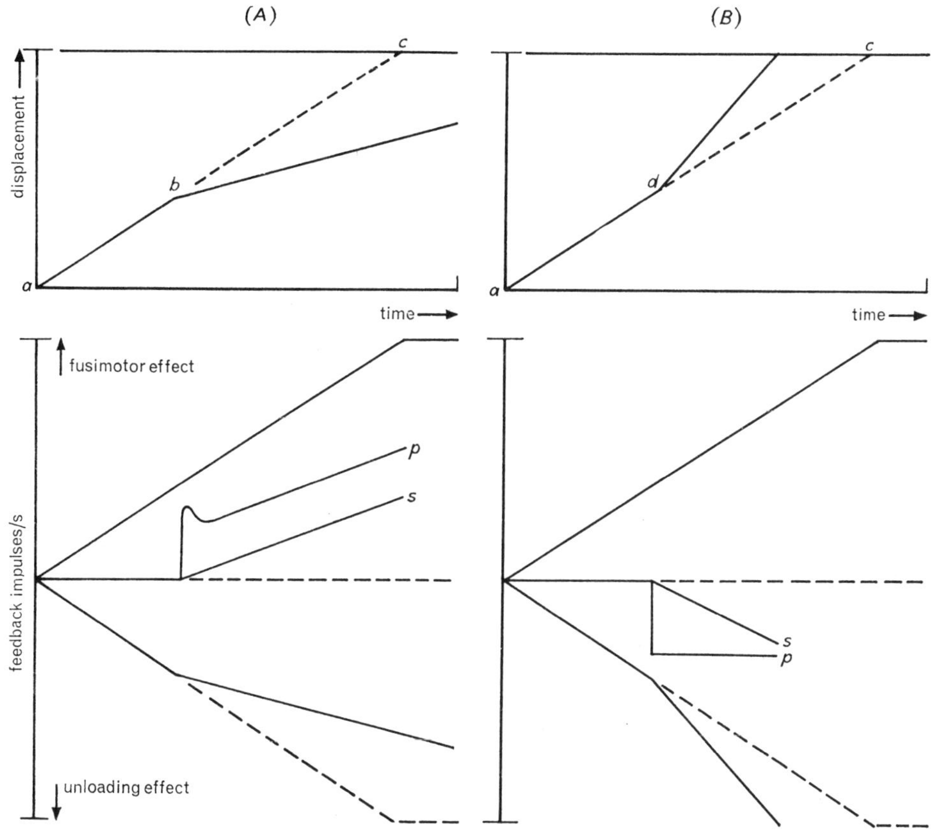

FIGURE 10. α–γ coactivation. Explanation in text.

from the 'expected' unloading effect (figure 10*A*, interrupted line). Since the feedback from the primary endings excites the α motoneurones monosynaptically, its abrupt increase at *b* will tend to produce a prompt increase in force at the instant of deviation, and its further increase will only be checked if the movement is brought back to its proper course towards the target *c*. (The compensated responses are not shown in the diagram.) The secondary endings (figure 10*A*, *s*) are also supposed to be under static fusimotor influence, and would signal the increasing discrepancy between the 'intended' and actual movements, but would differ from the primaries in transmitting no information of the velocity of the deviation. The secondaries do not excite the α motoneurons, but their information, as well as that supplied by the primaries, may well be important at cerebral and cerebellar levels (see below, p. 228).

Figure 10*B* presents the consequences of a speeding-up of the movement by an 'unexpected' reduction of resistance at time *d*. At this instant, the feedback from primary endings under dynamic fusimotor influence would fall abruptly to a lower level (figure 10*B*, *p*; cf. Crowe & Matthews 1964, figure 5), and monosynaptic excitation of the α motoneurons would be abruptly reduced, so that the movement would tend to slow down and resume its approach to the target *c*. The behaviour of primaries under static fusimotor influence would resemble that of the secondaries (figure 10*B*, *s*); intermediate types of response would be expected from primaries under both static and dynamic fusimotor influence.

In the special case of the movements of respiration, Corda, Eklund & Euler (1965) have shown experimentally that the α and fusimotor neurons of intercostal muscles are indeed coactivated, and that respiratory obstruction causes the expected increase in spindle feedback which leads in turn to a compensatory increase in the α discharge.

If the system is to operate in the special case of the baboon's hand, it should be possible to show (1) that the neural apparatus of the cortical hand area includes a projection (direct or polysynaptic) to fusimotor neurons, in addition to the powerful monosynaptic (CM) projection to α motoneurons; (2) that the hand muscles contain spindles which are unloaded by movement and made to discharge by fusimotor activation; and (3) that the spindles supply excitatory feedback to the α motoneurons. The distribution of this feedback would not, however, need to be as narrow as that required by the follow-up length servo hypothesis in its original form. For precision of movement could depend on selection of motor units by the CM projection: the spindle feedback, maintained by fusimotor coactivation, could have an auxiliary action on the membranes of the selected α motoneurons, increasing the discharge of the firing cells and recruiting fresh cells from the 'selected' subliminal fringe if the movement was obstructed by a peripheral load.

Muscle spindles of EDC and their responses to cortical stimulation

Koeze, Phillips & Sheridan (1968) recorded the discharges of single spindle afferents of EDC from baboon's cervical dorsal root filaments. These discharges were silenced during the rising phase of a maximal isometric twitch of EDC evoked by a single stimulus to its motor nerve (figure 12*d*). They were accelerated by linear stretch of the muscle, the responses showing dynamic and static components (figure 12*e*, *f*). The ventral roots were intact and there was plenty of evidence of 'spontaneous' fusimotor activity: in light nitrous oxide anaesthesia, supplemented by small doses of short-acting barbiturate, the frequency of firing altered from time to time when the muscle was kept at the same length (figure 12, beginning of *a* and *f*), and was sometimes irregular. In these conditions of high background fusimotor activity we could not classify our sample of receptors into pure primary and pure secondary endings. All were connected to axons conducting at 27 to 70 m/s; within this range, higher dynamic sensitivity tended to go with higher conduction velocity (Eccles *et al.* 1968). By contrast, primary afferents from cat's soleus conduct at 80 to 120 m/s (Matthews 1963).

To establish the significance of any spindle as a component of a servo-apparatus, it must be shown to be unloaded by the motor units engaged in a specific motor response; it would also be necessary to show that the feedback from this spindle is directed to the α motoneurons innervating its 'own' motor units. Figures 11 and 12 (Koeze *et al.* 1968) illustrate the responses to cortical stimulation of an EDC spindle for which the first of these criteria was satisfied. The second criterion has not yet been satisfied experimentally.

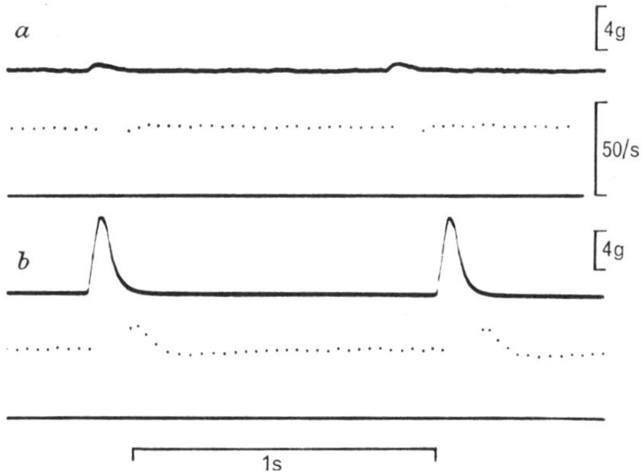

FIGURE 11. Upper records: isometric 'twitches' of EDC, each evoked by six 0·1 ms surface-anodal cortical pulses at 500/s; strength in *a*, 2·0 mA; in *b*, 4·2 mA. Lower records: frequency of impulses discharged by a single spindle receptor and recorded from a dorsal rootlet. Each dot represents one impulse; its height above the base-line measures the frequency of discharge at that instant (Koeze *et al.* 1968).

For figure 11 the cortical stimulus was a burst of six 0·1 ms S+ pulses at 500/s, roughly equivalent, in its action on α motoneurons, to the repetitive firing set by a single 5·0 ms pulse. When below α threshold this stimulus had no effect on the discharge of the spindle. In figure 11*a* it fired a single motor unit, which unloaded the spindle. Since the isometric myograph must have severely restricted the shortening of the motor unit, the mechanical coupling between the unit and the spindle must have been intimate. Stronger stimulation (figure 11*b*) excited many motor units (note tension scale); but if there was any fusimotor co-excitation, this was insufficient to offset the unloading effect of the contraction. Collateral experiments by J. D. Sheridan (to be published) discovered that brief corticospinal bursts discharged single impulses in single α and γ axons in the nerve to EDC, the cortical thresholds being of the same order. The γ impulses arrived at the entry to the muscle a few milliseconds later than the α impulses. It is not known whether these impulses were in static or dynamic γ axons. It is evident from figure 11 that any static impulses arriving early in the rising phase of a cortically evoked 'twitch' did not prevent the unloading of the spindle. The acceleration during the relaxation phase (in figure 11*b*) can be explained by the passive re-lengthening of the spindle, but an effect of dynamic γ axons is not excluded. The existence of the early γ discharges implies a close, possibly monosynaptic connexion between large corticospinal axons and fusimotor neurons of EDC (Clough & Sheridan 1968). But a single impulse in a single static or dynamic fusimotor axon produces an intrafusal contraction which hardly disturbs the background firing of a primary ending (Bessou, Laporte & Pagès 1968).

In figure 11, a mixed population of corticospinal cells would have been fired by the stimulus in unnatural combination, and the contribution of the powerful CM component may have been unnaturally preponderant. For this reason, clonic after-discharges set up by conventional repetitive ('faradic') stimulation are of great interest; the corticospinal populations involved in these may be associated in natural combinations by intrinsic cortical networks, so that a balance in favour of a functionally independent cortico-fusimotor projection might be established. Of ten spindles observed during clonic after-discharges, six were accelerated during the brief α contractions, or gave bursts of spindle acceleration independently of those contractions. Three of the six spindles were also tested in cortical 'twitches' (as in figure 11) and were unloaded (Koeze *et al.* 1968). They therefore satisfied the criterion of close parallel coupling.

Figure 12 shows responses of the spindle of figure 11 to conventional 'faradic' stimulation of the cortex by 3·3 s trains of 0·1 ms S+ pulses delivered at 50/s. At threshold (figure 12*a*), the spindle is unloaded: if there is any static fusimotor activity it is insufficient to prevent the silencing of the spindle. During the larger α responses of figure 12*b* and *c*, the initial unloading is followed by acceleration of the spindle, which continues until the end of the α contraction and subsides with it. The spindle bursts in figure 12*b* and *c* suggest the activity of static fusimotor fibres (Crowe & Matthews 1964), but since the myograph record has gone off the screen one cannot be certain that these bursts were not due to a tremulous α response. The experiment demonstrates that the cortex has access to a fusimotor system that is powerful enough to offset the unloading effect of 'isometric' contraction, and that its response is roughly coextensive in time with the α response. It could therefore be related accurately to phasic movements. The corticofugal pathway has not been defined and may be corticorubrospinal, corticointerneuronal, etc. Its functional independence of the CM projection has been shown by experiments in which the level of anaesthesia was deliberately varied, when the sequence of α and γ responses could sometimes be reversed, and the γ response could sometimes be elicited alone. When the γ response occurred in advance of the α, the latter might be thought to have been excited by feedback from the activated spindles; but identical patterns of response were seen in preparations in which the cervical dorsal roots had been cut (Koeze *et al.* 1968).

In a few unpublished experiments T. H. Koeze, C. G. Phillips and Wu Chien-ping stretched EDC during cortical stimulation and looked for an increase in the dynamic sensitivity of spindle afferents (Vedel 1966). In the changing background spindle excitation in our lightly anaesthetized preparations we could find no unequivocal evidence of evoked dynamic fusimotor activity.

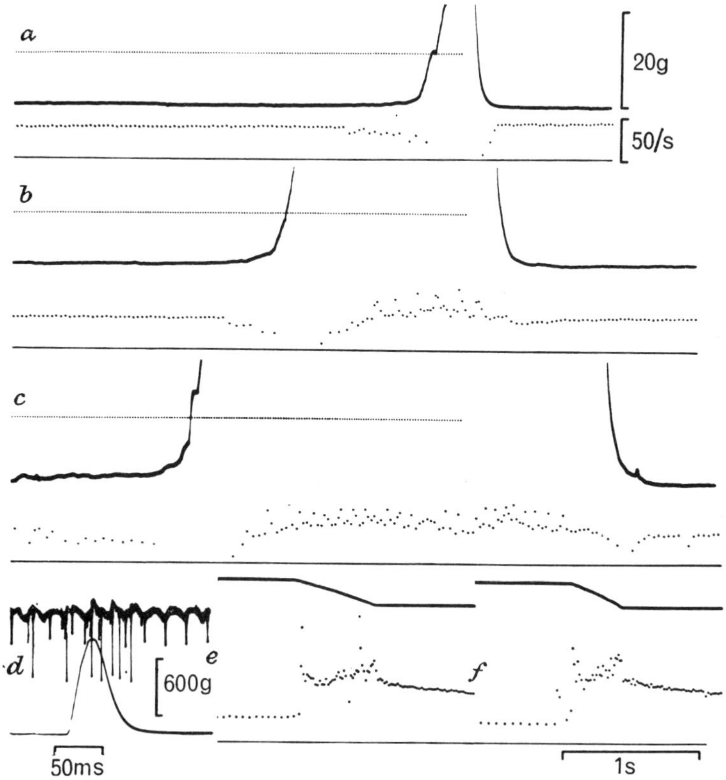

FIGURE 12. *a*, *b*, *c*, upper records: isometric myographic responses of EDC to stimulation of opposite precentral arm area with 0·1 ms surface-anodal pulses at 50/s for 3·3 s at strengths 4·2 mA in *a*; 4·35 mA in *b*; and 4·7 mA in *c*. Lower records: discharge of same spindle receptor as in figure 11; dorsal roots cut. *d*, maximal twitch elicited by stimulation of motor nerve. *e*, *f*, responses of spindle to passive lengthening of the muscle at 15 and 24 mm/s respectively (Koeze *et al.* 1968).

Central distribution and quantity of spindle feedback

It has already been stated that the spindle feedback shown in figure 12*b*, *c* ought (if the afferent axon had been intact) to have been directed to the motoneurons of its 'own' motor units, and that no experiment has yet shown that the feedback is thus accurately focused. But it has at least been possible to investigate whether the feedback from one muscle is confined to motoneurons of the same muscle (homonymous feedback), and to measure the quantity of monosynaptic excitatory action by recording intracellularly from a sample of α motoneurons of the same and of other muscles (Clough *et al.* 1968). The muscle afferents were electrically stimulated in the motor nerves from single muscles or functionally related groups of muscles. This method was introduced by Eccles *et al.* (1957) and R. M. Eccles & Lundberg (1958) in their classical studies of the cat's hind limb, and the work on the baboon was initiated in Gothenburg in 1966 by T. Hongo, C. G. Phillips, A. Lundberg & R. Thompson (to be published) in an extensive quantitative study of the homonymous and heteronymous feedback from most of the muscles of the hindlimb, undertaken with the same problem in mind.

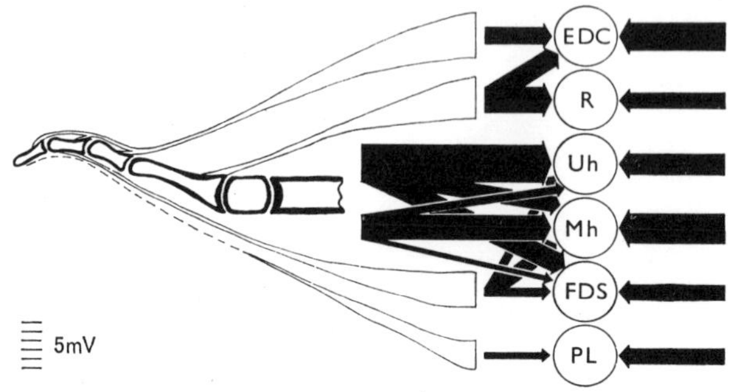

FIGURE 13. Diagram showing distribution and mean quantities of spindle feedback to motoneurons of different muscles and muscle-groups of the baboon's hand. Abbreviations as in figure 8. PL = palmaris longus. Arrows from right of figure show mean quantities of corticomotoneuronal excitatory synaptic action on motoneurons of the same muscles and muscle-groups. Breadths of arrow-shafts measure mean quantities of monosynaptic excitatory action (millivolt scale at left of figure) (Clough *et al.* 1968).

Figure 13 summarizes the results of the investigation of the hand by Clough *et al.* (1968). It shows that the spindles of EDC do not send any significant monosynaptic excitation to motoneurons of other muscles. This is consistent with the expectation of the follow-up-length-servo theory. Such an expectation is also satisfied in the case of palmaris longus (PL, figure 13). But it is certainly not fulfilled in the case of flexor digitorum sublimis (figure 13, FDS), whose spindles send more heteronymous excitation to motoneurons of the intrinsic muscles of the hand (figure 13, Uh and Mh) than homonymous excitation to the motoneurons of FDS itself. The remaining muscle afferent inputs illustrated in figure 13 (R, Uh, Mh) were from functionally related groups of muscles, not from single muscles. The distribution of these inputs was not confined to motoneurons of the same groups. Thus the spindles of wrist extensors (R, figure 13) sent about as much heteronymous monosynaptic excitation to motoneurons of EDC as those motoneurons received homonymously from the spindles of EDC itself. The spindles of intrinsic hand muscles (figure 13, Uh and Mh) supplied much more heteronymous monosynaptic excitation to FDS motoneurons than those motoneurons received homonymously from FDS spindles. Such diffuse distributions of spindle feedback would hardly be expected on the follow-up length servo theory; it is difficult to see how such a segmental motor apparatus could allow the cortex to initiate precise movements of the hand, employing only the fusimotor route (cf. Lundberg 1959). As will be shown, these findings are easier to reconcile with the theory of $\alpha\ \gamma$ coactivation.

The average quantities of monosynaptic feedback, which are measured by the widths of the arrows on the left side of figure 13, are obviously too small to discharge the α motoneurons unless these have already been brought by some other input to within 2 or 3 mV of the firing level. Granit (1968) has emphasized that tonic stretch reflexes were described and studied in decerebrate animals and are difficult to demonstrate under other conditions, for example in healthy man, and has discussed the possible reasons for this. The circumstances in which the feedback from the spindles of EDC could actually fire its motoneurons await full investigation. Wu Chien-ping and I (unpublished) made preliminary experiments on two baboons. We stretched the muscle from minimum to maximum physiological length at 24 mm/s, recording its responses both myographically and electromyographically.

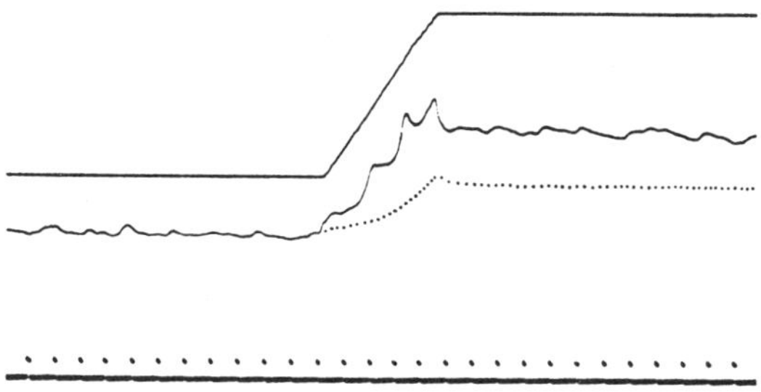

FIGURE 14. Myographic response of baboon's EDC to passive lengthening from minimum to maximum physiological length, 10 mm at 24 mm/s, at different levels of anaesthesia. Upper trace: signal of lengthening. Middle trace: isometric tension developed in light anaesthesia; note waves due to 'spontaneous' activity of α motoneurons, and small zero error with respect to upper trace; peak tension at end of phasic stretch is about 315 g wt., maintained static tension about 215 g wt. Dotted trace: three superimposed responses recorded when anaesthesia was deeper; peak tension 120 g wt., maintained tension 100 g wt. approx. Lower trace: time 10/s. (Wu Chien-ping & C. G. Phillips unpublished observations.)

Figure 14 shows mechanical responses recorded from one baboon in very light anaesthesia with nitrous oxide, oxygen and small doses of short-acting barbiturate. When anaesthesia was a little deeper the dotted curve was recorded. In lighter anaesthesia there was an irregular 'spontaneous' discharge of α motoneurons, revealed by the waves on the record at minimum length. There was now a myographic response to stretch (full curve). One of Granit's (1968) suggestions was that a high level of background fusimotor activity is essential if the monosynaptic bombardment of α motoneurons by spindle afferents is to be adequate to ensure their discharge. At comparable levels of anaesthesia the spindles of EDC are certainly highly excited (Koeze *et al.* 1968), but there is, as yet, no evidence whether such high levels of spindle feedback, even if aided by autogenous post-tetanic potentiation (cf. Granit 1956), can fire the α motoneurons monosynaptically unless they are simultaneously facilitated by some other input. Such background facilitation might be due to polysynaptic actions of the spindle input, or to excitation originating within the CNS. The absence of any obvious reflex effect at the peak of the dynamic phase of stretch in the dotted trace of figure 14, when some EDC spindles would be discharging at about 300/s, does suggest that the monosynaptic input needs to be supported by some other source of facilitation at the motoneuron membrane.

During the responses of which the middle trace of figure 14 is typical, simultaneous electrical records were made from three motor units of EDC (figure 15). The responses of the three α motoneurons differed unexpectedly. Two of them were discharging 'spontaneously'. One of these (figure 15*a*) was inhibited by the stretch, and resumed when the muscle was returned to its minimum length; the other (figure 15*b*) was unaffected by stretching or releasing the muscle. The third (figure 15*c*) behaved in the manner expected; it began to discharge during phasic stretch, continued as long as the muscle was held at its maximum length and ceased on release. In a second baboon no response to stretch could be obtained from EDC, either myographically or electromyographically.

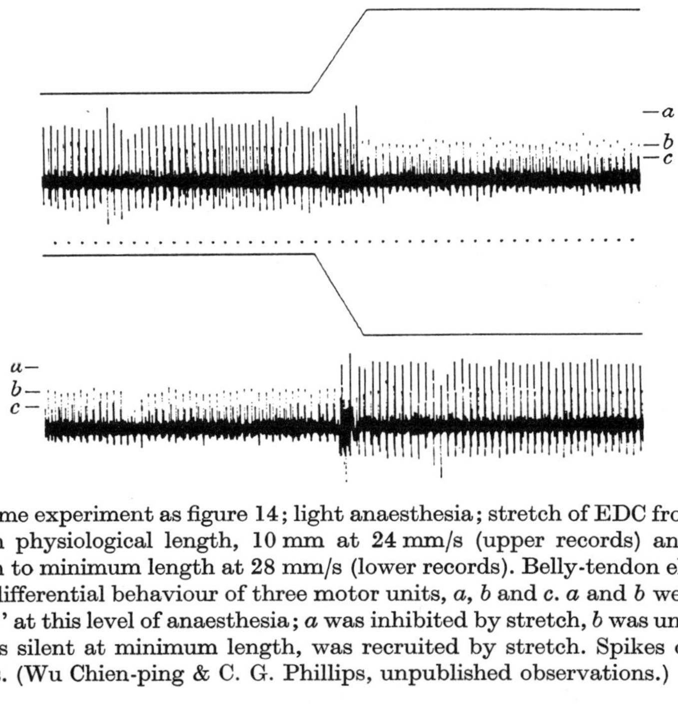

FIGURE 15. Same experiment as figure 14; light anaesthesia; stretch of EDC from minimum to maximum physiological length, 10 mm at 24 mm/s (upper records) and release from maximum to minimum length at 28 mm/s (lower records). Belly-tendon electromyogram showing differential behaviour of three motor units, *a*, *b* and *c*. *a* and *b* were firing 'spontaneously' at this level of anaesthesia; *a* was inhibited by stretch, *b* was unaffected, and *c*, which was silent at minimum length, was recruited by stretch. Spikes of *a* retouched. Time 10/s. (Wu Chien-ping & C. G. Phillips, unpublished observations.)

Interaction between spindle feedback and cortico-motoneuronal excitation at the motoneuron membrane

Another problem concerns the extent to which the CM system itself could supply the α motoneurons with the 'depolarizing activity from sources other than stretch afferents' which may be required to make the monosynaptic spindle feedback effective in maintaining a 'reflex' motoneuron discharge (Granit 1968).

The arrows to the right of figure 13 show that in the case of the single muscles (EDC, FDS, PL) the CM input produces, on average, more depolarization than the spindle feedback from the muscles' own spindles (the 'homonymous' input: the larger arrows to the left of Uh and R of course include heteronymous as well as homonymous spindle inputs to the sampled Uh and R motoneurons). The disparity between the homonymous spindle and the CM inputs to the α motoneurons is further increased by a property of CM synapses which is not shared by spindle afferents' synapses, namely that their transmitting potency is increased when a few impulses are delivered at frequencies of 200/s and over (Landgren *et al.* 1962*a*). Figure 16 (Phillips & Porter 1964) contrasts the lack of effect of such a brief period of repetition on the 'spindle' synapses with its powerful effect on the CM synapses at the membrane of the same motoneuron of the median nerve. It is this effect which gives the CM actions their overwhelming power when CM volleys are delivered at frequencies of the order of 200 to 300/s. Such frequencies are often seen in records from single spindle afferents, but in short bursts they have no central potentiating action on their own synapses.

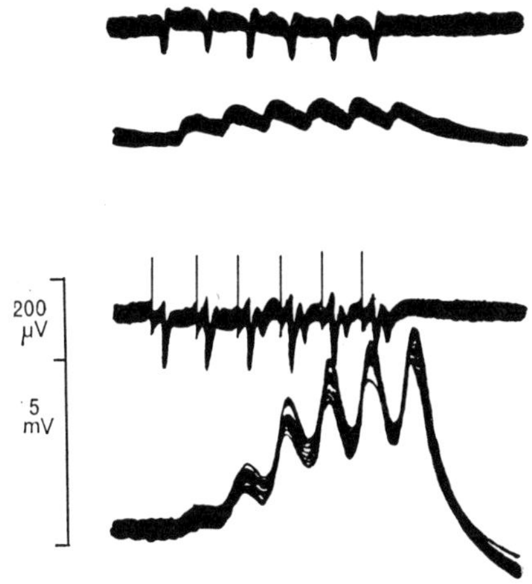

FIGURE 16. Superimposed intracellular records from motoneuron of median nerve. Upper pair: simultaneous records of dorsal root volleys set up by submaximal stimulation of muscle afferent axons at 200/s (upper record) and monosynaptic EPSPs evoked in the motoneuron (lower record). Lower pair: simultaneous records of corticospinal volleys at 200/s (upper record) and monosynaptic EPSPs evoked in the motoneuron (lower record); the strength of cortical stimulation was adjusted so that the response to the first volley was similar in size to the first muscle afferent volley in the upper set of records (Phillips & Porter (1964), in J. C. Eccles and J. P. Schadé (eds.), *Physiology of spinal neurons*, *Prog. Brain Research*, **12**, 222–242. Reproduced by kind permission of the publishers, Elsevier, Amsterdam).

The CM input provides a second monosynaptic input to the α motoneurons which may be used to test for any changes in excitability that might be brought about by alterations in the level of spindle input. In unpublished experiments Wu Chien-ping and I compared the cortical thresholds for minimal contractions of EDC in response to a brief, high-frequency burst when the muscle was held at its maximum and minimum physiological length (in six baboons) and before and after acute section of the dorsal roots (in four baboons). An idea of the large difference in quantity of spindle input in the different conditions can be gained by comparing the beginning and end of the frequency record of figure 12*f*; but we could detect no effect on the threshold of the testing CM input. (It must be stressed that these experiments are testing a simpler system than those of Gellhorn (1949), in which motor responses to strong and prolonged repetitive stimulation of the macaque's cortex were altered when different muscle groups were lengthened by changing the initial angles of the joints.) Such a lack of effect of changing the background spindle activity on the response to a brief testing CM input led us, in one preliminary experiment, to test the motoneurons at the peak of dynamic stretch, when the total spindle input should attain its highest value. The result is shown in figure 17, from the same preparation as figures 14 and 15. In figure 17*a* a control stretch had no effect on the motor unit with the large spike, but a second motor unit began to fire. (The e.m.g. electrodes had been moved, but these two units may have been the same as *b* and *c* in figure 15.) Figure 17*c* shows one of

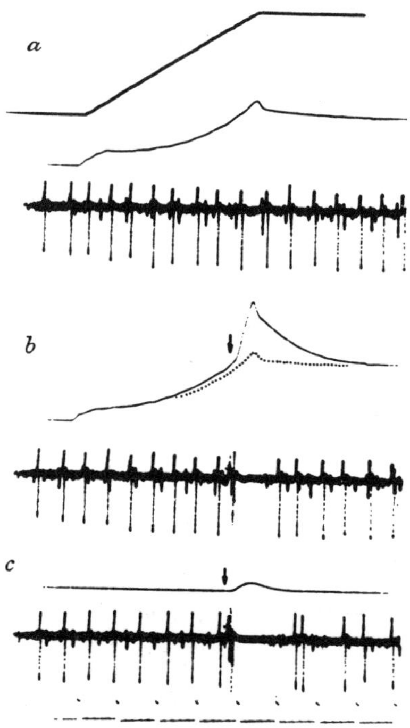

FIGURE 17. Same experiment as figures 14 and 15. *a*, Control stretch, 10 mm from minimum to maximum physiological length at 24 mm/s. The motor unit with the larger spike is unaffected. One with a smaller spike is recruited. *b*, Testing corticospinal volleys (five surface-anodal cortical pulses, 500/s, 2·3 mA) injected at arrow. Myographic response from *a* dotted for comparison. Peak tension exceeds that in *a* by 63 g wt. Both motor units show silent period. *c*, Control response to cortical stimulus with muscle held at maximum physiological length, Peak tension 11 g wt. (several controls ranged from 8 to 23 g wt.) Time 10/s (Wu-Chien-ping & C. G. Phillips, unpublished observations).

several control responses to the testing CM stimulus. This fires the unit that was unaffected by stretch and the impulse is followed by a 'silent period'. When the testing CM was injected just before the peak of phasic stretch (figure 17*b*), both units showed a 'silent period'. The peak tension exceeded that of the control stretch by about 60 g wt; the control CM 'twitches' ranged between 8 and 23 g wt. (11 g wt in figure 17*c*). The myogram shows that additional motor units had been recruited by the CM from a pool subliminally excited by the stretch. Control experiments have shown that the myographic effect was not an artifact of increased viscosity of the muscle (Gasser & Hill 1924), which was still being forcibly lengthened at the beginning of the cortically evoked contraction.

These experiments all suggest that even large totals of monosynaptic spindle input at high frequencies can have, at best, a marginal or auxiliary effect on the α motoneurons of the baboon's EDC. Such a conclusion favours the hypothesis of $\alpha\ \gamma$ coactivation against the hypothesis of the follow-up length servo in its original form. Clough *et al.* (1968) found that the largest spindle inputs are directed mainly to those α motoneurons which also receive the largest CM inputs. This property of the motor apparatus should improve the selectivity of control by $\alpha\ \gamma$ coactivation. In cortically driven movements of the hand, any spindle feedback directed to motoneurons other than those selected by the CM projection could well remain subthreshold, and need not blur the precision of the movements. The output of the CM-selected motoneurons could be governed automatically by marginal changes in a cortically sustained spindle feedback, reinforcing or withdrawing support from the CM input in response to changes in peripheral load (cf. figure 10).

Is the CM projection itself the efferent limb of a transcortical spindle circuit?

Up to this point the possibility of servo-control has been discussed solely in terms of the spindle–motoneuron–motor unit–spindle circuit. It remains to consider the possibility that in the case of the primate's hand, this segmental circuit, which has its most obvious importance in postural reactions (Eldred *et al.* 1953), has been overlaid in the course of evolution by a transcortical circuit of which the powerful and 'recent' CM projection is the efferent limb. There is evidence of spindle-evoked activity in cells in the precentral gyrus of the macaque (Albe-Fessard & Liebeskind 1966), but it is not yet known whether there is a 'direct' spindle input from third-order neurons to the corticospinal cells of the hand area in any primate. (No such connexion has been found in the cat by Oscarsson, Rosén & Sulg (1966)).

Figure 18 illustrates one of the very important experiments of Evarts (1967). It shows the discharges of a large pyramidal tract (PT) cell (conduction velocity 80 m/s) that occurred in relation to a movement of flexion of the contralateral wrist, for which the monkey was always rewarded. The middle record shows its discharge when the flexion movement was unhindered. In the top record the flexion was opposed by a weight. The response of the PT cell was greatly increased; in these conditions it was active also during extension of the wrist, which was now brought about by waning activity of the flexors, paying out the cord on which the weight was hung. In the bottom record the extensors were opposed by the weight. The rewarded movement of flexion was now brought about by the extensors

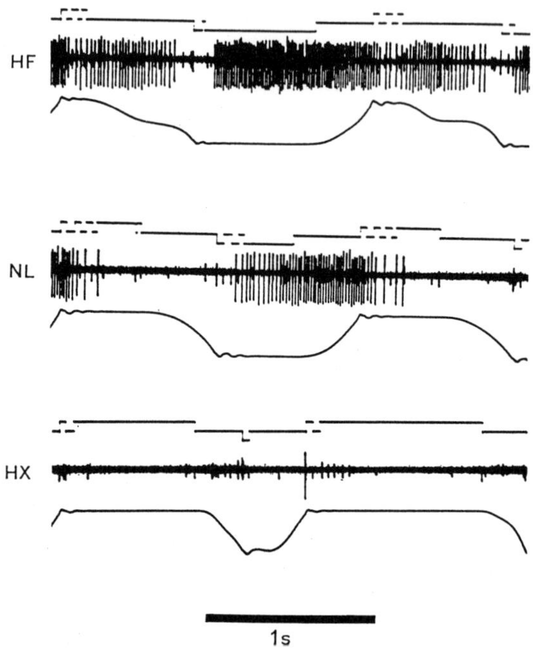

FIGURE 18. Discharges of PT cell of monkey (axonal conduction velocity 80 m/s) recorded during a movement of flexion of the wrist that was rewarded by grape juice. NL, the movement of wrist flexion was unresisted; the PT cell discharged in relation to the flexion movement (upward movement of lower trace). HF, flexion resisted by weight; PT cell discharge increased during flexion and continued during extension (in which paying-out of the weight was controlled by flexor force). HX, extension resisted by weight; flexion movement now controlled by extensor force; PT cell silent (Evarts (1967), reproduced by kind permission of the author and publishers, from *Neurophysiological basis of normal and abnormal motor activities* (ed. M. D. Yahr and D. P. Purpura), New York, Raven Press).

paying out the weight, and the PT cell remained silent. The output of this PT cell was thus related to flexor force rather than flexor displacement. One may hazard the speculation that the increased discharge of the PT cell in the top record was in response to a signal of mismatch between 'intended' and actual displacement. Whether this signal is a crude one from the muscle spindles, or whether the mismatch has been computed by the cerebellum is still unknown; nor, in this experiment, can the contributions of joints, skin and vision be assessed. But however 'instructed', the CM projection would transfer the 'instruction' for increased force to the α motoneurons with maximum directness. If the CM projection is indeed part of a control loop, new sense is made of the old observation that 'voluntary' movements of a monkey's arm are grossly impaired by deafferentation (Knapp, Taub & Berman 1963), when responses to cortical stimulation are unaffected (Mott & Sherrington 1895; Denny-Brown & Sherrington in Sherrington 1931). It may well be that the most important function of fusimotor co-activation in the case of the hand is to maintain the inflow of information of muscle length to the cortex and cerebellum.

Spindle feedback and willed movement of the human hand

The foregoing considerations lead naturally to the question of the part played by the spindle feedback in willed movements of the human hand. There are three types of approach to this problem.

An experiment desirable in principle would deprive a normal subject of his spindle feedback temporarily and reversibly, leaving all other inputs intact, and would compare the force, velocity and accuracy of goal-directed movements of the hand in the normal and in the abnormal condition. Less crucial but nevertheless very valuable observations could be made on patients whose cervical dorsal roots had been cut, in spite of the disadvantage that the inputs from the skin, joints, etc., would also be lost, and that many of the patients would have some prior neurological disability. Rhizotomy for relief of intractable pain would offer the most favourable opportunities. Foerster (1927) often performed the operation for the relief of spasticity, cutting the roots from C_3 to D_3. He found that even in the acute stage, and without the aid of vision, the patients could voluntarily flex, extend, raise, lower and rotate their arms, and pronate, supinate, extend and flex their hands. The start of these simple movements was delayed. The patients were often emphatic that they could not carry out the movements required of them, and said they were unaware of the movements that they had actually made. These proximal movements could be performed more promptly and accurately under visual guidance. But the loss of isolated movements of the fingers was much more severe, and was not helped when vision was allowed. An instruction to move an individual finger could rarely be obeyed. The wrong finger would be moved, or if the correct finger were moved there would be associated movements of other fingers or of the whole hand. With time and practice there was some improvement.

Visual information should be able to compensate in some degree for the lost information of position and movement from skin and joints: such compensation would be expected to be least effective in the case of the fingers, deprived of a cutaneous innervation so much richer than that of the proximal segments of the arm. It is clear that the movements of the fingers are the worst hit by dorsal root section. But it is hard to conceive that vision could substitute for the flow of precise measurements normally supplied by length receptors associated with motor units in $\alpha\ \gamma$ linkage. It may be suspected therefore that an important part of the disability is due to loss of the spindle feedback. Foerster (1927) thought that the prompt and correct addressing of the cortical message depended on afferent impulses. Presumably man has a CM projection to the fingers at least as powerful as the baboon's. Normal 'addressing' could well depend in part on spindle input to the CM colonies, and would also be helped at the segmental level by the mutually reinforcing action of CM and spindle impulses at the membranes of the target α motoneurons (cf. Clough *et al.* 1968).

A second type of approach is to deprive a normal subject of all inputs other than the spindle feedback and this experiment has been performed by Butt, Davies & Merton (Merton 1964). A sphygmomanometer cuff was kept inflated at the wrist until all sensation was lost from the hand. Skin and joints and intrinsic hand muscles were thus reversibly denervated, but the spindle feedback from the long flexors and extensors of the fingers remained intact. In this condition the subject was totally unaware of passive movements of his thumb. The input from the spindles of the long muscles does not therefore reach consciousness (even if it does reach the 'motor' cortex). But the subject was still able to move his thumb voluntarily, and even to move a pointer attached to it through a prescribed angle. He could still make the prescribed movement with his eyes closed. If then the movement was prevented by the examiner holding the thumb, 'the subject believed he had moved it just the same' (Merton 1964). Thus, when a voluntary effort has been put forth, the spindle feedback does not enter consciousness with information of what, if any, movement has actually occurred; such information must therefore normally be supplied by skin, joints, etc. It has yet to be shown whether the spindle feedback plays any part in the measuring-out of such a willed movement, that is to say, in halting it automatically when it reaches its goal. Merton concludes that Helmholtz was correct in supposing that we possess a 'sense of effort'. The movement would have been trained initially under visual guidance, and it was presumably vision that 'set' the 'sense of effort'.

A third type of approach is provided by vibrating the tendon of a muscle during the course of a movement, the important new method which K.-E. Hagbarth and his colleagues have introduced into clinical neurology. In the cat, longitudinal vibration is a highly selective stimulus to the primary endings of the muscle spindles (Brown, Engberg & Matthews 1967), and it is probable that transverse, percutaneous vibration of an intact tendon would act similarly. If the spindle feedback, under precise adjustment by fusimotor control, is necessary for the automatic governing of the course of a movement, then disorganization of the feedback by vibrating the tendon should seriously disturb the performance. Thus, in an experiment by Eklund & Hagbarth (1966), a normal subject was instructed to try to maintain a steady isometric force, flexing his elbow against a restraining strain-gauge dynamometer at the wrist. When the biceps tendon was vibrated there was a large increase in force, of which the subject was quite unaware. Presumably the motor output determined by his 'sense of effort' continued undisturbed by the vibration, which merely added its own quota of reflex force. Only if he was allowed to see the dial of the strain-gauge could he correct the force: there was then only a transient change in force at the beginning and end of the period of vibration. Again, the re-setting of the 'sense of effort' required visual control: the artificially disturbed spindle feedback provided no information that enabled the subject to correct the output of force (cf. Evarts 1967), although, in some of Eklund & Hagbarth's (1966) isometric experiments, it did enter consciousness as an illusion of the movement that would have taken place if the muscle had been allowed to shorten.

Conclusion

These new experiments uphold the old idea that the cerebrum is the head-ganglion of the distance receptors, and that its motor pathways, whose study was pushed forward by Ferrier's pioneer work, are internuncial pathways putting the body segments at the disposal of the distance receptors (Sherrington 1906). The work of the first century after Ferrier has been largely preoccupied with the structure of the motor apparatus, although for technical reasons it has had to neglect the distribution and actions of its very numerous fine nerve fibres. The first part of this lecture has reviewed some recent evidence that the spinal end of a very important component of the cortical motor projection has acquired a specially direct and powerful connexion with the motoneurons innervating the muscles of the primate's hand. High-frequency discharges of these corticomotoneuronal colonies would tend to forestall and to dominate the segmental inputs playing upon these motoneurons. It is almost as if these particular spinal motoneurons had been transplanted into the cerebral cortex: as if the dendrites of the corticospinal pyramids were acting vicariously for the dendrites of the α motoneurons as antennae sensitive to intracortical synaptic activities.

The next attack should be at the headward end of the pathway, to determine the range of peripheral inputs to the corticomotoeuronal colonies of the hand. Work of the type now in progress on inputs to PT cells in the cat (for review see Phillips (1966) needs to be extended to this special case. An important question is whether, in virtue of their dominance over the segmental circuits, these corticomotoneuronal colonies have acquired from the muscle spindles an input from third-order neurons sufficiently dense to function as a transcortical servo-loop which could dominate the segmental servo loop. It seems likely that the work of the second century after Ferrier will come to closer grips with the problems of cerebral motor function.

The lecture owes more than I can well acknowledge to my collaborators named within it, to discussions we have had with one another and with our colleagues in Gothenburg, Oxford and Stockholm.

References

Albe-Fessard, D. & Liebeskind, J. 1966 Origine des messages somatosensitifs activant les cellules du cortex moteur chez le singe. *Expl Brain Res.* **1**, 127–146.

Bernhard, C. G., Bohm, E. & Petersén, I. 1953 Investigations on the organization of the corticospinal system in monkeys. *Acta physiol. scand.* **29** (Suppl. **106**), 79–105.

Bessou, P., Laporte, Y. & Pagès, B. 1968 Frequencygrams of spindle primary endings elicited by stimulation of static and dynamic fusimotor fibres. *J. Physiol.* **196**, 47–63.
Brown, M. C., Engberg, I. & Matthews, P. B. C. 1967 The relative sensitivity to vibration of muscle receptors in the cat. *J. Physiol.* **192**, 773–800.
Burke, R. E. 1967 Motor unit types of cat triceps surae muscle. *J. Physiol.* **193**, 141–160.
Clough, J. F. M. & Sheridan, J. D. 1968 A fast pathway for cortical influence on cervical gamma motoneurones in the baboon. *J. Physiol.* **195**, 26–27*P*.
Clough, J. F. M., Kernell, D. & Phillips, C. G. 1968 The distribution of monosynaptic excitation from the pyramidal tract and from primary spindle afferents to motoneurones of the baboon's hand and forearm. *J. Physiol.* **198**, 145–166.
Cooper, S. & Denny-Brown, D. E. 1927 Responses to stimulation of the motor area of the cerebral cortex. *Proc. Roy. Soc.* B **102**, 222–236.
Cooper, S. & Eccles, J. C. 1930 The isometric responses of mammalian muscles. *J. Physiol.* **69**, 377–385.
Corda, M., Eklund, G. & Euler, C. von 1965 External intercostal and phrenic and motor responses to changes in respiratory load. *Acta physiol. scand.* **63**, 391–400.
Crowe, A. & Matthews, P. B. C. 1964 Further studies of static and dynamic fusimotor fibres. *J. Physiol.* **174**, 132–151.
Denny-Brown, D. E. 1929 The histological features of striped muscle in relation to its functional activity. *Proc. Roy. Soc.* B **104**, 371–411.
Devanandan, M. S., Eccles, R. M. & Westerman, R. A. 1965 Single motor units of mammalian muscle. *J. Physiol.* **178**, 359–367.
Eccles, J. C., Eccles, R. M. & Lundberg, A. 1957 The convergence of monosynaptic excitatory afferents on to many different species of alpha motoneurones. *J. Physiol.* **137**, 22–50.
Eccles, R. M. & Lundberg, A. 1958 Integrative pattern of 1a synaptic actions on motoneurones of hip and knee muscles. *J. Physiol.* **144**, 271–298.
Eccles, R. M., Phillips, C. G. & Wu Chien-ping 1968 Motor innervation, motor unit organization and afferent innervation of m. extensor digitorum communis of the baboon's forearm. *J. Physiol.* **198**, 179–192.
Eklund, G. & Hagbarth, K-E. 1966 Normal variability of tonic vibration reflexes in man. *Expl Neurol.* **16**, 80–92.
Eldred, E., Granit, R. & Merton, P. A. 1953 Supraspinal control of the muscle spindles and its significance. *J. Physiol.* **122**, 498–523.
Evarts, E. V. 1967 Representation of movements and muscles by pyramidal tract neurons of the precentral motor cortex. In *Neurophysiological basis of normal and abnormal motor activities* (ed. Yahr, M. D. and Purpura, D. P.). New York: Raven Press.
Ferrier, D. 1873 Experimental researches in cerebral physiology and pathology. *West Riding Lunatic Asylum Med. Rep.* **3**, 1–50.
Ferrier, D. 1876 *The functions of the brain.* London: Smith, Elder.
Ferrier, D. 1886 *The functions of the brain* (2nd ed.). London: Smith, Elder.
Ferrier, D. 1888 Discussion on cerebral localization. *Trans. Congr. Amer. Phys. Surg.* **1**, 337–340.
Foerster, O. 1927 Schlaffe und spastische Lähmung. In *Handbuch der normalen und pathologischen Physiologie* (ed. Bethe, A., Bergmann, G. v., Embden, G. and Ellinger, A.), **10**, pp. 900–901. Berlin: Springer.
Fritsch, G. & Hitzig, E. 1870 Uber die elektrische Erregbarkeit des Grosshirns. *Arch. Anat. Physiol. Wiss. Med.* **37**, 300–332.
Gasser, H. S. & Hill, A. V. 1924 The dynamics of muscular contraction. *Proc. Roy. Soc.* B **96**, 398–437.
Gellhorn, E. 1949 Proprioception and the motor cortex. *Brain* **72**, 35–62.
Gorman, A. L. F. 1966 Differential patterns of activation of the pyramidal system elicited by surface anodal and cathodal cortical stimulation. *J. Neurophysiol.* **29**, 547–564.
Granit, R. 1955 *Receptors and sensory perception,* p. 268. Newhaven and London: Yale University Press.
Granit, R. 1956 Reflex rebound by post-tetanic potentiation. Temporal summation-spasticity *J. Physiol.* **131**, 32–51.

Granit, R. 1968 Sherrington Memorial Lecture: The functional role of the muscle spindle's primary end organs. *Proc. R. Soc. Med.* **61**, 69–78.

Hammond, P. H., Merton, P. A. & Sutton, C. G. 1956 Nervous gradation of muscular contraction. *Br. med. Bull.* **12**, 214–218.

Hern, J. E. C., Landgren, S., Phillips, C. G. & Porter, R. 1962 Selective excitation of corticofugal neurones by surface-anodal stimulation of the baboon's motor cortex. *J. Physiol.* **161**, 73–90.

Hern, J. E. C., Phillips, C. G. & Porter, R. 1962 Electrical thresholds of unimpaled corticospinal cells in the cat. *Q. Jl exp. Physiol.* **47**, 134–140.

Jackson, J. H. 1873*a* On the anatomical and physiological localization of movements in the brain. *Lancet* **1**, 162. Reprinted in *Selected writings* (ed. Taylor, J.), **1**, 68 (1931).

Jackson, J. H. 1873*b* On the anatomical investigation of epilepsy and epileptiform convulsions. *Br. med. J.* **1**, 531. Reprinted in *Selected writings* (ed. Taylor, J.), **1**, 116 (1931).

Kernell, D. & Wu Chien-ping 1967*a* Responses of the pyramidal tract to stimulation of the baboon's motor cortex. *J. Physiol.* **191**, 653–672.

Kernell, D. & Wu Chien-ping 1967*b* Post-synaptic effects of cortical stimulation on forelimb motoneurones in the baboon. *J. Physiol.* **191**, 673–690.

Knapp, H. D., Taub, E. & Berman, A. J. 1963 Movements in monkeys with deafferented forelimbs. *Expl Neurol.* **7**, 305–315.

Koeze, T. H., Phillips, C. G. & Sheridan, J. D. 1968 Thresholds of cortical activation of muscle spindles and α motoneurones of the baboon's hand. *J. Physiol.* **195**, 419–449.

Kuypers, H. G. J. M. 1964 The descending pathways to the spinal cord, their anatomy and function. *Prog. Brain. Res.* **11**, 178–200.

Kuypers, H. G. J. M. & Lawrence, D. G. 1967 Cortical projections to the red nucleus and the brain stem in the rhesus monkey. *Brain Res.* **4**, 151–188.

Landau, W. M., Bishop, G. H. & Clare, M. H. 1965 Site of excitation in stimulation of the motor cortex. *J. Neurophysiol.* **26**, 1206–1222.

Landgren, S., Phillips, C. G. & Porter, R. 1962*a* Minimal synaptic actions of pyramidal impulses on some alpha motoneurones of the baboon's hand and forearm. *J. Physiol.* **161**, 91–111.

Landgren, S., Phillips, C. G. & Porter, R. 1962*b* Cortical fields of origin of the monosynaptic pyramidal pathways to some α motoneurones of the baboon's hand and forearm. *J. Physiol.* **161**, 112–125.

Lewis, R. P. & Brindley, G. S. 1965 The extrapyramidal cortical motor map. *Brain*, **88**, 397–406.

Liddell, E. G. T. & Phillips, C. G. 1950 Thresholds of cortical representation *Brain* **73**, 125–140.

Liddell, E. G. T. & Phillips, C. G. 1951 Overlapping areas in the motor cortex of the baboon. *J. Physiol.* **112**, 392–399.

Lundberg, A. 1959 Integrative significance of patterns of connexions made by muscle afferents in the spinal cord. *Symp. XXI Congr. int. Ciencias, Fisiol. Buenos Aires*, pp. 100–105.

Matthews, P. B. C. 1962 The differentiation of two types of fusimotor fibre by their effects on the dynamic response of muscle spindle primary endings *Q. Jl exp. Physiol.* **47**, 324–333.

Matthews, P. B. C. 1963 The response of de-efferented muscle spindle receptors to stretching at different velocities. *J. Physiol.* **168**, 660–678.

Matthews, P. B. C. 1964 Muscle spindles and their motor control. *Physiol. Rev.* **44**, 219–288.

Matthews, P. B. C. 1966 The reflex excitation of the soleus muscle of the decerebrate cat caused by vibration applied to its tendon. *J. Physiol.* **184**, 450–472.

Matthews, P. B. C. 1967 The reflex response to muscle vibration in the decerebrate cat. *J. Physiol.* **192**, 18–19*P*.

Merton, P. A. 1953 Speculations on the servo-control of movement. *Ciba Foundation Symposium on the spinal cord*, pp. 247–255. London: Churchill.

Merton, P. A. 1964 Human position sense and sense of effort. *Symp. Soc. exp. Biol.* **18**, 387–400.

Mott, F. W. & Sherrington, C. S. 1895 Experiments upon the influence of sensory nerves upon movement and nutrition of the limbs. Preliminary communication. *Proc. Roy. Soc.* **57**, 481–488.

Oscarsson, O., Rosén, I. & Sulg, I. 1966 Organization of neurones in the cat cerebral cortex that are influenced from group I muscle afferents. *J. Physiol.* **183**, 189–210.

Patton, H. D. & Amassian, V. E. 1954 Single- and multiple-unit analysis of cortical stage of pyramidal tract activation. *J. Neurophysiol.* **17**, 345–363.

Patton, H. D. & Amassian, V. E. 1960 The pyramidal tract: its excitation and functions. In *Handbook of physiology—neurophysiology II* (ed. Field, J.), pp. 837–861. Washington: American Physiological Society.

Phillips, C. G. 1956 Cortical motor threshold and the thresholds and distribution of excited Betz cells in the cat. *Q. Jl exp. Physiol.* **41**, 70–84.

Phillips, C. G. 1961 Some properties of pyramidal neurones of the motor cortex. In *The nature of sleep*, pp. 4–24. *Ciba Fdn Symp.* London: Churchill.

Phillips, C. G. 1966 Changing concepts of the precentral motor area. In *Brain and conscious experience* (ed. Eccles, J. C.), pp. 389–421. New York: Springer-Verlag.

Phillips, C. G. 1967 Corticomotoneuronal organization. Projection from the arm area of the baboon's motor cortex. *Archs Neurol. Psychiat., Chicago*, **17**, 188–195.

Phillips, C. G. & Porter, R. 1962 Unifocal and bifocal stimulation of the motor cortex. *J. Physiol.* **162**, 532–538.

Phillips, C. G. & Porter, R. 1964 The pyramidal projection to motoneurones of some muscle groups of the baboon's forelimb. *Prog. Brain Res.* **12**, 222–242.

Preston, J. B. & Whitlock, D. G. 1960 Precentral facilitation and inhibition of spinal motoneurons. *J. Neurophysiol.* **23**, 154–170.

Preston, J. B. & Whitlock, D. G. 1961 Intracellular potentials recorded from motoneurons following precentral gyrus stimulation in primate. *J. Neurophysiol.* **24**, 91–100.

Sherrington, C. S. 1906 *The integrative action of the nervous system.* New York: Scribners. (Reprinted 1947, Cambridge University Press.)

Sherrington, C. S. 1928 Sir David Ferrier, 1843–1928. Obituary Notice. *Proc. Roy. Soc.* B **103**, viii–xvi.

Sherrington, C. S. 1931 Quantitative management of contraction in lowest-level coordination. *Brain* **54**, 1–28.

Trotter, W. 1934 A landmark in modern neurology. *Lancet* **227**, 1207–1210.

Vedel, J-P. 1966 Mise en evidence d'un contrôle de l'activité des fibres fusimotrices dynamiques chez le Chat par la voie pyramidale. *C. r. hebd. Séanc. Acad. Sci., Paris* **262** *D*, 908–911.

Walshe, F. M. R. 1943 The mode of representation of movements in the motor cortex, with special reference to 'convulsions beginning unilaterally'. *Brain* **66**, 104–139.

Walshe, F. M. R. 1963 *Diseases of the nervous system*, 10th ed., p. 21. Edinburgh: Livingstone.

Wuerker, R. B., McPhedran, A. M. & Henneman, E. 1965 Properties of motor units in a heterogeneous pale muscle (M. gastrocnemius) of the cat. *J. Neurophysiol.* **28**, 85–99.

Review Lecture

Novel organic chemical processes

By H. M. Stanley, F.R.S.

(*Lecture delivered* 22 *June* 1967—*MS. received* 17 *August* 1967)

Proceedings of the Royal Society, A, *volume* 303, *pp*. 259–273, 1968

1. Introduction

The purpose of this paper is to review some recent developments in industrial organic chemistry which have led, or seem likely to lead, to the establishment of new industrial processes for the production of bulk organic chemical products. The significance of some of the new and cheap, or potentially cheap, intermediates produced in such processes is pointed out in a few selected cases.

In this review I shall concentrate on major new processes leading to large-scale organic chemical production. This invariably means continuous processes using as raw materials either petroleum hydrocarbons, including natural gas, or their cracked products. Because many important organic chemicals are oxygenated compounds, it is quite natural that the reaction of hydrocarbons with oxygen or air should have become quantitatively the most important area of industrial organic chemistry, apart that is from the field of polymerization. Certain processes involving the simultaneous reaction of hydrocarbons with oxygen and with a third reactant (such as hydrogen chloride, ammonia, etc.) have recently become very important industrially and will be reviewed in some detail. The potentialities of reactions involving hydrocarbons and carbon monoxide, cheap and readily available raw materials, will also be considered.

Many of the reactions described are of great scientific, as well as commercial interest, and would seem to deserve more basic study from the viewpoint of mechanism, and also of possible extensions to other and more complex systems.

Broadly speaking it may be said that advances in methods of manufacture of existing bulk organic chemicals have come about in two ways, namely:

(i) by improvement in existing processes, and

(ii) by the development of entirely new processes, often based on alternative, and cheaper, raw materials and involving fewer reaction stages.

The type of highly integrated chemical process involved in such bulk manufacture is almost invariably capital intensive, the economics thus favouring very large plants. However, for any particular scale of operation the commercial attractiveness of a new, or improved, process as compared with an established one will depend on the achievement of one or more of the following desiderata:

(i) a reduction in raw materials cost,

(ii) a reduction in capital cost, and

(iii) a reduction in utilities consumption.

A lower raw material cost can be achieved for an existing process by improvements in chemical efficiency, but there are obvious limits to this approach.

Alternatively this desirable result can be achieved by a switch to an entirely new raw material basis. An example is the switch from acetylene to ethylene as a raw material for the production of vinyl chloride and acetaldehyde, and from acetylene to propylene for the manufacture of acrylonitrile.

Capital costs and associated depreciation and maintenance charges tend to increase with process complexity, and especially with an increasing number of chemical stages. Hence, generally speaking, it is advantageous to carry out reactions in the minimum number of stages, and preferably in a single reaction stage. There are many examples of the value of such reduction in the number of stages including the direct conversion of ethylene to acetaldehyde, rather than by the two-stage process based on initial conversion of ethylene to ethanol and its subsequent dehydrogenation or oxidation to acetaldehyde. An even more striking example is provided by processes which produce acetic acid directly from the lower paraffin hydrocarbons in a single stage reaction, the alternative route involving cracking to ethylene and oxidation of this hydrocarbon in two separate reaction stages to acetic acid through acetaldehyde as intermediate.

The recovery and purification sections of a continuous chemical process are obviously an essential part of the overall process, and indeed often represent a major part of the total capital involved and an appreciable proportion of the operating costs of the process. Standards of purity now demanded in respect of most bulk organic chemicals are high and are being constantly raised. This is particularly so with polymerizable monomers and intermediates for polycondensation reactions where certain types of impurity are often highly undesirable even at very low concentrations. In the development of entirely new processes for such polymer intermediates, quite new impurities, not found in the product from the established process, are often encountered and some of these are undesirable down to very low concentrations.

2. Oxidation of saturated hydrocarbons

(a) *Production of acetic acid*

Acetic acid is a bulk organic chemical of major industrial importance in the production of cellulose acetate, vinyl acetate, solvent esters, etc. For decades it has been produced industrially by the oxidation of acetaldehyde with air or oxygen according to the equation

$$CH_3.CHO+\tfrac{1}{2}O_2 \longrightarrow CH_3.COOH. \qquad (1)$$

This process, which is still used industrially on a very large scale, may be modified so as to yield acetic anhydride with some acetic acid. However, in the last decade processes based on the direct reaction of hydrocarbons with molecular oxygen have become increasingly important in the production of acetic acid.

The liquid phase oxidation of the paraffin hydrocarbons with molecular oxygen at temperatures above 100 °C gives rise to the production of a complex mixture of products, including ketones, esters, acids, etc., of a relatively wide carbon number range. Such oxidation reactions proceed by a free radical mechanism, the initial products being RO_2 radicals and alkyl hydroperoxides; these undergo further reactions whereby carbon chains are ruptured, with further oxidation of the resulting fragments. In the case of the lower paraffins, the carbon number range of the products is obviously limited, and since the lower aliphatic acids, particularly acetic acid, tend to be the more stable end-products of oxidation of these hydrocarbons in the liquid phase, it is not surprising that the oxidation of the lower paraffins in the C_4 to C_8 range in the liquid or dissolved phase has been applied industrially for the manufacture of acetic acid.

Acetic acid is produced in the U.S.A. on a very large scale by the oxidation of *n*-butane under pressures of about 70 atm and at temperatures of 170 to 200 °C. This temperature range is above the critical, and the oxidation takes place in solution in the products of oxidation. In this continuous process, butane and air, which can be enriched in oxygen, are fed to a stainless steel tower reactor, the vaporous reaction products and unreacted butane being cooled to low temperatures, and the resulting condensate separated. Unreacted *n*-butane and intermediate esters and ketones are recycled to the oxidation system. The yield of acetic acid appears to be in the region of 50 to 60 % of the theory, assuming that one molecule of butane can yield two molecules of acetic acid. Small quantities of formic acid are also formed.

In this country, The Distillers Company Ltd in its Chemicals and Plastics Group (now B.P. Chemicals (U.K.) Ltd), has developed and applied industrially a similar process for the production of acetic acid, but based on the oxidation of a mixture of the lower paraffins in the C_4 to C_7 range. Here again, the reaction proceeds in solution and intermediate oxidation products are recycled and oxidized further to organic acids. In addition to acetic acid substantial quantities of formic and propionic acids are also produced together with some succinic acid. A large plant for the production of around 75000 tons per annum of acetic acid has just been commissioned in this country, and will make available at the same time substantial quantities of formic and propionic acids for chemical utilization and for industrial purposes.

It is convenient to mention here two other methods of producing acetic acid which have been developed in the last few years. One of these processes involves the reaction of *n*-butenes with acetic acid at about 100 °C and 15 atm pressure to yield secondary butyl acetate which is thereafter oxidized at 190 °C and 60 atm pressure with air to yield acetic acid (with some formic acid) in an overall yield of about 60 %. This process which can be ideally represented by the equations:

$$\left.\begin{matrix} CH_3.CH{=}CH.CH_3 \\ CH_2{=}CH.CH_2.CH_3 \end{matrix}\right\} + CH_3.COOH \longrightarrow CH_3.CH(O.COCH_3).CH_2.CH_3, \qquad (2)$$

$$CH_3.CH(O.COCH_3).CH_2.CH_3 + 2O_2 \longrightarrow 3CH_3.COOH \qquad (3)$$

has been developed to the large pilot scale by Farbenfabriken Bayer (Krönig 1967). No plans for its industrialization have yet been announced.

A second process not involving hydrocarbon oxidation has been developed by the B.A.S.F. in Germany on the basis of the combination of carbon monoxide with methanol in accordance with the equation

$$CH_3OH + CO \longrightarrow CH_3.COOH. \tag{4}$$

In this process which has just been industrialized on a large scale in the U.S.A., carbon monoxide under high pressure reacts with methanol at about 200 °C in the presence of a catalyst consisting of cobalt iodide (Hohenschutz, Kutepow & Himmele 1966). Though the reaction medium is highly corrosive and necessitates special equipment, notably reactors lined with Hastelloy C, the process operates in a single stage from basically cheap intermediates, and would seem to be commercially attractive, if in addition to methanol a supply of cheap carbon monoxide is available. The reaction mechanism proposed involves the formation of a methyl cobalt carbonyl complex followed by insertion of the carbon monoxide between the methyl group and the cobalt atom to yield an acetyl cobalt carbonyl complex, which is thereafter hydrolysed with water.

(b) *Secondary alcohols from n-paraffins and cyclohexane*

As already noted the oxidation of the paraffin hydrocarbons by molecular oxygen is generally unselective (except at very limited degrees of oxidation) and gives rise to a complex mixture of products including acids, esters, ketones, etc. The favourable influence of boric acid in directing the oxidation of the *n*-paraffin hydrocarbons to the production of mixed secondary alcohols of the same carbon number as the raw material used was noted in 1943 by Nobori *et al.* but the discovery seems to have lain dormant until it was revived in the late 1950's by Bashkirov and his co-workers in Russia (Bashkirov 1956; Bashkirov, Kamzolkin, Sokova & Andreeva 1958).

In several industrial processes based on this effect, the hydrocarbon is oxidized in the liquid phase in the presence of metaboric acid under conditions favouring the removal of any water formed in the reaction. The products are esters of boric or metaboric acids which are stable against further oxidation provided they are not hydrolysed by water—hence the necessity for continuous removal of reaction water.

The process has been industrialized in the U.S.A. for the production of mixed secondary alcohols from the C_{12} to C_{16} *n*-paraffins, the resulting mixed secondary alcohols being useful in the production of biodegradable detergents. It has also been applied in the oxidation of cyclohexane to cyclohexanol notably in this country, in Japan, and also in Australia. The oxidation of cyclododecane to the corresponding alcohol is practised in Germany by Chemische Werke Hüls as one step in the production of laurolactam. In the case of the oxidation of cyclohexane, selectivities to cyclohexanol and cyclohexanone of 90 % are achievable in place of the usual 75 % and at higher percentage conversions (Alagy, Asselineau, Busson, Cha & Sandler 1967).

The specific effect of metaboric acid in directing the oxidation to the production of secondary alcohols (in the form of their esters) has been the subject of study by Broich and his co-workers of Chemische Werke Hüls (Broich & Grasemann 1965) and also by Bashkirov and his collaborators in Russia (Novak, Kamzolkin Talysenkow & Bashkirov 1967). It appears that the metaboric acid has little influence on the chain mechanism, but operates by selectively decomposing the intermediate hydroperoxides. This interesting and industrially important reaction would appear worthy of further and more detailed mechanistic study.

(c) *Co-oxidation of paraffins with inorganic reactants*

A number of co-oxidation reactions of paraffin hydrocarbons with inorganic reactants are known and specially interesting from the industrial point of view is the sulphoxidation reaction illustrated by the equation

$$RH + SO_2 + \tfrac{1}{2}O_2 \longrightarrow R.SO_3H. \qquad (5)$$

This chain reaction gives a mixture of all the possible isomeric sulphonic acids, together with some disulphonic acids and is likely to be of considerable technical importance in the production of detergents.

One reaction of cyclohexane of substantial scientific and technical interest is its photochemical combination with nitrosyl chloride in accordance with the equation:

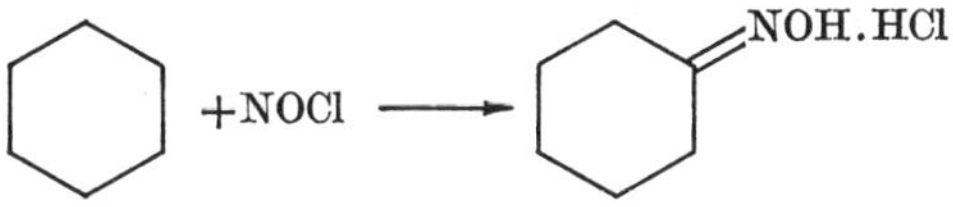

(6)

This low quantum yield photochemical process is operated industrially on a large scale by the Toyo Rayon Company in Japan as a step in the synthesis of caprolactam, the intermediate for Nylon 6. Its successful industrialization has involved the development of very large mercury light sources of over 20 kW each. The electric power consumption of this process is thus fairly high, being somewhat below 2 kWh per pound of oxime produced (Aikawa 1964).

3. Oxidation of unsaturated hydrocarbons

(a) *Olefin oxides by direct oxidation*

The direct oxidation route used commercially and so successfully in the case of ethylene, i.e. vapour phase reaction over a supported silver catalyst, has not been found applicable to the manufacture of homologues from the corresponding higher olefins.

A great deal of effort is being concentrated on methods for the direct oxidation of propylene to propylene oxide, which is a major industrial intermediate produced commercially at the moment only indirectly by way of the chlorohydrin. While the liquid phase non-catalytic oxidation of propylene and its immediate homologues can give rise to fair yields, around 50 %, of epoxides, no practicable commercial process has so far been developed on this basis, mainly because of the difficulty of recovering the epoxide from the complex mixture of other products formed.

A more satisfactory approach to the problem of directly oxidizing propylene and similar olefins to the corresponding epoxide is provided by a new process developed by Halcon International (Landau, Brown, Russell & Kollar 1967) and being industrialized in the U.S.A. and Spain. This involves oxidation with a hydroperoxide, such as ethylbenzene hydroperoxide, in solution at 100 to 130 °C in the presence of soluble catalysts based on molybdenum, tungsten and other elements. The preferred catalyst appears to be molybdenum naphthenate. The reactions involved are illustrated by the equations

$$C_6H_5.CH_2.CH_3 \xrightarrow{O_2} C_6H_5.CHOOH.CH_3, \tag{7}$$

$$\begin{aligned} &C_6H_5.CHOOH.CH_3+CH_3.CH{=}CH_2 \\ &\qquad \longrightarrow CH_3.\underset{\diagdown\,O\,\diagup}{CH{-}CH_2}+C_6H_5.CHOH.CH_3, \end{aligned} \tag{8}$$

$$C_6H_5.CHOH.CH_3 \longrightarrow H_2O+C_6H_5.CH{=}CH_2. \tag{9}$$

Reaction (8) bears a resemblance to the well-known Prilezaev reaction, represented by the equation

$$CH_3.COOOH+\gt C{=}C\lt \longrightarrow \gt \underset{\diagdown\,O\,\diagup}{C{-}C}\lt +CH_3.COOH, \tag{10}$$

which is used commercially for the production of higher epoxides such as those based on unsaturated natural oils.

Yields are said to be high, e.g. 80 to 90 % of propylene oxide based on hydroperoxide consumed. A disadvantage of this type of process is that it involves the co-production of quite large amounts of a second product, in this case styrene, or isobutene if tertiary butylhydroperoxide is used. However, the process is an interesting one both technically and scientifically.

(*b*) *Vapour phase heterogeneous oxidation processes*

It has long been known that acrolein can be formed from propylene and air or oxygen in the presence of certain heterogeneous catalysts. Large-scale production of acrolein by this technique has been established in the U.S.A., France and elsewhere for use as an intermediate in the production of allyl alcohol, methionine, etc.

More recently it has been found possible to develop catalysts capable of catalysing the reaction of propylene with air to yield a mixture of acrylic acid and acrolein. Since the latter can be recycled and ultimately converted into acrylic acid over the same catalyst, this represents a method of producing acrylic acid (and thence its commercially useful esters) directly from propylene in what is essentially a single stage operation. The reactions involved are represented by the equations

$$CH_3.CH{=}CH_2+O_2 \longrightarrow \left.\begin{array}{c} CH_2{=}CH.CHO \\ + \\ CH_2{=}CH.COOH \end{array}\right\} \tag{11}$$

Such a process, developed by the Distillers Company Ltd is now being industrialized in the U.S.A. on a large scale.

Attempts to apply this type of process to the *n*-butenes have been found to yield butadiene rather than the expected crotonaldehyde or methylvinylketone. A method of producing butadiene from *n*-butenes by oxidative dehydrogenation over an oxide catalyst has been developed by The Distillers Company Ltd on a large pilot plant scale but has not yet been industrialized. Alternative processes using halogen compounds as catalysts are in course of development or industrialization.

(c) *Oxidation and co-oxidation of olefins in the presence of palladium catalysts*

During the last few years, a direct process for the production of acetaldehyde from ethylene has been developed in Germany and applied in plants in many countries of the world. It involves the oxidation of ethylene directly to acetaldehyde in the presence of aqueous solutions of palladium chloride containing also cupric chloride (Smidt *et al.* 1959; Smidt 1962). The reaction may be pictured as involving a number of stages, in the first of which an ethylene palladium complex is decomposed by water to form acetaldehyde and palladium. Palladium is reoxidized by a cupric salt back to palladium chloride, and the cuprous salt resulting from this reduction is reoxidized to the cupric condition by oxygen. The series of reactions involved together with the overall reaction (15) are illustrated by the following equations:

$$C_2H_4+PdCl_2+H_2O \longrightarrow CH_3.CHO+Pd^0+2HCl, \tag{12}$$

$$Pd^0+2CuCl_2 \longrightarrow PdCl_2+2CuCl, \tag{13}$$

$$2CuCl+2HCl+\tfrac{1}{2}O_2 \longrightarrow 2CuCl_2+H_2O, \tag{14}$$

$$C_2H_4+\tfrac{1}{2}O_2 \longrightarrow CH_3.CHO. \tag{15}$$

It is possible to carry out this reaction in two ways, namely (i) in a single-stage process in which a gas containing ethylene and oxygen is reacted with the aqueous catalytic medium or (ii) in a two-stage process in which ethylene is first reacted and the aqueous reaction medium is thereafter re-oxidized with air in a separate stage. Both types of operation have been applied industrially.

A similar process has also been developed for the manufacture of acetone directly from propylene. Higher olefins can also be directly oxidized to ketones, i.e. 4-methyl-pentene-1 to methyl-isobutyl ketone but there is an increasing tendency as one ascends the series for chlorinated by-products to be produced.

An important subsequent development in the use of palladium complexes in olefin oxidation reactions has arisen as the result of the observation by Moiseev and his co-workers in Russia (Moiseev, Vargaftik & Syrkin 1960; Moiseev & Vargaftik 1965) that ethylene reacts with palladium chloride in acetic acid containing sodium acetate to produce vinyl acetate; the precipitation of elemental palladium can be avoided by the presence of a suitable redox system.

Vinyl acetate

Arising out of this Russian work, several processes have been developed for the production of vinyl acetate from ethylene, oxygen and acetic acid in the presence of dissolved palladium salts, or of elemental palladium suspended on a suitable support. The two main types of process so far evolved have been recently described by Krekeler & Krönig (1967) and the following details are taken from their paper:

Liquid catalytic system

In this system, vinyl acetate and acetaldehyde are produced simultaneously by bringing a gas mixture containing ethylene and molecular oxygen (usually about 5 % concentration) at a pressure of about 30 atm into contact with a solution of palladium salt in acetic acid containing other catalyst salts particularly copper salts, lithium acetate and lithium chloride, and at temperatures of around 120 °C. Recycled acetic acid and fresh acetic acid are fed to the reaction system. Reaction products are carried forward in the exit gas, separated by cooling and the excess ethylene and oxygen recycled. Because of the limitation of oxygen concentration imposed by explosion limits, conversions per pass of the ethylene are low, but the productivity is high, i.e. 1½ mole of vinyl acetate per hour per litre of medium together with up to a similar amount of acetaldehyde.

The palladium concentration in the medium must be closely controlled, while the chlorine loss arising from the formation of volatile chlorinated organic products must be compensated by continuous addition of a suitable source of chloride ion, e.g. hydrochloric acid. The ratio of acetaldehyde to vinyl acetate in the product depends on the water content of the medium and can be varied within certain limits, i.e. (in moles) from 1:1 to about 1 of vinyl acetate to 0·2 of acetaldehyde. The yield of acetaldehyde and vinyl acetate is about 90 % on the ethylene and 95 % on the acetic acid consumed. Because of the necessity of limiting the concentration of salts in the reaction medium, the use of a separate oxidizing zone is not practicable in this type of process, which has been industrialized in the U.K. and in the U.S.A. and elsewhere.

Heterogeneous catalytic system

An alternative system which has been carried to large pilot plant scale in Germany and elsewhere consists in passing a mixture of ethylene with up to 5 % of oxygen, together with the vapour of acetic acid, over a solid catalyst consisting of elemental palladium on a support, e.g. 0·1 to 2·0 % of palladium on alumina or silica in a multi-tubular reactor system operating at 160 °C and 5 to 7 atm. The catalyst may advantageously contain other noble metals such as gold and, for long life, must contain also 0·5 to 5 % of a metal acetate such as sodium acetate. The hot gases are cooled to 0 °C to condense the products and the gases, after freeing from carbon dioxide, are recycled. In contradistinction to the liquid phase process no appreciable amount of acetaldehyde is formed; the yield of vinyl acetate is about 91 to 94 % with 1 % of acetaldehyde on the ethylene consumed, the yield on acetic acid being virtually quantitative. Conversions per pass are 10 to 15 % on ethylene and 60 to 90 % on oxygen, and a satisfactory catalyst life has been achieved.

It is interesting to note that the catalytic effect of palladium salts in homogeneous aqueous phase are closely simulated by heterogeneous systems involving the element itself supported on a suitable carrier. Presumably the mechanisms involved are similar.

Compared with the alternative acetylene-based process, ethylene-based operations involve lower raw material cost per ton of vinyl acetate though capital costs and energy costs tend to be higher. The vapour phase operation is attractive where acetaldehyde is not needed (that is for the production of acetic acid), and also because it avoids problems of handling very corrosive liquid media.

(*d*) *Co-oxidation of olefins with inorganic reactants*

The following equations show some industrially significant reactions of olefins with oxygen or air and a third reactant in the vapour phase in the presence of specific heterogeneous catalysts:

$$CH_3.CH{=}CH_2+1\tfrac{1}{2}O_2+NH_3 \longrightarrow CH_2{=}CH.CN+3H_2O, \quad (16)$$

$$(CH_3)_2C{=}CH_2+1\tfrac{1}{2}O_2+NH_3 \longrightarrow CH_2{=}C(CH_3).CN+3H_2O, \quad (17)$$

$$CH_2{=}CH_2+\tfrac{1}{2}O_2+2HCl \longrightarrow CH_2Cl.CH_2Cl+H_2O. \quad (18)$$

The manufacture of acrylonitrile from propylene by the so-called ammoxidation process (equation (16)) is now firmly established, while the production of methacrylonitrile by a similar reaction with isobutene (equation (17)) is now practised in the U.S.A. (Anon. 1966). Hydrogen cyanide is a useful by-product of these processes.

The oxychlorination of ethylene to produce ethylene dichloride (equation (18)) is now utilized industrially as a key step in the manufacture of vinyl chloride monomer on the basis of ethylene in a sequence of reactions represented by the following equations:

$$\left.\begin{array}{l} C_2H_4+Cl_2 \longrightarrow CH_2Cl.CH_2Cl, \\ CH_2Cl.CH_2Cl \longrightarrow CH_2{=}CHCl+HCl, \\ C_2H_4+2HCl+\tfrac{1}{2}O_2 \longrightarrow CH_2Cl.CH_2Cl+H_2O, \end{array}\right\} \quad (19)$$

the overall effect being

$$2C_2H_4+Cl_2+\tfrac{1}{2}O_2 \longrightarrow 2CH_2{=}CH.Cl+H_2O. \quad (20)$$

The oxychlorination process is normally carried out by passing a mixture of ethylene, anhydrous hydrogen chloride and air over a heterogeneous catalyst consisting of cupric chloride on a support. Because of the highly exothermic nature of the process, fluidized bed reactors are normally used. There is much evidence that the ethylene dichloride is formed by the chlorination of ethylene by cupric chloride, which is continuously regenerated by the oxygen–hydrogen chloride mixture.

An interesting new approach to oxychlorination is the development by the M. W. Kellogg Co. of a liquid phase oxychlorination catalytic system consisting of an aqueous solution of cupric chloride containing some cuprous chloride (Spector, Heinemann & Miller 1967; Anon. 1966). A gas mixture containing ethylene, oxygen and hydrogen chloride is passed through the medium at about 185 °C and at 10 to 80 atm pressure to yield ethylene dichloride at conversions per pass of ethylene of 8 to 24 % and of oxygen 29 to 84 %, the space time yield being about

4 moles per hour per litre of catalytic liquid. The unreacted gas is recycled and yields of ethylene dichloride approach 96 %, while about 1 % of chloral is produced. The catalytic medium is very corrosive and may demand materials of construction such as tantalum or brick lining. No plans for industrializing this process have so far been announced.

4. Reactions involving carbon monoxide

Reactions of organic compounds, particularly hydrocarbons, with carbon monoxide, are of considerable industrial importance, because of the relative cheapness and ready availability of the reactants and the high value of some of the products produced including aldehydes, acids, etc. Many of these reactions are also of great scientific interest from the viewpoint of their mechanism. The following equations show some of these reactions, the first four of which are of established industrial importance:

$$C_nH_{2n}+CO+H_2 \longrightarrow C_nH_{2n+1}.CHO, \tag{21}$$

$$CH_3.CH{=}CH_2+3CO+2H_2O \longrightarrow CH_3.CH_2.CH_2.CH_2OH+2CO_2, \tag{22}$$

$$CH_3OH+CO \longrightarrow CH_3.COOH, \tag{23}$$

$$C_nH_{2n}+CO+H_2O \longrightarrow C_nH_{2n+1}.COOH, \tag{24}$$

$$C_2H_4+CO+\tfrac{1}{2}O_2 \longrightarrow CH_2{=}CH.COOH, \tag{25}$$

$$R.NO_2+3CO \longrightarrow R.NCO+2CO_2. \tag{26}$$

Perhaps the most important of these reactions is the so-called Oxo process (equation (21)) which is applied commercially for the production of aldehydes, and thence of primary alcohols, from olefins. The catalyst is cobalt carbonyl hydride and the reaction takes place at elevated pressures and temperatures of around 150 °C. Important advances have recently been made in modifying the activity of the catalyst and these have rendered possible operation at lower pressures and with greater selectivity to the more desirable linear product, e.g. to yield a higher ratio of *n*-butyraldehyde to isobutyraldehyde from propylene.

The reaction of olefins with carbon monoxide and water to produce carboxylic acids (equation (24)) has been practised for many years in Germany for the production of propionic acid from ethylene using an aqueous solution of nickel salt (furnishing nickel carbonyl) as catalyst at elevated temperatures and pressures, e.g. 250 °C and 200 atm. More recently a different type of reaction, the so-called Koch reaction, has become industrially important. This involves the reaction of an olefin with carbon monoxide in the presence of a strong mineral acid catalyst such as phosphoric acid–boron trifluoride at 40 to 80 °C and pressures of 50 to 150 atm followed by hydrolysis of the resulting product to produce highly branched aliphatic carboxylic acids, which are now finding some considerable interest in a variety of industrial applications. The process is operated commercially in Europe and the U.S.A. to produce branched chain acids mainly in the C_9 to C_{11} range.

Equation (25) shows a novel reaction described in Belgian Patent 664782 to the

Union Oil Company of California, whereby ethylene is reacted with carbon monoxide and oxygen in an acetic acid solution of a palladium salt resulting in the production of acrylic acid and β-acetoxy propionic acid. This would seem to open up the possibility of a relatively simple method of synthesizing acrylic acid directly from ethylene and carbon monoxide.

Yet another extremely novel reaction is that illustrated in equation (26), whereby nitrobenzene or its intermediate reduction products are reacted with carbon monoxide in the liquid phase, e.g. at 200 °C and 500 atm pressure in the presence of a noble metal catalyst and a Lewis acid to yield phenyl isocyanate (Hardy & Bennett 1967). Such a process would seem to have industrial possibilities, particularly if it can be applied to the production of the industrially important aromatic diisocyanates.

5. The dehydrogenation of paraffin hydrocarbons

The catalytic dehydrogenation of the lower paraffins, particularly butane to butenes and butadiene, has been practised for many years, but certain advances have recently been made in this type of reaction, particularly by the development of the so-called oxydehydrogenation route, which is illustrated by the equation

$$n-C_4H_{10}+\tfrac{1}{2}O_2 \longrightarrow H_2O+C_4H_8, \tag{27}$$

$$n-C_4H_8+\tfrac{1}{2}O_2 \longrightarrow H_2O+C_4H_6. \tag{28}$$

In contradistinction to the catalytic dehydrogenation process, namely

$$C_nH_{2n+2} \rightleftharpoons H_2+C_nH_{2n}, \tag{29}$$

these oxidative dehydrogenation processes are irreversible and exothermic.

A commercial plant for the oxidative dehydrogenation of n-butane to butadiene using iodine and oxygen (King 1966), and a catalytic medium involving lithium iodide and lithium oxide, is said to have been constructed in southern France, though no details of its operation are available. The oxidative dehydrogenation of butane with oxygen at elevated temperatures in the presence of ammonium bromide has also been developed by the Petro-Tex Corporation in the U.S.A.

Great interest attaches to methods for dehydrogenating the higher normal paraffins in the carbon number range 10 to 16, now readily available from petroleum fractions by new separation techniques, based for example on molecular sieves. To produce olefins in the same carbon number range for use in detergent manufacture, it has not so far been possible to apply catalytic dehydrogenation or oxidative dehydrogenation to such hydrocarbons because of the ease of chain rupture giving rise to products of lower carbon numbers. Accordingly the conversion of a higher normal paraffin to an olefin or to an olefin precursor has hitherto been accomplished by chlorination followed by dehydrochlorination in accordance with the following equations:

$$C_nH_{2n+2}+Cl_2 \longrightarrow C_nH_{2n+1}.Cl+HCl, \tag{30}$$

$$C_nH_{2n+1}.Cl \longrightarrow C_nH_{2n}+HCl. \tag{31}$$

This process, which has been practised industrially in Germany, involves a substantial usage of chlorine and produces quantities of by-product hydrogen chloride.

Very recently, however, the Universal Oil Products Company has developed a process for the catalytic dehydrogenation of C_{12}–C_{16} paraffins to the corresponding olefins in the presence of a specific catalyst based on one of the noble metals (Bloch 1966). The corresponding olefins are produced in yields of 95 %, the remainder being mainly diolefins. The double bond in the product is distributed uniformly over the carbon chain, and cracking and dehydrocyclization is very low. Such a process would seem to offer many advantages over the alternative in the production of olefins for detergent manufacture. There is no evidence so far of a successful development based on oxidative dehydrogenation for the production of these higher olefins.

6. Potentialities of New Cheap Intermediates

As a result of improvements in existing processes, as well as of the industrialization of entirely new processes, a number of important intermediates are now available or are likely to be available soon at price levels which will render them of considerable importance in further chemical transformations of industrial importance. Among these intermediates may be mentioned butadiene, isoprene, acrolein, acrylonitrile, methacrylonitrile, 1:4-hexadiene and oligomers of butadiene and isoprene. A great deal of effort has been, and is being, devoted to exploring the chemistry of many of these intermediates. In some cases, e.g. acrolein, the results so far have been rather meagre from the industrial point of view.

Some recent developments will be discussed briefly in relation to the chemical utilization of butadiene and acrylonitrile.

(a) *Butadiene derivatives*

With the increasing production of butadiene as a co-product of naphtha cracking, the price of this hydrocarbon has fallen and will probably fall considerably more in the future, a factor which will undoubtedly stimulate its utilization, not only in the field of synthetic rubbers, in which it plays a dominant role, but also in a range of synthetic chemical processes, some of which are represented by the following equations:

$$\left.\begin{array}{lcl} CH_2{=}CH.CH{=}CH_2+Cl_2 & \longrightarrow & CH_2Cl.CH{=}CH.CH_2Cl \\ & & \quad\rightleftarrows \\ CH_2{=}CCl.CH{=}CH_2 & \xleftarrow{-HCl} & CH_2{=}CH{=}CHCl.CH_2Cl, \end{array}\right\} \qquad (32)$$

$$3CH_2{=}CH.CH{=}CH_2 \longrightarrow \text{cyclododecatriene} \longrightarrow C_{12}\ \text{Lactam}, \qquad (33)$$

$$CH_2{=}CH.CH{=}CH_2+C_2H_4 \xrightarrow{RhCl_3} CH_2{=}CH{-}CH{-}CH{=}CH.CH_3, \qquad (34)$$

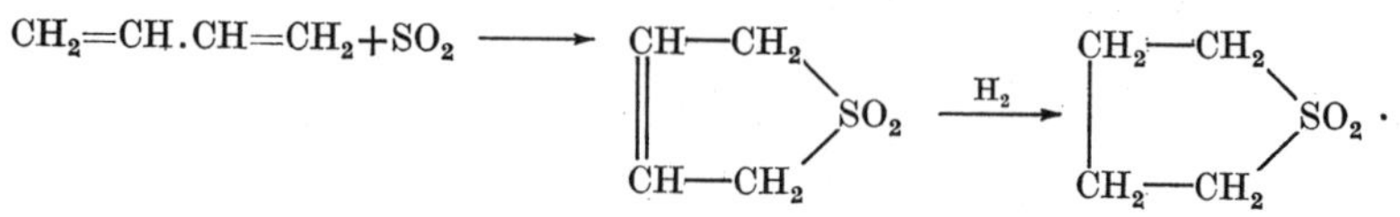

(35)

The production of chloroprene monomer (equation (32)) is now practised industrially in France on the basis of a process developed by my colleagues and involving the chlorination of butadiene and dehydrochlorination of 3:4-dichlorobutene-1.

The trimerization of butadiene to its cyclic trimer, namely, 1,5,9-cyclododecatriene can be accomplished with certain complex catalysts, particularly those based on ethyl aluminium sesquichloride with minor proportions of titanium tetrachloride, in high yields with high productivities at relatively low temperatures and near atmosphere pressures (Wilke 1957, 1960, 1963; Weber, Ring, Hochmuth & Franke 1965; Ring & Gaube 1966). This cyclic trimer, together with the corresponding dimer, have excited a great deal of interest and much research has been concentrated on the potentialities of these materials. One important industrial process based on the cyclic trimer is the manufacture of laurolactam (a monomer for a high duty polyamide), by the Chemische Werke Hüls on a scale of several thousand tons per annum (Franke & Müller 1964). The synthesis goes through the stages indicated in the following equations:

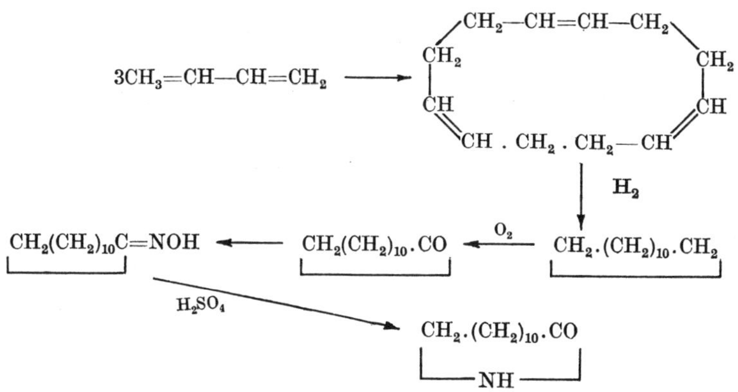

(36)

Another interesting product is the co-dimer of butadiene and ethylene, namely 1,4-hexadiene, which can be produced very selectively by polymerization in the presence of rhodium chloride (Alderson, Jenner & Lindsey 1965), as well as by other catalytic systems (equation (34)). This material is said to be a useful diene in the production of ethylene-propylene terpolymer rubbers.

(*b*) *Acrylonitrile derivatives*

The new process for making acrylonitrile from propylene by the ammoxidation route has largely displaced the alternative methods and has resulted in a substantial reduction in price, which in turn has stimulated interest in the use of this material as an intermediate for other syntheses.

One such synthesis is that of adiponitrile, an intermediate for the production of hexamethylene diamine by reductive dimerization sometimes called 'hydrodimerization' of acrylonitrile. An electrolytic process for carrying out the reaction has now been industrialized in the U.S.A. by the Monsanto Company and is represented by the equation

$$2CH_2{=}CH.CN \xrightarrow[H_2O]{2e} NC.CH_2.CH_2.CH_2.CH_2.CN.$$

In this process, acrylonitrile in aqueous solution solubilized by aromatic sulphonate salts is electrolysed in a cathode compartment using tetraethyl ammonium *p*-toluene sulphonate as electrolyte (Baizer & Anderson 1964). These cations are chosen because they can be discharged at cathode potentials much greater than those required to effect the hydrodimerization reaction. The pH of the cathode liquid must be controlled within the range 7 to 9·5 by addition of acid which can be either hydrochloric acid or carbon dioxide and the electrolysis is carried out at temperatures of 20 to 30 °C. The cathode may be made of lead or mercury and anodes of platinum or lead are suitable. Yields of adiponitrile up to 90 % on the actylonitrile converted are quoted.

As an alternative, this hydrodimerization may be effected by using sodium amalgam (Knunyants & Vyazankin 1957; Matsuda 1966) and also by the use of diaphragm cells with a mercury cathode in which the amalgam is generated continuously. The reaction is carried out in a slightly acid medium and preferably in polar organic solvents such as dimethylformamide containing a little water. Yields of adiponitrile on acrylonitrile consumed of up to 90 % are quoted. Yields on sodium can be up to 70 %. So far as is known, no commercial process based on this method has yet been industrialized.

An alternative approach to the problem of converting acrylonitrile efficiently into adiponitrile is based on the possibility of dimerizing and then hydrogenating the dimer to the desired end-product, namely hexamethylenediamine. At least three dimers of acrylonitrile can be produced (depending on the conditions and particularly on the catalyst), including 1:2-dicyanocyclobutane, 1:4-dicyanobutene-1 and 1:3-dicyanobutene-3. Of these the first two are capable of being hydrogenated to adiponitrile and thence to hexamethylenediamine.

Conclusion

This brings me to the end of my review, in which I have attempted to outline some of the more important newer developments in industrial organic chemical processes. Though I fear I have been able to say little that is new to those intimately concerned with research and development in this field, I hope that I have succeeded in giving to those not so familiar with this subject an impression of the great effort being applied on a world-wide basis in process research and development, and what intense, and therefore stimulating, competition exists to achieve the most economic processes.

Finally, I would stress again that many of the newer processes I have described raise theoretical questions of profound interest, and would therefore seem to be well worthy of a great deal more basic investigation.

References

Aikiwa, Koki 1964 *Hydrocarb. Process. Petrol. Refin.* **43** (10), 157.

Alagy, J., Asselineau, L., Busson, C., Cha, B. & Sandler, H. 1967 Paper presented at Seventh World Petroleum Congress, Mexico City.

Alderson, T., Jenner, E. L. & Lindsey, R. V. Jr. 1965 *J. Am. Chem. Soc.* **87**, 5638.

Anon. 1966 *Chem. Engng News* **76**.

Anon. 1966 *Chem. Week*, p. 93.

Baizer, M. M. & Anderson, J. D. 1964 *J. Electrochem. Soc.* **111**, 215, 223, 226.

Bashkirov, A. N. 1956 *Khim. Nauka Prom.* **1**, 273.

Bashkirov, A. N., Kamzolkin, Sokova, K. M. & Andreeva, T. P. 1958 *Dokl. Akad. Nauk S.S.S.R.* **119**, 705.

Bloch, H. S. 1966 Paper presented to Symposium on N-paraffins, Manchester.

Broich, F. & Grasemann, H. 1965 *Erdöl Kohle* **18** (5), 360.

Franke, W. K. & Müller, K. A. 1964 *Chemie-Ingenieur-Technik* **36**, 960.

Hardy, W. B. & Bennett, R. P. 1967 *Tetrahedron Lett.* **11**, 961.

Hohenschutz, H., Kutepow, N. von & Himmele, W. 1966 *Hydrocarb. Process. Petrol. Refin.* **45** (11), 141.

King, R. W. 1966 *Hydrocarb. Process. Petrol. Refin.* **45** (11), 189.

Knunyants, I. L. & Vyazankin, N. S. 1957 *Dokl. Akad. Nauk, Otdel. Khim. Nauk* **113**, 112.

Krekeler, H. & Krönig, W. 1967 Paper presented to Seventh World Petroleum Congress, Mexico City.

Krönig, W. 1967; Paper presented at Seventh World Petroleum Congress, Mexico City; see also British Patent 1072399 to Farbenfabriken Bayer A.G.

Landau, R., Brown, D., Russell, J. L. & Kollar, J. 1967 Paper presented at Seventh World Petroleum Congress, Mexico City.

Matsuda, Fujio 1966 *Tetrahedron Lett.* **49**, 6193.

Moiseev, I. I. & Vargaftik, M. N. 1965 *Izv. Akad. Nauk S.S.S.R. Ser. Khim.* no. 4, 759.

Moiseev, I. I., Vargaftik, M. N. & Syrkin, Ya. K. 1960 *Dokl. Akad. Nauk S.S.S.R.* **133** (2), 377.

Nobori *et al.* 1943 *J. Soc. Chem. Ind. Japan* **46**, 195, 765, 895, 1208, 1252.

Novak, F. I., Kamzolkin, V. V., Talysenkov, Ya. A. & Bashkirov, A. A. 1967 *Neftekhim.* **7** (2), 248.

Ring, W. & Gaube, J. 1966 *Chemie-Ingenieur-Technik* **38**, 1041.

Smidt, J. *et al.* 1959 *Angew. Chem.* **71**, 176.

Smidt, J. 1962 *Chem. Ind.* **54**.

Spector, M. L., Heinemann, H. & Miller, K. D. 1967 *Ind. Engng Chem. Process Design Develop.* **6** (3), 327.

Weber, H., Ring. W., Hochmuth, U. & Franke, W. K. 1965 *Liebigs Ann. Chem.* **681**, 10.

Wilke, G. 1957 *Angew. Chem.* **69**, 397.

Wilke, G. 1960 *Angew. Chem.* **72**, 581.

Wilke, G. 1963 *Angew. Chem.* **75**, 10.

Review Lecture

Living corals

By C. M. Yonge, F.R.S.*
University of Glasgow

(*Lecture delivered* 2 *March* 1967—*MS. received* 20 *September* 1967)

Proceedings of the Royal Society, B, *volume* 169, *pp*. 329–344, 1968

Coral reefs have a dramatic impact owing to their frequent vast extent; above sea level they may form the basis of inhabited islands while below the surface they constitute major hazards to navigation. Perhaps for this reason the precise nature of the animals primarily responsible for their creation tends to be overlooked. Literature on the 'coral reef problem', a matter of primarily geological and geographical concern, is far greater than that on corals themselves. It is as though these animals have become buried under the vast mass of the skeletons they secrete, of the calcium carbonate they extract from the sea to convert into aragonite. Perhaps even the beauty of living coral colonies distracts from consideration of the animal itself.

In the hope of in some measure correcting the balance, this lecture is concerned solely with the animal, in particular with the functional aspects of structure together with development and ecology, all viewed from a broadly evolutionary standpoint.

Coral is a loose term which must immediately be restricted by confining it to stony corals, that is to coelenterates secreting a massive calcareous skeleton. While noting that this includes hydrozoans such as *Millepora* and octocorallines such as *Heliopora* and *Tubipora*, the two first of which are extremely important reef-builders, attention must be further confined to members of the Order Scleractinia (Madreporaria) of the class Anthozoa. It is their skeletons which constitute the vast proportion of coral reefs. At the same time the supreme importance of calcareous red algae, of nullipores or 'lithothamnia' in the consolidation of the reef surfaces on the seaward side cannot but be mentioned.

Basically, as the name of the order indicates, scleractinian corals have the same structure as sea anemones (Actiniaria); they are restricted to the polyp form with radial mesenteries and unbranched tentacles in multiples of six. Below the level of the stomodaeum the mesenteries are bounded with filaments which, not only within the coelenteron but in certain cases extending far outside this, provide a digestive mechanism of the greatest efficiency. The obvious difference lies in the possession of a calcareous skeleton with an accompanying reduction of the thick mesogleal layer so characteristic of the Actiniaria. This skeleton consists of aragonite which, in the case of *Pocillopora damicornis*, forms up to 99·9 % by weight with only 0·01 to 0·1 % of organic matter, this largely consisting of a spongework of

* Now Sir Maurice Yonge.

fibrils of chitin as described by Wainwright (1963) who infers a relationship between the orientation of the fibrils and that of the crystals of aragonite. The living tissues form no more than a most delicate skin over the skeleton, rising higher and higher as this is continually added to by the secretory activity of the underlying skeletogenic epithelium. The precise mode of secretion, within or outside the cells, remains obscure. This skeleton is cemented to the substrate, in other words corals are primitively sedentary and epifaunistic.

In its simplest form, in early stages of development and throughout life in a simple 'cup coral' such as *Caryophyllia smithii* amd *Balanophyllia regia* of British coasts, the skeleton consists essentially of a first formed basal plate from which arises a thecal cup, the calix, with internally converging septa, also in multiples of six. The narrow spaces between the septa form a series of deep, radially disposed compartments into which, following contraction of their muscles, the mesenteries with the all-important filaments are withdrawn, the disk tissues and the retracted tentacle being stretched tightly over the surface of the skeleton. The prime function of the skeleton is thus protection. Only when the polyp is expanded and rises well above the skeleton can it function. This follows relaxation of the muscles with accompanying intake of water by ciliary action. With the mouth closed, hydrostatic pressure is then generated within the coelenteric cavity.

The great majority of scleractinian corals form colonies, some of great size and elaboration. They assume many forms, rounded and massive, branching, foliacious and encrusting. It is in some cases difficult to know whether a particular colony is a species or a growth form because the final form of certain types of corals is certainly influenced by the environment. Colony formation as such does not occur in the Actiniaria although great aggregations do occur, such as in *Bunodactis elegantissima* which, as a close carpet of individuals, covers wide intertidal areas along the Pacific coast of North America. The biological advantage of such an extensive 2-dimensional 'colony' is somewhat debatable, but the scleractinian 3-dimensional colony, formed by repeated asexual division of the original polyp and calyx, has the obvious dual advantage of raising the colony above the bottom. and so clear of falling sediment, and also of greatly extending the feeding surface represented by the tentacles which extend from each of the now numerous polyps. These are secondary, but major, advantages arising out of the possession of a skeleton.

The smothering action of particles of sediment continually raining upon them is a problem common to all sedentary animals and is met by possession of some form of cleansing mechanism. In the Scleractinia this consists of a copious secretion of mucus* associated with the cilia borne on the coenosarc which covers the colony. Particles embedded in mucus are conveyed off the surface of the colony as first described and illustrated in a series of photographs for *Favia pectinata* and *Fungia danai* by Marshall & Orr (1931). The mechanism is simple but highly effective. Sediment does not represent the absolute bar to the establishment of coral growth it was formerly considered to do. This was indeed demonstrated by the

* Recently suggested as a possible source of food for zooplankton off coral reefs (Johannes 1967).

observations of Umbgrove and other Dutch workers on the presence of corals on the muddy bottom of the Bay of Batavia.

Feeding in corals, as in coelenterates generally, is primitively and most usually by means of tentacles armed with nematocysts which seize and pass food into the mouth. Again, as in the great majority of coelenterates, the Scleractinia are exclusively carnivorous, neither tentacles nor mesenterial filaments react to the presence of vegetable matter. They will accept any form of animal matter, and react to any extract of this, but in life must feed largely on zooplankton with any fragments of dead or moribund animal material that may come within the range of the tentacles. When fully extended these may obscure the form of the colony and constitute an extent of feeding surface in relation to the actual volume of living tissue unparalleled in the animal kingdom. This is undoubtedly one of the major reasons for the supreme success of this group.

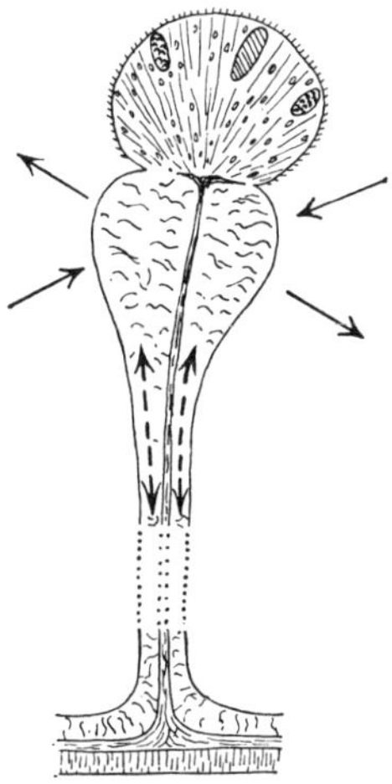

FIGURE 1. Diagram of mesentery with terminal filament in transverse section. The filament is ciliated and has enzymic and mucous glands. Both intake and excretion of particulate matter, as indicated by arrows, occurs in the region of the mesentery adjacent to the filament.

Digestion immediately follows introduction of food into the coelenteron. In *Fungia* (Yonge & Nicholls 1931), the pH of the contained fluid drops from a previous 7·8 to around 7·1 due to secretion by the gland cells in the epithelium of the mesenterial filaments which immediately wrap round the prey. These also produce a powerful protease which, with optimum activity at pH 7·1, breaks down proteins to polypeptides with resultant speedy liquefaction of the food mass. The soluble or finely particulate material is then absorbed or ingested phagocytically within the cells of the thickened 'absorptive' zone of the mesenteries immediately adjacent to the filaments (see figure 1). Here digestion proceeds intracellularly with final conversion of polypeptides to amino acids, some digestion of fats and also of glycogen but of no other carbohydrate. Digestion is extremely rapid, small crustaceans, such as copepods, mysids or amphipods of up to 13 mm in length being digested in from 4 to 13 h by species of *Fungia*, *Favia* and *Symphyllia* at an ambient temperature of around 25 °C. Thus zooplankton captured after nightfall, when the majority of corals expand (and when the plankton ascends in the water) will have been digested and the empty skeletons ejected by the following morning. Essentially the same digestive process occurs in the Actiniaria as most recently described by Nicol (1959).

Ecologically it is possible, indeed altogether necessary, to divide the Scleractinia into two distinct groups, the ahermatypic or non-reef-building and the hermatypic or reef-building, corals. (Of the hydrozoan corals, *Millepora* is hermatypic, so are the octocorallians, *Heliopora* and *Tubipora*, but not the red coral of commerce, *Corallium*). Structural differences between hermatypic and ahermatypic corals are slight, most notably the much greater spacing out of the polyps in the colonial ahermatypes such as *Lophelia* (*Lophohelia*) a member of the British offshore deep-water fauna. The really significant distinction is the invariable presence within the hermatypes of symbiotic algae, unicellular brown spheres of around 10 μm in diameter, known as zooxanthellae, which occur within the endodermal tissues. Associated with their presence is a profound difference between the distribution of the two groups. The ahermatypes are widely distributed in all latitudes and from the surface to a present known maximum depth of 5870 m. They are most numerous between 180 and 550 m (Vaughan & Wells 1943). The three British genera already mentioned, *Balanophyllia*, *Caryophyllia* and *Lophelia*, are all ahermatypic.

The reef-builders, on the other hand, have closely packed polyps and are confined to waters where the surface temperature never drops significantly below 18·5 °C, i.e. are effectively restricted to the tropics, and do not occur below a greatest depth of around 90 m with the majority of species occurring in less than half that depth. In other words, hermatypic corals demand light, and this undoubtedly in connexion with their contained zooxanthellae. In the very extensive collections made by him in the Marshall Islands, Wells (1954) records the presence of 240 corals, 229 of these being Scleractinia. Of the latter only 15 were ahermatypic. The essentially tropical distribution of the reef-builders is emphasized by the fact that the greatest number of species occur in the mid-tropics (in terms of temperature if not always of latitude) with diminution in numbers both to the north and the south. No new species appear as temperature falls.

Comparison between the distribution of hermatypic and ahermatypic corals is indicated in figure 2. It is apparent that association with zooxanthellae is not essential for the formation of the aragonite composing the skeleton; this occurs in ahermatypes in cold waters both near the surface and in profound depths. Nor are symbiotic unicellular algae confined to warm seas. They occur in temperate waters in Protozoa and in a variety of invertebrates, notably Porifera, Coelenterata and Turbellaria, all of which digest intracellularly (Yonge 1957). Amongst coelenterates they are present in some species of British sea anemones, notably *Anthopleura ballii*, *Aiptasia couchii* and *Anemonia sulcata*. But what is of occasional occurrence in temperate seas is universal in tropical seas where effectively every species of coelenterate—hydrozoans, scyphozoans and anthozoans—with the exception of the few ahermatypic corals, harbours vast numbers of zooxanthellae usually confined to the endoderm. In the mid-tropics also occur the bivalve

Tridacnidae, producing exceptionally large and massive shells, in which the entire anatomy is profoundly modified owing to the presence of zooxanthellae in the greatly hypertrophied siphonal tissues (Yonge 1936, 1953) and the very differently modified heart cockle, *Corculum* (also a member of the Superfamily Cardiacea) in which the laterally extended shell is exceptionally thin, permitting passage of light for the maintenance of much smaller algal populations (Kawaguti 1950; Yonge 1967).

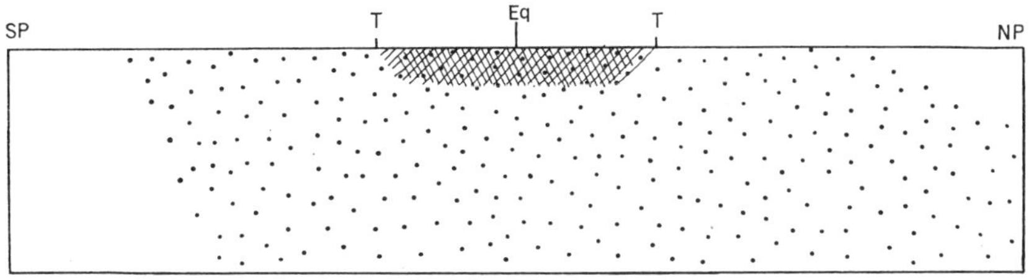

FIGURE 2. Diagram indicating the wide distribution of ahermatypic Scleractinia (indicated by dots) and the concentrated occurrence of hermatypes confined to shallow tropical waters.

It is initially clear that association with zooxanthellae is not primarily connected with calcium metabolism because this occurs equally well in coelenterates and bivalves without algae, while many invertebrates harbouring algae do not produce calcareous shells. However, there is some indication that the exceptional powers of calcium metabolism possessed by reef-building corals and which represents a major difference between them and the non-reef-builders, may be associated with the presence of zooxanthellae. Only in shallow (i.e. illuminated) tropical waters does this exceptional capacity for calcium metabolism exist and always in animals with enormous populations of zooxanthellae (this could include the Tridacnidae as well as the Scleractinia). Here only can reefs be formed and maintained against a diversity of physical and biological agents of destruction. What is surprising and still impossible to explain is the sharp delimitation between hermatypes and ahermatypes.

Zooxanthellae are now known to represent vegetative stages of dinoflagellates. Kawaguti (1944) provided the first direct proof by cultivating algae from *Acropora corymbosa* in Miquel-Allen's solution and finding that some changed from spheres into motile organisms with the characteristic two flagella. Detailed study by workers from the Haskins Laboratory, New York, on algae from a scyphozoan and from actinians (essentially similar to, if not identical with, zooxanthellae from corals) culminated in the description by Freudenthal (1962) of *Symbiodinium adricaticum* (from the scyphozoan, *Cassiopeia*). From observations made on his cultures he was able to deduce a life history involving the development of 'the vegetative single thin-walled autotrophic cell, commonly recognized as the zooxanthella'. These appear sometimes to become thick-walled cysts which either undergo binary fission to form autospores or else change into gymnodinioid zoospores, both later becoming zooxanthellae. Swarms of spherical bodies each with a single flagellum also appear, both in culture and in the field, and could be isogametes which fuse to form zooxanthellae.

The precise significance of the association has long been a matter of major controversy. It may be considered first in terms of the algae, then of the corals. There is little problem about the former. They obtain protection and a source of inorganic food for carbohydrate and protein synthesis. They are in effect imprisoned phytoplankton. Aquarium experiments reveal that they remove all available phosphorus and nitrogen which would otherwise be excreted by the corals (Yonge & Nicholls 1931; Kawaguti 1953) while in cultures they have been shown capable of utilizing urea, uric acid, guanine, adenine or any of the twelve amino acids, also a range of phosphoric acids (McLaughlin & Zahl 1959). Their numbers within any coral colony appear dependent on the 'health' of the animal. If corals be exposed to low oxygen tensions or to sublethal temperatures the 'excess' zooxanthellae are ejected from the tissues and always by way of the 'absorptive' zone in the mesenteries, which is revealed as also 'excretory', and so as the one region where particulate matter can pass into or out of the body (figure 1). The algae are carried within wandering cells. This process also occurs in nature where, following exposure to high temperatures during low day tides coral colonies, although still alive, may become white owing to the expulsion of the great majority of the algae. This has been observed both on the Great Barrier Reef of Australia (Yonge & Nicholls 1931) and in Jamaica (Goreau 1964). It remains to be discovered what stimulus leads to this mass expulsion; but it happens so quickly that it presumably follows death of the algae. Coral colonies would normally seem to possess the maximum algal population that the waste products of their metabolism can support. Certainly if additional nutrients be added to sea water in which corals are contained these are quickly removed, indicating that the algal population within the corals is controlled by the supply of available nutrients just as is that of the 'free' phytoplankton in the surrounding sea. But intimate association with supplies of available nutrients represented by the excrement of the animals in which zooxanthellae live permits an extremely high population. Single planulae of *Porites*, about 1 mm long and 0·5 mm wide, were recorded by Marshall (1932) to possess an algal population of up to 7400.

The algae, as recently demonstrated, can live free in the sea as well as, in the vegetative form, indefinitely within corals. The latter, as individual colonies, are also capable of independent life. In nature they may be found in regions of deepest shade growing without zooxanthellae, these can also be removed by exposing colonies to darkness; the algae are literally starved to death and then expelled (Yonge & Nicholls 1931; Goreau 1964). But in nature, except in these rare conditions of perpetual darkness, the corals are always densely packed with zooxanthellae, their numbers only diminishing with increasing depth and corresponding reduction of light intensity.

It has been suggested that corals, by which should be understood growing colonies concerned with the formation of reefs, may be dependent on the contained algae for (1) food, (2) oxygen, (3) removal of excrement, (4) aid in calcification. These may be considered separately although they are not mutually exclusive.

(1) *Food*

It has already been noted that, like the vast majority of coelenterates, Scleractinia are specialized carnivores. They certainly do not, and cannot, digest zooxanthellae. Presence of these algae in a degenerating state in the 'absorptive' zone, once advanced as evidence for this, is now known as due to the expulsion of dead algae through this equally 'excretory' area (Yonge 1931). Experiments in which specimens of *Fungia* were starved for up to 10 weeks in light revealed that the tissues almost immediately began to recede, finally uncovering the greater part of the upper surface of the septa (Yonge & Nicholls 1931). This was considered sound proof that the animal obtains no nutriment from the algae. However Kawaguti (1964) has since kept colonies of *Oulastrea crispata* alive for as long as 15 months without food; they remained healthy and grew. Species of other genera survived equally well. These were all kept in full daylight, whereas personal experiments on *Fungia* had, perforce, to be conducted in dim light or the temperature would have risen. In other words, the algae may well have been in some measure starved whereas those in *Oulastrea* were more actively photosynthesizing. It is now known that in cultures unicellular algae *do* release quantities of organic matter into the surrounding medium, also that radio-active labelled material passes from zooxanthellae into the tissues both in Actiniaria (Muscatine & Hand 1958) and in *Tridacna* (Goreau, Goreau & Yonge 1966). In the freshwater hydrozoan, *Chlorohydra* there is clear evidence that the animal gains nutriment from symbiotic algae, not in this case zooxanthellae (Muscatine 1961). Further work is certainly necessary, but it does now appear that there is a probably significant interchange, notably, in the opinion of T. F. Goreau (personal communication), of amino acids, between the zooxanthellae and the 'host' coral.

(2) *Oxygen*

The presence of so much chlorophyll within the coral tissues involves an excess of oxygen production over oxygen consumption over the hours of bright light, at any rate near the surface. This has been frequently demonstrated, probably most thoroughly in an experiment involving eight hermatypic corals run for a period of 27 h (Yonge, Yonge & Nicholls 1932). The greatest excess of oxygen was produced some time before noon indicating that either the light became too intense (the experiment was run in the open sea with the sun almost vertically overhead) or accumulation of the products of photosynthesis slowed this process down. Other data indicated that, certainly in the mid-tropics, optimal photosynthetic action is at a few fathoms below the surface. However, there is no evidence that the additional oxygen produced within the coral colonies by the zooxanthellae has any significance. The respiratory needs of corals are low—high figures reported for some species, notably of *Acropora*, in the past have been shown due to oxidation of the great quantities of mucus secreted by them.

(3) *Removal of excrement*

The effect of the algae, as noted above, is to remove at source the CO_2 with nitrogen, phosphorus and doubtless sulphur, produced by metabolic breakdown within the coral. Although perhaps difficult of precise proof, there is obvious advantage in the presence of these automatic agents of excretion. Immediate removal of the end products of metabolism is an obvious aid to efficiency. This view has long been held by the writer and is supported more recently by Goreau.

(4) *Aid in calcification*

The most important recent work on reef-building corals, that of T. F. Goreau in Jamaica, has been concerned with this. In the course of a series of papers (Goreau 1959, 1961*a*, *b*; Goreau & Goreau 1959, 1960*a*, *b*) he has clearly demonstrated, using calcium-45 as radio-active tracer and conducting experiments *in situ* on the reef by diving, that calcification in a series of representative corals is on the average *ten times greater* in light than in darkness. The rate of uptake of calcium was even reduced by about a half on cloudy as compared with sunny days. No such differences were found in ahermatypic corals or in hermatypes which had been deprived of zooxanthellae by prolonged exposure to darkness (when they are expelled as already described). An absolute correlation between the presence of photosynthesizing zooxanthellae and rate of calcification was established. In the briefest outline, Goreau postulates the process to be somewhat as follows. Calcium is taken up direct from the sea and is considered to pass through the cells to be adsorbed on a mucopolysaccharide-like layer on the surface of the skeleton. This he considers to form part of the organic matrix (the existence of which has been noted) of the skeleton and to act, as he describes it, 'as the template upon which the initial stages of skeleton mineralization occur'. Within this layer Goreau further postulates that calcium combines with bicarbonate produced in metabolism forming first calcium bicarbonate and then calcium carbonate. The later process, depending on the removal of H_2CO_3, is catalysed by carbonic anhydrase, the presence of which he demonstrated by treatment with the inhibitor, Diamox, when calcification was strikingly reduced, by about 50 % in light and 75 % in darkness The effect of photosynthesis by the zooxanthellae in the tissues is to increase the speed with which CO_2 produced in metabolism is removed and so to assist in the production of calcium carbonate.

While there can now be no doubt whatever as to the effect of the zooxanthellae on the rate of calcification, Goreau does also feel that some of this may be a consequence of the speed with which not only CO_2 but also other end products of metabolism are removed, in other words that the speeding up of excretion is a highly significant factor. This is supported by the fact that calcification in darkness is greater when zooxanthellae are present; they cannot then photosynthesize but protein synthesis continues with consequent removal of available phosphorus and nitrogen.

The work of Goreau gives clear experimental proof to the general conclusion earlier stated, that the possession of zooxanthellae is an essential factor in reef formation. Broadly speaking, as he points out, the effect of photosynthesis *within* the tissues is to increase the free energy of the community through primary production as well as to create optimum conditions for skeleton formation in corals (and in the fundamentally no less important calcareous algae). It is the unique powers—above all speed—of calcareous deposition possessed by hermatypic corals which have made reef formation possible and permitted them in shallow water to grow out against the greatest forces of wind driven seas. Ahermatypic corals have no such powers, even the extensive banks of *Lophelia* found in the bottom of Norwegian fjords form only an open meshwork which could not exist in shallow water because it would immediately be crushed by surf. It is noteworthy that in other tropical coelenterates (apart from the Alcyonacea which require careful study) the presence of zooxanthellae appears no more than incidental, the gorgonids, for instance, attain far greater size in cold seas.

Consideration of the role of the zooxanthellae, without which individual coral colonies can exist but certainly not form reefs, leads to a brief examination of the consequences of reef formation on hermatypic corals. The Scleractinia, of dubious connexion with the palaeozoic rugose corals, appear within the Middle Triassic. While there is much subsequent variation in the amount of reef formation nevertheless hermatypic corals do soon appear (Vaughan & Wells 1943) and it is reasonable to assume that they possessed zooxanthellae. The modern coral fauna becomes established in the Tertiary (the Eniwetok boring descends into the Eocene) with hermatypic and ahermatypic corals apparently evolving separately.

The evolution of the reef-builders can only adequately be considered in relation to the reefs their assisted powers of calcium metabolism enable them to construct and, above all, the range of environments brought about when these reefs rose towards the surface and so presented an obstacle in the way of the steady force of seas driven by the Trade Winds. There is immediate separation between an exposed and a sheltered environment, on the weather and lee sides, a further environment is provided by the reef crest over which the surf rushes and which is consolidated by the calcareous algae which grow best in these highly oxygenated waters. A soft substrate is provided by the calcareous sand which accumulates between coral masses in the lee. There are many gradations between these major environments, all brought into being by the upward and outward growth of reefs. There is a zonation of corals associated with depth and with degrees of exposure on the upper surface of reefs, as particularly well shown by Manton (1935). Above extreme low-water level corals will be exposed to varying degrees of desiccation as well as to sudden variations in temperature and salinity, also to mechanical shock. Only corals possessing the physiological adaptations of intertidal animals can exist in such an environment. Other corals are always sublittoral. No less then seventeen major environments may be distinguished passing across an atoll from the weather side (against the N.E. or S.E. Trades) across the reef surface to the lee side of the lagoon, then across this with its numerous contained patch reefs to the weather side of the lagoon, then again over the surface of the reef to the seaward slopes on the lee (see Yonge (1963) for figure and references.)

Light which, almost certainly through the intermediary of the zooxanthellae, has such a profound effect on the growth of hermatypic corals (see Yonge 1963), is the factor influencing their distribution in depth. In the West Indies, Goreau (1963*a*) has shown that 90 % of all the species of hermatypes occur in depths of between 5 and 20 m but only 25 % at 60 m, although nine of the ten species there found occurred in shallower depths. The differences between the coral population in shallow and greater depths lay not in the species, because nine of the ten species at 60 m also occurred in shallower water, but in the smaller size of individual colonies and their greater spacing at that depth. They have no significance as reef builders, whereas in the shallower optimal depths the same species are among a population which occupies 'the whole of the available area with tier on tier of very large colonies in closed pavement growth'.

With the exception of the few corals to be mentioned later, the position occupied by any coral colony indicates where the initial planula (or planulae) settled. The small amount of work so far conducted on the behaviour of the liberated larvae indicates initial positive phototaxis with negative geotaxis and some positive rheotaxis with final positive thigmotaxis resulting in the settlement of the planula. Differences in the pattern of behaviour of planulae from different species may be assumed to be responsible for differences in the regions where they settle and so for zonation, the adult colonies being adapted for life in the zones where the planulae settle. There is no later opportunity for movement (although there may be much modification in form). Here again light is of supreme importance, planulae of ahermatypic corals develop in its absence, those of hermatypes do not.

Planulation is usually periodic, in most cases at some phase in the lunar cycle. In the case of *Pocillopora bulbosa*, however, it does not occur at the same phase in every lunar month, depending on the time of the year (see Yonge (1963) for details). Planulae settle in numbers together, in part, no doubt, owing to the great quantities liberated at any planulation, but they are also presumably influenced by factors causing the aggregation observed in other sedentary invertebrates. On settlement they may fuse to form colonies of precisely the same form as those produced by growth from a single planula. In meandrines such as *Manicina* (*Meandra*) *areolata* the origin of the oval or rounded colony from one, two or three (possibly more) planulae is revealed by the number of separate 'valley' systems. Colonies of *Pocillopora bulbosa* and other corals consisting of large numbers of fused planulae have been reared experimentally. The form of the colony is unaffected by the number of planulae from which it arose. There would seem to be some means here of distinguishing between species and growth forms; planulae from different individuals of the same species, whatever their form, would presumably blend in this way, but not planulae from different species.

The wide variation in the form of corals is well known. There is a range from branching colonies such as those of *Acropora* and *Pocillopora* to the massively rounded Favidae and the numerous meandrines; from the folacious species of *Montipora* and *Tridacophyllia* to encrusting growths of *Siderastrea*, *Pavona* and other agaricids. Growth of corals must be considered from the dual standpoints of tissue increase and skeletal formation, the former most conveniently measured by nitrogen content, the latter by increase in calcium (Goreau 1959, 1961*b*, 1963*a*). Living tissue is related directly to surface area and the mass of the skeleton to the volume, relationship between the two being constant for any species. The form of any coral species depends on local differences in growth gradients; in hemispherical colonies there is little difference from one area to another while in branching colonies growth rates in apical areas may be as much as ten times greater than those in lateral or basal areas.

Such are the determining factors between species. But the environment plays a major part in determining the final form of any colony. A long controversy has ranged on the relative importance of growth forms as against species in these sedentary colonies which are so obviously subject to the major effects of wind, weather, exposure—and also depth. A good instance of the effect of exposure on form is provided by the Atlantic agaricid *Siderastrea radians* which grows as flat encrustations in its usual habitat on beach rock fully exposed to surf but as hemispherical colonies in the complete shelter of the moat at Fort Jefferson, Dry Tortugas (Yonge 1935). Similar effects have been noted following division of colonies and subsequent growth of the halves in different shallow water environments; the eventual form of these frequently varied considerably. It has been left to Goreau (1963*a*) to demonstrate the effect of diminished light, whether this be due

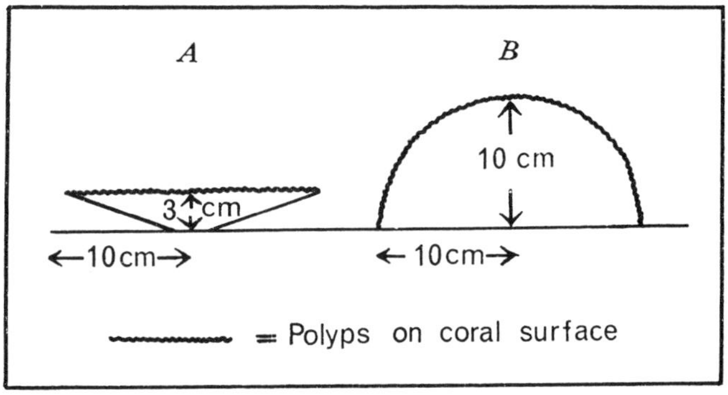

FIGURE 3. Diagrams of colonies of the same species of coral, *A* with flattened form typical of poorly illuminated regions, *B* with hemispheroidal form found in shallow, well illuminated waters. In the latter, the living polyps extend to the base to form a firm attachment. (From Goreau.)

to increasing depth or to the effect of shade in shallow water. The effect, as shown in figure 3, is to change the form of a hemispherical colony into a flat plate, i.e. to return from a three-dimensional to a two-dimensional symmetry. There being no reason to assume significant reduction in the planktonic food of the corals at these minor depths (25 to 50 m) the change in form can reasonably be attributed to reduction in calcification due to lessening of photosynthetic activity by the zooxanthellae. No such effect is observed in any ahermatypic coral. Colonies of hermatypes such as *Porites astreoides* and *Montastrea annularis* which are hemispherical near the surface where they are major reef builders, at greater depths become flattened and plate-like, projecting from the reef slope like bracket fungi from the trunk of a tree. The change in form and consequent greater lightness of the skeleton has the incidental but not insignificant effect of rendering them less liable to be carried into greater depths when their attachments are eroded by the action of boring sponges (Goreau & Hartman 1963).

The final form of any coral is clearly the consequence both of genetic constitution and of environment, But many corals are very highly specialized. Apart from the form of the colony, there are the wide differences in the size of the polyp and the associated ciliary currents (figure 4). Those with large polyps (e.g. *Favia* or

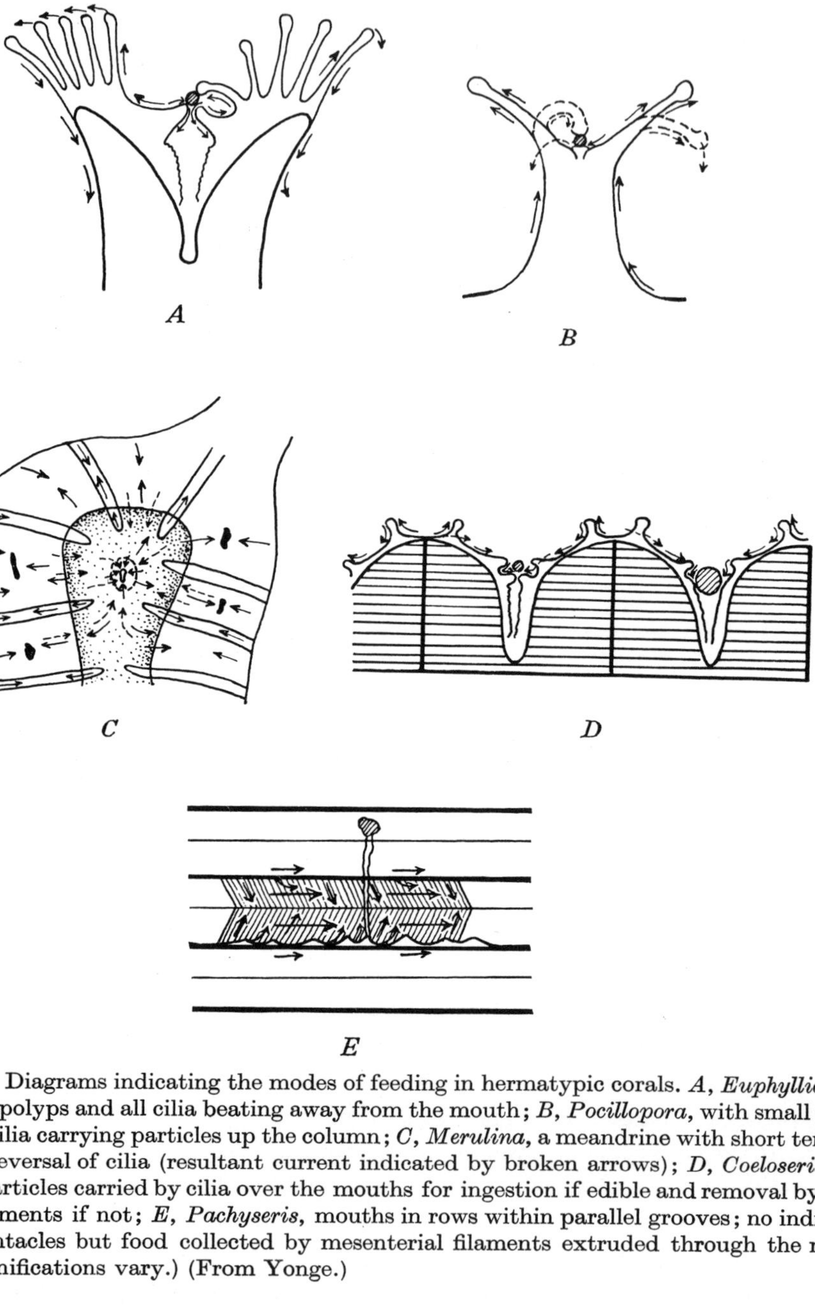

FIGURE 4. Diagrams indicating the modes of feeding in hermatypic corals. *A*, *Euphyllia*, with large polyps and all cilia beating away from the mouth; *B*, *Pocillopora*, with small polyps and cilia carrying particles up the column; *C*, *Merulina*, a meandrine with short tentacles and reversal of cilia (resultant current indicated by broken arrows); *D*, *Coeloseris*, with all particles carried by cilia over the mouths for ingestion if edible and removal by water movements if not; *E*, *Pachyseris*, mouths in rows within parallel grooves; no indication of tentacles but food collected by mesenterial filaments extruded through the mouth. (Magnifications vary.) (From Yonge.)

Euphyllia) feed essentially as do actinarians by means of the tentacles, ciliary currents carrying waste material away from the disk, down the column and so eventually clear of the colony. In some species with smaller polyps, such as *Pocillopora*, the ciliary currents carry material up the column, in other words the general surface of the coenosarc appears to assist in the collection of at any rate minute fragments of organic matter. Where tentacles are reduced in length, the cilia on the disk, which normally beat away from the mouth, temporarily reverse on the stimulus of animal matter and carry this to the mouth, later returning to their normal direction of beat. This is true of species of *Merulina*, *Tridacophyllia* and many fungids (but not in *Fungia actiniformis* which alone has long tentacles). As in *Pocillopora*, the changes represent modifications in the cleansing mechanism (analogous changes occur in ciliary feeding Gastropoda). Certain corals, notably *Coeloseris* which lives on the surface of reefs, have all the superficial cilia beating towards the polyps and rely entirely on water movements for cleansing. Finally in *Pachyseris* there appear to be no tentacles, food being captured directly by the extruded mesenterial filaments. This also occurs in many corals possessing tentacles and may well prove to be widespread and most important. A great deal remains to be ascertained about the feeding of corals but the main impression is of a group of animals specialized, among the constituent species, for collection of all available animal material, alive, moribund or disintegrated, from the most minute to the size of medium crustaceans and small fishes.

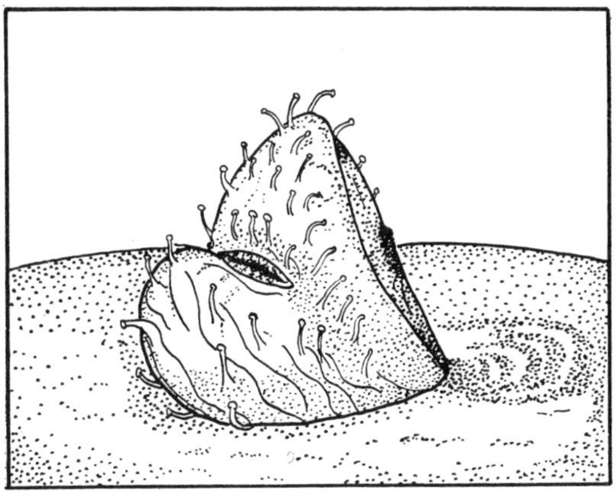

FIGURE 5. *Fungia actiniformis*, righting movement with disk tissues distended with water. (From Abe.)

Undoubtedly the most obviously specialized corals are those that, although settling initially on hard substrates, eventually become free. They exploit the areas of sand and loose fragments found in the lee of reefs and also occur in depressions on the reef surface. They include the very numerous Fungiidae of the Indo-Pacific and *Manicina* (*Meandra*) *areolata* in the Atlantic. After settlement the former develop into an anthocaulus-stage from which, from time to time, a free disk, the anthocyathus, becomes detached (see Wells 1966). These are carried by water movements to regions of still water. They are capable of uncovering when buried and of righting themselves (figure 5) when turned over—both natural hazards. *M. areolata* is also initially attached but eventually breaks away owing to withdrawal of tissues from the restricted basal region. It has the same capacity for uncovering and for righting itself. In both groups these activities are the result of ciliary action associated with swelling of the polyp following stomodeal intake of water. Of very different origins, therefore, the fungids and this meandrine have adapted themselves very similarly to meet the hazards and exploit the possibilities of life on unstable substrates.

In general the success of the hermatypic corals may be attributed to the wide adaptibility of some species combined with high degrees of specialization in others.

Hermatypic corals are of compelling interest because of their powers of skeletal formation, so great in shallow waters as to form, in association with the essentially no less important calcareous algae (nullipores) a vast cemented mass of interlocking skeletons continually being overgrown by the products of subsequent secretion. This is the primary reef framework with secondary or detrital reefs in the deeper, calmer but poorly illuminated regions around their base (Goreau 1963*a*). The primary framework provides a series of environmental niches for a host of associated animals and plants. Considered as a community, a coral reef thus constitutes an elaborate ecosystem some of the major constituents of which have been ennumerated and illustrated by Gerlach (1960). They include a variety of organisms which penetrate the skeletal mass and some which actually prey on living corals. The former include the diverse boring invertebrates, sponges (the great significance of which is emphasized by Goreau & Hartman 1963), various polychaetes and sipunculids and many bivalve and a few gastropod molluscs. All confine their activities to the skeleton and only incidentally affect the living corals.

Predators are another matter. A number of parrot fishes (Scaridae) and file fishes (Monacanthidae) scrape or nibble the surface of living corals doing no more than superficial damage. At least one polychaete, *Hermodice carunculata*, is able to feed on coral polyps, swallowing those of *Porites porites* when these expand at night. But again damage must be negligible. It is otherwise with the attacks of the large spiny starfish, the so-called Crown of Thorns starfish, *Acanthaster planci*, which was originally described as eating corals by Mortensen (1931). It spreads widely over the surface, everting the stomach and liquefying and absorbing the tissues of corals. This species appears to have increased in numbers spectacularly of recent years. A single specimen only was contained in the collections of the Great Barrier Reef Expedition of 1928–29. The author does not personally remember seeing a specimen of this most striking species during that year but it is now extremely common. Over 27000 of these animals were taken in 15 months for one area alone of Green Island reef off Cairns, N.Q. (Barnes 1966). The author has observed great numbers of these starfishes on reefs off the north-east of Borneo. Goreau (1963*b*) has given the fullest account of their biology in a description of specimens obtained off Eritrea in the Red Sea.

These predators would seem to present a major hazard to the continuance of coral reefs. These are only maintained by the presence of an excessively thin covering of living matter which is able—with the assistance of the zooxanthellae—to secrete an altogether disproportionate skeletal mass. Any serious break in this covering and erosion, by a combination of boring organisms and the forces of the sea, would inevitably prevail and the reef mass rapidly be breached. The corals obviously have no protection against these predators. Of course such attacks may have been of frequent occurrence in the past with some compensating factor, of parasitism for instance, coming later to control the numbers of the starfish. But their present abundance and their effect on living corals most strikingly emphasize the essentially precarious nature of coral reefs dependent for their maintenance on the secretory activities of the skeletogenic epithelia of hermatypic Sceleractinia.

The author wishes to thank his research assistant, Miss J. I. Campbell, for preparation of the figures.

References

Barnes, J. H. 1966 The crown of thorns starfish as a destroyer of coral. *Aust. nat. Hist.* **15**, 257–261.

Freudenthal, H. D. 1962 *Symbiodinium* gen.nov. and *Symbiodinium microadriaticum* sp.nov., a zooxanthella: taxonomy, life cycle and morphology. *J. Protozool.* **9**, 45–52.

Gerlach, S. A. 1960 Über das tropische Korallenriff als Lebenstaum. *Zoo. Anz.* (Suppl.) **23**, 356–363.

Goreau, T. F. 1959 The physiology of skeleton formation in corals. I. A method for measuring the rate of calcium deposition by corals under different conditions. *Biol. Bull. mar. biol. Lab. Woods Hole* **116**, 59–75.

Goreau, T. F. 1961*a* Problems of growth and calcium deposition in reef corals. *Endeavour* **20**, 32–39.

Goreau, T. F. 1961*b* On the relation of calcification to primary production in reef-building organisms. In *The biology of hydra etc.*, pp. 269–285. (Ed. H. M. Lenhoff and W. F. Loomis.) University of Miami Press.

Goreau, T. F. 1963*a* Calcium carbonate deposition by coralline algae and corals in relation to their roles as reef-builders. *Ann. N.Y. Acad. Sci.* **109**, 127–167.

Goreau, T. F. 1963*b* On the predation of coral by the spiny starfish *Acanthaster planci* (L.) in the southern Red Sea. *Israel South Red Sea Exped.*, 1962, *Reports*, no. 2.

Goreau, T. F. 1964 Mass expulsion of zooxanthellae from Jamaican reef communities after Hurricane Flora. *Science* **145**, 383–386.

Goreau, T. F. & Goreau, N. I. 1959 The physiology of skeleton formation in corals. II. Calcium deposition by hermatypic corals under various conditions in the reef. *Biol. Bull. mar. biol. Lab. Woods Hole* **117**, 239–250.

Goreau, T. F. & Goreau, N. I. 1960*a* Distribution of labelled carbon in reef-building corals with and without Zooxanthellae. *Science* **131**, 668–669.

Goreau, T. F. & Goreau, N. I. 1960*b* The physiology of skeleton formation in corals. III. Calcification rate as a function of colony weight and total nitrogen content in the reef coral *Manicina areolata* (Linnaeus). *Biol. Bull. mar. biol. Lab. Woods Hole* **118**, 419–429.

Goreau, T. F., Goreau, N. I. & Yonge, C. M. 1966 Evidence for a soluble algal factor produced by the zooxanthellae of *Tridacna elongata*. *Intern. Conf. Trop. Oceanography*, 17–24 Nov., 1965, Miami. (Abstract only.)

Goreau, T. F. & Hartman, W. D. 1963 Boring sponges as controlling factors in the formation and maintenance of coral reefs. In *Mechanisms of hard tissue destruction*, pp. 25–54. Publication no. 75, Am. Ass. Adv. Science, Washington, D.C.

Johannes, R. E. 1967 Ecology of organic aggregates in the vicinity of a coral reef. *Limnol. Ocean.* **12**, 189–195.

Kawaguti, S. 1944 On the physiology of reef corals. VII. Zooxanthellae of the reef corals in *Gymnodinium* sp., Dinoflagellata; its culture *in vitro*. *Palao trop. biol. Stud.* **2**, 675–679.

Kawaguti, S. 1950 Observations on the Heart Cockle, *Corculum cardissa* (L.), and its associated zooxanthellae. *Pacif. Sci.* **4**, 43–49.

Kawaguti, S. 1953 Ammonium metabolism of the coral reefs. *Biol. J. Okayama Univ.* **1**, 171–176.

Kawaguti, S. 1964 An electron microscope proof for a path of nutritive substances from zooxanthellae to the reef coral tissue. *Proc. Jap. Acad.* **40**, 832–835.

Manton, S. M. 1935 Ecological surveys of coral reefs. *Sci. Rep. Gr. Barrier Reef Exped.* 1928–29, *Br. Mus.* (*Nat. Hist.*) **3**, 273–312.

McLaughlin, J. J. A. & Zahl, P. A. 1959 Axenic zooxanthellae from various invertebrate hosts. *Ann. N.Y. Acad. Sci.* **77**, 55–72.

Marshall, S. M. 1932 Notes on oxygen production in coral planulae. *Sci. Rep. Gr. Barrier Reef Exped.* 1928–29, *Br. Mus.* (*Nat. Hist.*) **1**, 253–258.

Marshall, S. M. & Orr, A. P. 1931 Sedimentation on Low Isles Reef and its relation to coral growth. *Sci. Rep. Gr. Barrier Reef Exped.* 1928–29, *Br. Mus.* (*Nat. Hist.*) **1**, 93–132.

Mortensen, T. 1931 Contributions to the study of larval forms of echinoderms. *Kong. Dansk. Vidensk. Selsk. Skrifter* (9), **4**, 1–39.

Muscatine, L. 1961 Symbiosis in marine and freshwater coelenterates. In *The biology of hydra etc.*, pp. 255–268. (Ed. H. M. Lenhoff and W. F. Loomis.) University of Miami Press.

Muscatine, L. & Hand, C. 1958 Direct evidence for the transfer of materials from symbiotic algae to the tissues of a coelenterate. *Proc. nat. Acad. Sci. Wash.* **44**, 1259–1263.

Nicol, J. A. C. 1959 Digestion in sea anemones. *J. mar. biol. Ass. U.K.* **38**, 469–476.

Vaughan, T. W. & Wells, J. W. 1943 Revision of the suborders, families, and genera of the Scleractinia. *Spec. Pap. geol. Soc. Am.* no. 44.

Wainwright, S. A. 1963 Skeletal organization in the coral, *Pocillopora damicornis*. *Q. Jl microsc. Sci.* **104**, 169–183.

Wells, J. W. 1954 Recent corals of the Marshall Islands, Bikini and nearby atolls. *U.S. geol. Surv. Pap.* **260**–I, pp. 385–486.

Wells, J. W. 1966 Evolutionary development in the scleractinian family Fungiidae. In *The Cnidaria and their evolution*, pp. 223–246. London: Academic Press.

Yonge, C. M. 1931 Studies on the physiology of corals. III. Assimilation and excretion. *Sci. Rep. Gr. Barrier Reef Exped.* 1928–29, *Br. Mus.* (*Nat. Hist.*) **1**, 83–91.

Yonge, C. M. 1935*a* Studies on the biology of Tortugas corals. I. Observations on *Maeandra areolata* Linn. *Pap. Tortugas Lab.* **29**, 185–198.

Yonge, C. M. 1935*b* Studies on the biology of Tortugas corals. II. Variation in the genus *Siderastrea*. *Pap. Tortugas Lab.* **29**, 199–208.

Yonge, C. M. 1936 Mode of life, feeding, digestion and symbiosis with zooxanthellae in the Tridacnidae. *Sci. Rep. Gr. Barrier Reef Exped.* 1928–29, *Br. Mus.* (*Nat. Hist.*), **1**, 283–321.

Yonge, C. M. 1953 Mantle chambers and water circulation in the Tridacnidae. *Proc. Zool. Soc. Lond.* **123**, 551–561.

Yonge, C. M. 1957 Symbiosis. *Mem. geol. Soc. Am.* **67**, **1**, 429–442.

Yonge, C. M. 1963 The biology of coral reefs. *Advanc. Marine Biol.* **1**, 209–260.

Yonge, C. M. 1967 Form, habit and evolution in the Chamidae (Bivalvia) with reference to conditions in the rudists (Hippuritacea). *Phil. Trans.* B **252**, 49–105.

Yonge, C. M. & Nicholls, A. G. 1931 Studies on the physiology of corals. IV. The structure, distribution and physiology of the zooxanthellae. *Sci. Rep. Gr. Barrier Reef Exped.* 1928–29, *Br. Mus.* (*Nat. Hist.*) **1**, 135–176.

Yonge, C. M., Yonge, M. J. & Nicholls, A. G. 1932 Studies on the physiology of corals. VI. The relationship between respiration in corals and the production of oxygen by their zooxanthellae. *Sci. Rep. Gr. Barrier Reef Exped.* 1928–29, *Br. Mus.* (*Nat. Hist.*) **1**, 213–251.

The Rutherford Memorial Lecture, 1968

Some problems of the growth and spread of science into developing countries†

By J. M. Ziman, F.R.S.

H. H. Wills Physics Laboratory, University of Bristol

(*Delivered* 2 *December* 1968—*Received* 20 *January* 1969)

Proceedings of the Royal Society, A, *volume* 311, *pp.* 349–369, 1969

For a physicist brought up in New Zealand, it is a special honour, and especially humbling, to be asked to speak in memory of Ernest Rutherford. Whatever we may think of the subsequent applications of nuclear physics, we cannot fail to class him among the immortals for his brilliant voyages of discovery in that vast and unsuspected realm of Nature. How could I present my own little scholarly pastime —that agreeable game called the theory of the solid state—under the shadow of his achievement. He dominated the scientific life of his day; and there are too many distinguished scholars who remember him personally for me to dare to say anything about him at second hand.

The story of the poor farmer's son, from the backblocks of distant little New Zealand, who won a scholarship to Cambridge and became world famous, is still one of the inspirational epics of the life of the mind. Many another young man, from many another far-off land, has taken it as his model, and gone to seek his fortune thus, in the great intellectual metropolises of Europe and North America. The urge to contend and make ones mark in the great game of 'high science' is as strong as ever among the ideals and ambitions of gifted youth.

Yet times have changed; from the point of view of the patriotic New Zealander, Rutherford's permanent migration to England might now be seen as a serious case of 'brain drainage'. In speaking in his memory nearly a century after his birth, it seems appropriate to discuss the converse phenomenon—the growth and spread of basic science, from its original nuclei in the industrial countries of western Europe, into all corners of the Earth. We hold it almost as self-evident that currents of knowledge, skill, attitudes and techniques should diffuse the culture of scientific research throughout the world, so that eventually the conditions for the pursuit of mysterious Truth may be provided anywhere, from Timbuctoo to Tahiti, from Kamchatka to Kathmandu.

Now, of course, for a developing country, struggling desperately to provide a better life for millions of ordinary men, women and children, the problem of giving

† Lecture delivered on 2 December 1968 at the University of Delhi, during a tour of scientific institutions in India and Pakistan, as a guest of the Indian University Grants Committee and of the Pakistan Atomic Energy Commission.

an appropriate priority to this essentially exotic activity is not trivial. I am not speaking now of the techniques that pure science has uncovered and created—improved methods of agriculture, the control of birth and disease, rapid communication and transport, and so on—but of research without any more conscious aim than the understanding of how things are, or were, or might be. When money and men are scarce, it is by no means self-evident that an institute of theoretical physics, or molecular biology, or archaeology, should take precedence over a tractor factory, a hospital, or a school of civil engineering. Even in Britain, which is already as rich in material comforts as any sane man would wish, we argue about these priorities; in a country such as Indonesia, or Brazil, or the Congo, a very strong case indeed must be made for the diversion of any financial and human resources at all into such unproductive channels (cf. Blackett 1967).

As a first approximation this argument is compelling. Those of us, in advanced countries, who have the good fortune to be allowed to make our living solving the splendid puzzles of Natural Philosophy may be well aware that what we are doing is not likely to be very useful; but we can comfort ourselves with the thought that we live in a society where most of the basic human needs are adequately satisfied, and that we are no more parasitic than such admirable citizens as estate agents, jockeys, bar tenders or advertising executives. But this irresponsible attitude is not permissible in less fortunate societies; in a country such as India or Pakistan, I can see no justification for, shall we say, the neglect of practical animal husbandry for the sake of research in pure physiology, or for the study of the theory of superconductivity in the absence of good schools of electrical engineering. I want to say this at the beginning in case I seem to be yet another pure scientist who has his nose so deep in his silly little field of research that he cannot look out at the tragic real world around him. Indeed, I suspect that the solution of technological problems, which come already loaded with human cares and consequences, is more satisfying than finding the answers to those abstract impersonal puzzles which we dream up for ourselves in our ivory towers; and I simply cannot understand the intellectual snobbery of those silly people who give more credits for the discovery of another meson than for the design of a suspension bridge, and who cannot see that a zip fastener is a far more beautiful idea than a zeta function. Who knows: Rutherford might easily have earned his immortality for developing radio and radar rather than for splitting the atom.

But the utilitarian argument against basic research can be carried too far. In an interesting series of controversial articles Professor Harry Johnson (1965, 1966, 1967) has attempted to show quantitatively that pure science is essentially an extravagance whose applicable results may be bought from abroad far more cheaply than they can be produced at home. This follows from the rationale of free trade and the international division of labour. The developing country benefits more by the export of a few clever people, and by the subsequent import of technical knowhow, than it would by attempting to expand its own very inefficient and unproductive local research industry.

But as others have countered (Kidd 1968; Toulmin 1966) mere economic accountancy is not compelling in such a complex of subtle circumstances. A certain amount of fundamental research *must* be sponsored in a developing country, for a number of excellent reasons which have been clearly stated by Moravcsik (1964).

In the first place the education of technical experts—engineers, doctors, agricultural advisers, even government administrators—cannot be left entirely in the hands of technologists of their own practical kind. Modern engineering, for example, requires the exercise of skills, and the application of knowledge, acquired from pure physics, chemistry and mathematics. The rapid development of new techniques can only be exploited if the practitioners are adequately trained in these fundamental disciplines, by teachers who are themselves in close contact with the latest theoretical principles. If we have learnt anything in the past century about scientific education, it is that only those who are actively engaged in research can truly absorb and retransmit these new principles as they arise. The research work of university teachers is not an extravagant irrelevance to their professional task; it is absolutely essential to their efficiency in passing on useful, applicable knowledge to their pupils.

From a purely practical point of view also, it is impossible to import technical know-how, and to apply it successfully, if one does not have available, locally, a corps of learned men to whom one can appeal for guidance on matters of pure scientific principle. Suppose, as an imaginary example, that we have set up a factory for the production of simple radio sets. The beautiful machinery for making the transistors, imported from America or Japan, does not work properly because, perhaps, there is too much copper in the water supply. The engineers study the problem and eventually unearth a paper in the *British Journal of Applied Physics* referring to 'the diffusion of transition metal ions along dislocations in covalent semiconductors'. Their training is adequate to suggest the relevance of this paper to their practical problem—but only an 'academic' physicist, with a profound knowledge of the science of materials, could really tell them just what it meant, how much it could be relied on and whether it contained the answer to their questions. All our experience tells us that such queries cannot be dealt with by correspondence at a distance. As Crawford (1966) has pointed out, local experts can see more clearly the needs and wants of the country. Science-based technology is not a lusty crop of rules of thumb, perfected by evolution over long periods; it is a delicate plant, which thrives only when tended by mixed teams of experts, including those impractical specialists to whom the buck can be passed when fundamental principles are at stake.

Quite apart from utilitarian arguments, there are spiritual considerations. The pursuit of scientific knowledge for its own sake has become one of the major artistic enterprises of humanity. Even though we would deplore the supersession of all other forms of art, sport or religion by such a specialized activity, we must surely recognize that the opportunity to undertake scientific research is one of the important attributes of the Good Society. I cannot imagine any goal towards which a human society might claim that it was 'developing' that did not nowadays give some room for at least a few of its members to participate in this great, world-wide movement. The attempt to understand the genetic code may promise no immediate practical dividends, and may perhaps give direct pleasure to only a small proportion of the population. Nevertheless, like a ballet, a poem, a park or a stained glass window, it embellishes the society that has commissioned it, both for its own sake as a thing of beauty and as a symbol of the time to come when bread will not be our immediate lack. Failure to give at least token support to pure science would be as philistine as banning the public performance of orchestral music, or ploughing up Regent's Park to grow potatoes.

There is, therefore, a very good case indeed for modest encouragement and financial support for basic research in any country on the threshold of industrial development. On the other hand, these arguments should not be carried too far. They do not, for example, justify leaving the choice of basic sciences entirely to the accident of the availability of trained men, and the directions into which their enthusiasms happen to lead them. There are occasions when deliberate decisions of policy can be taken at a relatively general level. Shall we foster ecology or physiology? Should we strengthen plasma physics or materials science? Do we need an institute of molecular biology, or a department of radio astronomy? Surely one should then take account of the relevance of the subject to the economic, geographical, meteorological or social background of one's country. Even within pure science there are branches that are more natural, so to speak, to one region of the world than another, and which are therefore more likely to thrive in the appropriate context. In Alaska, for example, a physicist might find glaciology congenial—while the study of thunderstorms could be left to the fortunate inhabitants of the tropics: in Western Australia, with its vast new mining industry, he ought to become interested in mineralogy—while the Japanese steel industry might encourage him to study the mechanical properties of metals—and so on. This chameleon behaviour is not a matter for formal legislation, but quite apart from its technical consequences it is the sort of shrewd and practical policy that can counteract the brain drain. It gives the local scientist a better reason for staying where he is than mere patriotism. If, like one of the first real scientists I ever knew, you specialize in the physiology of the mammary gland, and if cow's milk is the life blood of New Zealand, then where better to live and work than beside the Waikato River! On the other hand, what could be more demoralizing than attempting to construct theories of superconductivity in a laboratory without low temperature facilities, or more frustrating than trying to look through a large astronomical telescope in the climate of the British Isles!

It would seem sensible, also, to put the main emphasis on what I should describe as 'potentially applicable' science. Of course we do not know in advance what may come of any particular scientific discovery. It is a trite argument, often used to justify the most esoteric activities, to point out that nobody could have known the practical uses of radioactivity, or electromagnetism, when they were first discovered '. . . and therefore, gentlemen, you must of course give me another million dollars in case I discover something equally useful up there in the stratosphere' etc., etc. I am afraid I do not see the necessity. Let us not stand in the way of utterly pure mathematics, astronomy, palaeontology, and other apparently useless disciplines; but do not let's fall into the trap of equating uselessness with snob value. There are numerous excellent, difficult, and rewarding branches of pure science which at least attempt some understanding of humanity and the everyday human environment and which therefore promise, some day, to be of real use. Although I have pretended to laugh at my own little game of solid state physics, I could easily trace the connexions between the basic problems we study and such practical activities as the design of radio sets or of bridges.

Indeed, the distinction that is here being made between 'applied' and 'basic', 'fundamental' or 'pure' science is not at all precise, and the attempt to make it sharp and clear is bound to fail (Reagan 1967). I am using a rough and ready conventional classification, which I should not want to defend in detail. The sooner we all face up to the fact that theory and practice are indissoluble, and that there is no contradiction between the qualities of usefulness and beauty, the better. If there is one feature of American science that we should all imitate—I speak for Britain, Europe, and most of the rest of the world—it is their willingness to have both fundamental and applied research going on together under the same roof. Their universities do not despise technological development work, while their industrial laboratories are well equipped with basic research groups. In a developing country, where short-term problems have much the higher priority, the conventional practice of keeping such work out of the universities is quite indefensible; as I have already said, the main function of long-term research without specific goals is to provide the best intellectual and educational background for well-informed technical innovation.

From these relatively sober considerations, it follows that any developing country, once it has emerged from the most primitive poverty, should foster a small amount of basic science. Of course this can be overdone; a research reactor can become as extravagant a 'status symbol' for a small country as its own international airline. But I do not wish to dwell upon the difficult problem of relative priorities and the criteria for choice between technology, applied science and fundamental research. For the remainder of this lecture I shall try to express some opinions concerning the establishment and maintenance of basic scientific work under these difficult circumstances. Although the obstacles are all too evident for those actually involved, very little attention seems to have been given to this problem in all the mass of writing and talking on industrial and technical development. In particular, I want to emphasize the intellectual and psychological aspects, which are neglected in the struggle for material resources.

At the outset, one must have a clear idea of the nature of science itself. It is all too easy to adopt an uncritical, imitative attitude. 'Science is what they do at the

Cavendish Laboratory, so we must try to build a laboratory that is just the same.' But any attempt to make a carbon copy of a social institution, under another sky, is bound to end in disappointment. If we are to foster new institutions that really work—productive and creative Indian science, or Pakistani science or Nigerian science, or Patagonian science—then we must have a proper understanding of the long-term goals and short-term functioning of the parent institution on which they are moulded. To give a ridiculous example: putting on a white coat and peering knowingly down a microscope no more makes a biologist than chanting metrical psalms makes a presbyterian. It is the inner spirit that counts, not the ritual.

For this purpose, it is essential to begin the analysis at the most general level; indeed, one must first try to define science itself. As I have argued at length, in a recently published book (Ziman 1968) the most useful definition comprehends the *social* dimension of science. The role of the pure scientist, however eccentric, imaginative, speculative or opinionated, is to make a contribution to 'public knowledge', to help construct a rational consensus. Many familiar dichotomies—experiment and theory, creation and criticism, accidental observation and deliberate search, logic and intuition, master and pupil, freedom and authority—can live together without contradiction in this definition. And, as I hope to show in this lecture, one can deduce from it some quite powerful guiding principles for the practical management of scholarly affairs.

This point of view implies, conversely, that there is no very special 'philosophy of science' that must already be latent in the culture of a developing country before a scientific community can be established. It is sometimes argued (Odhiambo 1967; Basalla 1967) that the philosophical, religious or cultural climate in certain regions of the world is so hostile to the scientific attitude that research could never thrive there without a complete psychological reorientation. 'In Jub-Jub Land', this argument runs, 'the study of botany would be quite impossible, for everybody is brought up to believe that oak trees spend the hours of darkness flying to the Moon', or 'How could you teach Euclid to a Zen Buddist, whose every utterance is deliberately irrational?'

We are not entitled to reject such assertions just because they are uttered with a hint of cultural superiority. Yet I do not think we need take them seriously into account, for several reasons. As Levi-Strauss has demonstrated, the supposed credulity and irrationality of primitive peoples is often over-emphasized in anthropological writings. Primitive systems of thought are not illogical. They respect such general principles as 'seeing is believing'; they know the difference between cause and effect; and they make regular categorizations of natural objects. From our point of view, the weakness of primitive thinking is the incorporation of incongruous and (to us) irrelevant elements in the chain of deduction—totem animals, ancestral spirits, ghosts and such like—which can, of course, lead eventually to the most bizarre consequences. But then many Western scientists seem not to be hampered by belief in the Virgin Birth, the infallibility of the Pope, the prophetic foresight of Karl Marx, the unwholesomeness of the flesh of the pig, or the superiority of the American Way of Life. I do not see the harm in a little totemism, white magic, or contemplation of Nirvana—provided it is out of office hours. The initial premises and modes of thought of science are essentially those of matter-of-fact, everyday life, and not those of the sort of metaphysician who feels that he has to prove that the tree is still in the Quad when nobody is about to see it. The primitive savage—*and* the plain man in the streets of London, Paris, Tokyo or Delhi—has many non-scientific beliefs, which another lifetime of education could scarcely correct; but he is, none the less, a naïve realist at heart. Only a very small proportion of the population, even of the most civilized societies, is susceptible to a genuinely *anti*-scientific, genuinely irrational, metaphysic. It takes a good dose of neurosis, or an altogether too literary education, to arrive at such a decadent intellectual impasse.

We must be careful here not to confuse the conditions required for the growth and spread of science from those required for it to appear spontaneously. However much we may admire the technical achievements of various other civilizations, we must concede that the science about which we are here talking came to birth in Western Europe in the seventeenth century, and has not had the occasion to appear independently at another time and place. One of the grand problems of cultural history is just why it did appear there; what were the origins and causes of this unique event. Of course, the whole thing is very well documented; but it still remains almost as baffling as the corresponding biological problem of the origin of life. This, indeed, is a good analogy, for the conditions envisaged for the original spontaneous generation of the very first living organisms are entirely different from those which could easily be colonized in the later stages of biological evolution. The sophisticated science of today could certainly not now be expected to appear all of a sudden out of the forests of the Amazon or the aboriginal wastes of Australia; but we know that it can be taught quite well, and can thus reproduce itself, in almost any civilized cultural environment.

But although the basic metaphysical foundations of science are relatively naïve, we all know what an immense superstructure of actual knowledge they carry. The unity of science—the fact that it is the product of innumerable hands, cooperating over the centuries—and the unique power of the dialectical processes of imaginative creation and critical reappraisal, make it by far the most elaborate intellectual construction that the world has ever seen. If one is to contribute to this vast system, one must become the master of what is already known, at least in some limited aspect. Occasionally an ignorant scientist may stumble accidentally upon an important new discovery but such serendipity must not become an excuse for lack of education as a policy. Self-taught geniuses, like the incomparable Srinivasa Ramanujan, are entirely exceptional and prove nothing against the necessity for the deliberate teaching of scientific knowledge.

This is, of course, fully recognized, and the first condition for the growth of science in a developing country is the establishment of centres of higher education. It is here that the child brought up in an advanced society has his major advantage.

Up to the age of ten or twelve he is still no better than a primitive savage, with only a material, instrumental, familiarity with the products of modern technology to offset his natural beliefs in magic and mystery. But the ordinary curriculum of secondary education introduces him rapidly and efficiently to the general principles of scientific thought, and his talents are guided by experienced teachers into the higher realms of knowledge. If he works hard, then he may have learnt enough by his early twenties to begin to do research on his own account.

For a child born into a developing country this process is far more erratic and uncertain. His whole environment may be entirely traditional, materially and intellectually, and it may only be a stroke of fortune that carries him beyond mere literacy. He will be lucky if he encounters good secondary teachers to give him the necessary grounding in scientific subjects, and it is more than likely that most of his university education will be at the feet of scholars who have not, themselves, fully mastered their subjects, and who can convey only the outward inessentials to their pupils. As I have already emphasized, university science teachers who are not in contact with research find great difficulty in keeping their knowledge fresh and up-to-date. The most serious obstacle to the creation of a genuine scientific community in a developing country is often the prior existence of a self-perpetuating academic system of low quality.

This problem is not confined to poor and technologically backward countries where higher education itself is relatively new. In a number of European countries where universities have existed for centuries, a tradition of uncritical book-learning, taught by lazy professors to a mass of students who are only seeking a nominal qualification for a white collar technical job, is the greatest impediment to scientific progress. It is instructive to notice that the best basic research workers in such countries often come through the 'Technical High Schools' or 'Institutes of Technology' which are supposed to be for the training of practical engineers. The reason is that these schools have developed outside of the old medieval university system, and as their qualifications have become more highly regarded they have built up very high standards of competitive entry and rigorous professional training. They make the most of the traditional 4 to 5 years course and give that thorough grounding in the basic sciences which is the hall mark of Continental European scholarship at its best. Whether or not it makes sense to copy such institutions in other countries with different cultural traditions, the value of a few such Centres of Excellence cannot be denied. The recent Indian decision in favour of such a policy can only be welcomed.

But a good basic education in the established principles of mathematics, physics, chemistry or biology is only half the training of a proper scientist. The longer this is drawn out, the more difficult it becomes to learn the other half—how to do research.

The harshness of the psychological transition from being a student to being a researcher is not appreciated in the world at large. Even in scientific circles—especially in the older Oxbridge tradition—it is sometimes assumed that a man who has shown that he has a 'first class mind' by answering the stereotyped riddles of the Tripos can be put down in a laboratory and safely left to get on with his research on his very own, with no more assistance than is appropriate to a junior colleague. That is what happened in Rutherford's day, so why should we act differently now.

Well, of course, Rutherford was a genius, and had enormous reserves of spiritual energy. The policy of 'throwing 'em into the deep end and larning 'em to swim' is not inappropriate to a strongly independent 'inner-directed' personality. Although I would not go all the way with Feuer (1963) in his attempt to show that modern science was founded by 'hedonist libertarians', and I am certainly not convinced that there is a direct intellectual link with the rise of capitalism or of Protestant theology, I am sure that this facet of the scientific attitude comes relatively easily and naturally to persons whose psychology has been moulded by the Puritan culture of northern Europe and the United States. The professional scientist must be able to shut off his ego from the opinions of other people, and 'press on regardless' with his own ideas (Merton 1968). In an age when bureaucratic conformism is said to have totally infected the human spirit, he must still assert the stubborn 'It still moves' of Galileo, which echoes Luther's 'I cannot do otherwise'. That same psychological type also arises from another more ancient variety of non-conformism—the ghetto. At its worst, the orthodox Jewish personality is pig-headed, intolerant and argumentative; but these are traits which can become the virtues of strong-mindedness, scepticism and intellectuality!

Unfortunately, these qualities are certainly not those that are encouraged by a system of mass education geared to the passing of book-work examinations (see, for example, Karve 1963). Even in the best of circumstances, one may be caught in a psychological trap. To become a scientist one must first master the current consensus, which demands many years of accepting the arguments and opinions of one's teachers without serious question; one must at the same time preserve, or acquire, the self-confidence to reject some of those arguments when, in due course, one uncovers evidence against them. There is no simple way of avoiding this dilemma; we must plunge into the educational system and endeavour to struggle out of it again without too much loss of impetus. In a strange way, some of the most brilliant scientific intellects have escaped the trap by never falling into it in the first place; either arrogantly, or in a dream, they simply opted out of the competition for early academic success, and preserved themselves for their imaginative, critical, creative role as research workers.

Again, we cannot legislate for the Einsteins and Darwins of this world. In the past, when science absorbed an infinitesimal proportion of the manpower and gross national product even of the most advanced countries, one could safely leave training in research to the rough and ready competition of 'sink or swim'. Now we need to organize and plan this stage in the creation of efficient scientific cadres with care and forethought.

By convention, this has now become the function of what the Americans have taught us to call a 'graduate school'. Unfortunately the very name—school—tends to put the emphasis on the quality of the advanced courses of instruction offered by the faculty. These are, of course, very important, especially in countries where the ordinary undergraduate curriculum is of a low standard, or too heavily weighted by old-fashioned 'classical' topics. In the ordinary way of things, advanced lectures both by the resident staff and by visitors from abroad, are the means by which living scientific attitudes and issues are communicated to the students. This is information that cannot be picked up by reading. The textbooks are inevitably a few years out of date, and review articles usually demand too much expertise for the beginner. Although the time of the student should not be overloaded with such courses, in the vain hope of covering all possible aspects of some very large subject, they are absolutely essential, both to those who listen to them and to those who must assemble the material to deliver them.

But a good graduate school is far more important as a training ground in research. This is not just a matter of picking up the practical techniques of the particular science—electronics, glass blowing, computing, plotting data and drawing logical conclusions: much depends upon the quality of the supervisors. For example, one of the attributes of the successful leader of research is his ability to set problems that will yield good doctoral theses for his students and teach them the art of research while making a positive contribution to the advancement of knowledge (Merton 1968). It is no good telling a young man to go away and solve the Riddle of the Universe, or find the Philosopher's Stone; unless he is a genius, he will fail, and become hopelessly discouraged; or even go off his head. On the other hand, putting him to work as a mere technical assistant, turning the knobs, reading the meters, and plotting the points in yet another set of measurements with the departmental, white elephant apparatus, will never make him 'self winding' as a scholar. One of the real disadvantages (besides its enormous cost!) of the very elaborate equipment now deemed essential to research is that a whole team of students and assistants is sometimes needed to mount an experiment. In neutron physics, in high energy physics, in space science and other sophisticated fields, the individual student may get an excellent training in advanced technology, but he is seldom called upon to defeat a real live problem of natural philosophy. Scientists in developing countries may not always be wise to deplore the poverty that cuts them off from these beautiful but extravagant playthings.

Out of this necessity one can scrape a little virtue; the experience of successfully conducting a modest investigation with limited apparatus may well be of greater value to the student than being one of a team of a hundred Ph.Ds manipulating a billion dollar 'facility' under the distant direction of powerful boss. Let me give a simple example. Following the lead of several other British universities, we have been experimenting recently in Bristol with research 'projects', in places of set piece laboratory work in the final undergraduate year. A little problem is set: could the sulphur dioxide emitted by a factory chimney be detected optically from a distance; design an electronic organ capable of imitating any type of musical instrument; what is the actual mechanism by which a steel wire passes through a block of ice? Under the supervision of a member of the staff, two students work together planning experiments, designing apparatus, and interpreting the results. Very little of this work is at all profound, but it has the very important characteristic that nobody—not even the supervisor—knows the answer in advance. In this respect, therefore, these little projects are true to the spirit of science, and provide a most valuable psychological introduction to professional research. The old tradition of a sharp transition from undergraduate course work to postgraduate research may have been quite wrong. The changeover should perhaps be quite gradual, over a period of years; research projects introduced into the undergraduate curriculum are, so to speak, a counterbalance to the further formal instruction now required at the graduate level. In this way, also, we, and our students early discover their natural inclination—or disinclination— towards a scientific career, and they can move away into more useful and practical professions without loss of face.

But a graduate school is more than a stock pile of scientific learning, technical advice, solvable problems and material equipment. The psychological transition from student to researcher is not merely a matter of strengthening one's personality and throwing away the crutches of bookish knowledge. Paradoxically, one must also learn to be 'other directed' in a special way: one must acquire the habits and conventions of a responsible member of the scientific community.

For example, the graduate student must learn to write scientific papers in the peculiar impersonal style which is now customary. He must learn to accept criticism without personal offence, and to offer it without rancour. He must learn to give due credit to other people for their prior discoveries, and not to claim too much for himself. He must learn to scour the literature for references and to keep abreast of all relevant work in his own subject. Above all, he must somehow acquire high standards of scientific accuracy, honesty, and judgement, so as to distinguish quickly between the true and the false, the meaningful and the trivial.

Now it often seems to be taken for granted that these desirable habits of mind and art will come quite naturally to any sufficiently clever young man who is set down to a good scientific problem in a well-equipped laboratory. Nothing could be further from the truth. They do not come from some inner source of soul, but are acquired by imitation of the current standards and conventions of the particular institution where he is trained. They cannot be mugged up from a textbook—which in any case, would be about as useful as an instruction leaflet on skiing or horse-riding. It is rightly assumed that the difficult art of research is learnt by apprenticeship, by direct personal contact with an experienced practising master.

This is of the very greatest consequence for the whole theme of my lecture. I have emphasized the world-wide unity of pure science. That unity is not achieved and maintained by such feeble strands as the publication of learned journals, nor by corporate junketings at conferences and congresses; it exists because modern science stems historically from a single source in Renaissance Europe, and has spread outwards by the natural process of an apostolic succession—by the fact that almost every practising scientist has at one time or another been the personal pupil of a scholar of a previous generation. In our own day we have institutionalized this tradition; the regulations for the Ph.D. now practically ensure that one cannot get registered as a professional research worker unless one has been supervised by someone who already has a Ph.D. As one who is often called upon to act as external examiner for Ph.Ds of other universities, I must admit that this ritual is becoming a little like a secular version of the 'laying on of hands' by which a priest is catechized and ordained by a Bishop.

The spread of science throughout the world has not been the haphazard distribution of wind blown seeds, encapsulated within the dry covers of reprints of scientific papers, but by runners and tendrils reaching out from the original institutions and establishing themselves in new soil. Ideas, standards and traditions travel around inside people, and are only transferred from one to another by prolonged contact.

Of course the major developing nations now have self-sustaining scientific communities where this tradition is preserved and retransmitted. But this is quite a recent development, even in some of the most advanced industrial societies (Basalla 1967). It is interesting to recall, for example, that the St Petersburg Academy, founded by Peter the Great in 1724, was largely staffed by non-Russian scholars until late in the nineteenth century (Vucinich 1963) while even the United States was dependent upon German graduate schools for the training of research workers in many scientific fields until a similar date. The stage of 'take-off' in pure science has been reached only in the past few years in so rich a country as Australia; while small nations, such as New Zealand, or Finland, still find very great difficulties in becoming self-sufficient in graduate studies over a wide range of the sciences.

It is easy enough, at this stage in the argument, to echo the points expressed very properly and concisely by Moravcsik (1966). Graduate schools, he says, should be established in developing countries because of the cost of sending people overseas, because of shortages of places in Western institutions, because graduate students are valuable technical assistants in research, and because once you have sent a man overseas for training it becomes difficult for him to re-adjust to more primitive conditions on his return. All these are valid reasons, and the various suggestions for greater technical assistance by way of buildings, equipment, and visiting experts would fulfil genuine needs, which should be given the highest priority.

It is also generally agreed (Kidd 1968; Maheshwari 1964; Blackett 1964) that one should try to give a future research worker the fullest possible preparation in his own country before sending him to an institution for advanced training overseas. There are many factors, of cost and morale, in favour of such a policy.

Yet there are real dangers in forcing the premature growth of graduate schools before sufficient human resources are available to man them. Only a policy of rigorous selectivity, by the encouragement of really first rate scholars, makes sense.

I know how difficult it is to achieve such a policy, against all the academic political arts of log-rolling, me-tooism, appeals to fair play, etc; but the consequences of flabbiness are all too sadly evident in all quarters of the globe—the proliferation of third-rate research which is just as expensive of money and materials as the best, but does not really satisfy those who carry it out, and adds nothing at all to the world's stock of useful or useless knowledge. The old adage should be stood on its head. Of applied science, it can reasonably be said 'If a thing is worth doing, it is even worth doing badly'; of pure science I am tempted to say 'If a thing is not worth doing, it is only worth doing well!'

The difficulty is that one cannot create good research groups just by setting up graduate schools: the two types of institution are strongly interacting, and come into existence simultaneously. An indigenous school of research is the *culmination* of the growth of pure science in a particular country, not a mere subsidiary agent.

We have got beyond the historical phase where Henry Cavendish or Michael Faraday could work away in his own little laboratory, make his own apparatus, carry out his own experiments and report his own interpretations and theories in communications to the *Philosophical Transactions of the Royal Society*. We find ourselves now with *groups* of professional scientists, organized in teams, departments, divisions, etc., working collectively. If such a group is to live for more than a few years, it must have a balanced composition, with a range of ages and experience and mechanisms for acquiring new members as the older ones retire or become administrators. These are the makings of a graduate school; the problem of tacking graduate students on at the bottom, and arranging for their supervision and formal instruction, is quite trivial by comparison with the problem of creating such a research group in the first place. All active scientific institutions are schools of apprenticeship for their junior members, whether we call them graduate students, research students, 'post-docs' or assistant scientific officers. The responsibility for producing well-formed research workers out of raw graduates should be the most serious burden on any senior scientist or laboratory head. These are the seed corn of the next crop, and the whole quality and success of the scientific activity of one's country depends on how they are treated in these vital years.

Until that happens, there is no substitute for post-graduate and post-doctoral study abroad by the most promising young scientists. Whatever the disadvantages, costs and difficulties of such a policy, they must be borne if a developing country is ever to acquire a worthwhile scientific establishment.

In any case, the notion that a country can ever become so self-sufficient scientifically that individual research workers need not travel is a gross fallacy, only pardonable in a Chancellor of the Exchequer, a cost accountant or a Commissar of Culture. The whole tradition of science feeds on the sacrifices of pilgrimage and exile. Even in Europe, we are learning that polite letters and occasional conferences do not constitute a network of communications: we must actually go and work for a while with our colleagues across the border.

The reasons for this are more fundamental than that 'travel broadens the mind' or that 'one needs the stimulus of a new environment' or even that one can only do good work under the ideal climatic conditions of California—presumably with a good fat Californian income to keep the wolf from the door. The universality of science is at stake.

Consider, for example, the choice of a research problem, and the strategy and tactics of one's attack upon it. The aim is to produce a piece of original, *publishable* research. What this really means is that it must be the sort of research that is of interest to others; in other words, it must seem to contribute to the great body of 'consensible' knowledge.

There is always a stimulating philosophical debate going on concerning the 'importance' or 'significance' of various branches of science, and of particular discoveries in particular fields. It is doubtful whether such questions of taste can be resolved by argument; but it is certain that the vast majority of scientific workers have only a single simple definition of publishable research: it should look as much like other published research as possible, with a few new features to make it seem adequately original.

This remark is not meant to be altogether sarcastic. The great power and philosophical authority of scientific knowledge stems from its cooperative character; each of us builds on and into the work of his predecessors and contemporaries. The rare geniuses are master masons, projecting tremendous extensions of the fabric, but the stones are put in place, one by one, by the numerous journeymen. A piece of research that does, indeed, add just a little to the work of others is more valuable than a disconnected, fruitless, speculation, on however grand a theme.

But this means that the topics and techniques of current research are determined by the actions and opinions of the whole scientific community, throughout the world. By its very nature, pure science is completely supernational, for it is essential that every new discovery should be communicated to every scientist whom it might interest, and that it be subject to the critical eye of every competent authority, whether he lives in Tobolsk, Medicine Hat, or the legendary hamlet of Waikikamukau. Everyone knows that science, like love, knows no frontiers; but I do not think it is adequately realized that this is of its very essence. To cut a scholar off from his potential audience (however critical) by administrative, financial, and political curtains and walls is to imprison his mind.

International art fashions or international political movements are signs either of impoverishment of imagination or crude powerseeking. It is sad and unnecessary, for example, that Japanese motor cars should look just like Italian ones, and that Mexican communists should use the same debased Leninist terminology as their Malayan comrades. But there is nothing out of the ordinary in my personal observation that the Mössbauer effect is being studied in Helsinki, and in Western Australia, as well as in its native Germany. Even in China, whose scholars are so unhealthily cut off from direct personal contact with foreign intellectuals, current scientific work is still recognizably international in its appeal (Orleans 1967). While there may be charming variations of *style* in research from country to country, the idea of a distinctive American, Russian, or Brazilian science, with a different *intellectual* content, obtaining different results, is a contradiction in terms: When, as sometimes happens in wartime, the usual channels of scientific communication and criticism between two countries are blocked, and divergences of opinion develop on purely scientific questions, the very first task of the scholars in both countries is to organize meetings at which the breach can be mended—not by force but by the presentation of the most convincing arguments and evidence until a minimum consensus is re-established. It is the analogy of such meetings that lies behind the Pugwash movement, which has tried to extend the range of consensible topics to include more serious questions of peace, war and politics.

Science in a developing country cannot, therefore, cut itself off from competition, and protect itself by tariffs like an infant manufacturing industry. If it is to be genuine science at all, it must, from the very beginning, be able to stand on its own feet. The international scientific community is not a very lenient examiner, and does not award even one mark out of ten for just attempting the question. Work that is not up to the standard set on the world market of ideas will simply be ignored; it will either not get published, or will appear in an obscure local journal which is not read outside its own country.

Now there is nothing more disheartening and debilitating to a scholar than to know that his work goes unrecognized and unread. His whole professional activity is directed towards the production of some contribution to knowledge; his labour is utterly wasted if that contribution is eventually judged to be negligible. If pure science is to exist at all, in any country, it must be adequate in quality, by this criterion. But only by keeping in close contact with his foreign contemporaries by word of mouth, hand-waving, and sketches on the backs of envelopes or lunch table napkins, can the active research worker maintain his critical standards. He must actually meet the authors of the papers he reads; he must have the opportunity to persuade them, face to face, over periods of weeks and months, that he has something of his own to contribute to the discussion. Without such intercourse, his work is liable to drift into a backwater, where he engages in mock debates with himself, quoting only his own papers and ignoring the progress made by other people. Until the day of the universal, free, intercontinental videophone, there is no alternative to carrying the body around when the mind needs to travel.

Even within the range of publishable science with an international audience there is a great deal of choice. The mistake that is often made by those who are a little out of touch is to follow fashion—not even current fashion but the styles of a few years back. This is natural enough, I suppose. Topics and techniques that are heavily emphasized in the literature have a high visibility, even from a distant continent. Imperfect communications will make them seem the only important topics for research; while the peculiar frenzy that excites some ambitious scientists when they see a problem half way to solution can seem the collective wisdom of the community at large. The phenomenon of fashion in science is of great interest and significance (see, for example, Hagstrom 1965) but its saddest consequence is the tail of ill-equipped research groups who have not quite succeeded in jumping on that gaudy band-wagon as it flashed by. The fact, of course, is that fashionable subjects are those in which the competition is fiercest, and often where too many good people are chasing too few ideas. It does not seem very wise for a small research group, with limited equipment and many distractions and difficulties, to take on the Bell Telephone Laboratories at their own game!

The optimum strategy for the organization and planning of pure scientific research in a developing country would thus seem to be to concentrate on a few solid scientific problems, not necessarily those that are currently fashionable, and to establish a sound reputation for good if unspectacular work in these particular fields. The free trade in knowledge, which is the key to the so-called scientific method, demands specialization and division of labour. It is trite to remark that no single person can now have the whole of knowledge as his province, that the era of the universal scholar is long past. I would go further, and say that the age of the universal university is over—that even such great institutions as M.I.T. and Imperial College cannot hope to have experts in every field of science on the faculty.

It is much more profitable to emphasize particular lines of research, allowing each research group to reach the critical size for continued viability. I have thrown a spanner at team research which reduces the graduate student to a mere pair of hands; on the other hand, most modern scientists work best if they have a number of close colleagues and contemporaries with whom to discuss problems and from whom to acquire the unwritten expertise of the subject. One cannot lay down an optimum size for such a group, but my own guess is that one needs three or four permanent, established staff, with perhaps a dozen juniors and graduate students, to generate enough intellectual heat, by their mutual interaction, to keep the pot boiling. There are, of course, many forces leading to the fragmentation of research—and there are some gross evils in the building of academic empires—but this degree of concentration and specialization should be the aim of any deliberate policy for the encouragement of pure science, however 'advanced' the country.

Such a policy, however, demands just as much attention to the machinery of communication and travel as trying to follow fashion. It is necessary, for example, to know, and be known to, the few other research groups in the world who are also interested in the same topics. The formal exchange of reprints and pre-prints is not enough (Moravcsik 1966). To ensure coordination, collaboration, and fruitful competition rather than sterile shadow boxing, the leaders need to be on terms of personal friendship, and there must be regular exchanges of persons between the groups. I cannot emphasize too strongly the difference that it makes to one's attitude to a paper when the author is not just an outlandish, unpronounceable name, but that clever young man one met last summer in Uppsala—or, conversely, that long-winded old fogey from Nebraska who was mooning about the Department a couple of years ago. Perhaps such emotional prejudices ought not to be, but they are to be reckoned with in the real world.

I am fully aware that this optimum strategy is a counsel of perfection which is not at all easily followed. The reality of pure scientific research in many developing countries is often tragically wasteful of training and talent. There is a regular pattern of failure, which continually frustrates the well-meant effort to construct viable scientific institutions. Let me draw attention to one typical phenomenon.

Let us assume that a young and able student has taken himself off to a good graduate school, in his native country or abroad, and has got his Ph.D. He has the makings of a competent scientist, and given a decade of experience in active research he might well have become the powerful nucleus of a good new research group. But for perfectly proper reasons—patriotism, availability of jobs, family and cultural ties etc.—he takes a relatively junior position in a small university away from any major scientific centre. What happens to him?

We do not know in detail, because no sociologist has yet had the inspiration to go and find out. But it does not take all the professional technique of sociology to ask a few questions, and to make a few modest enquiries. On this topic I can strongly recommend an excellent article by Amar Kumar Singh (1962), who discusses very perceptively the impact of foreign study on Indian students. He expresses very forcibly the dissatisfaction and disillusionment with such things as (I quote) 'nepotism and corruption in public organizations and government; poverty and low standards of living; waste and dishonesty; low morality in commerce; red-tape and bureaucratic delay; the discouragement and obstructive attitudes of senior persons in positions of authority; the general inefficiency, lethargy and disorganization permeating all spheres of social and political life, and the absence of social justice and individual dignity'. Stevan Dedijer, in an equally forthright article (1963) adds, for good measure, the lack of any general cultural sympathy for the scientific point of view, and remarks on certain pathological forms of careerism that flourish in this unhealthy soil. I have quoted evidence from India, but much the same would be said about many developing countries by those who know them best. One could even apply some of these remarks to a region such as Sicily, where the dishonesty and moral corruption of the Mafia is a formidable obstacle to the scientific spirit.

It is not surprising, in these circumstances, that many a promising young scholar abandons serious science at this stage. He loses enthusiasm for research, and becomes eventually one of those lazy, do-nothing politicking professors at whom so many brickbats are thrown—or else he resigns his post, and flees to a bigger centre at home or abroad, where he can continue to be a scientist. At the moment when he is trying at last to stand on his own feet, he is bowled over by the irresistible tides of an inimical cultural environment. It is asking too much to expect superhuman fortitude in the face of such circumstances.

I see little point in belabouring such a familiar topic and sermonizing on the ills of other people's cultures. One might as well address stern admonitions to the weather, and rage, like King Lear, at the elements! But I do believe that there is a very special factor which particularly affects the professional scientist, and about which something can be done. He becomes intellectually isolated.

As I have emphasized throughout this lecture the romantic picture of the scientist as a lonely hero, on a solitary expedition through a sort of starlit outer space of the mind, is applicable only to a few extraordinary geniuses; whatever *we* do, *they* look after themselves. The modern research worker is a highly trained professional, who has been taught to cooperate—and compete—with his contemporaries. The better his training, the more he will have to come to depend upon the face to face contact with his peers, for discussion, stimulation, criticism and technical information. He is no better prepared for doing research on his own, with no one near to whom he can speak about his work, than the inhabitant of a city is equipped for life in the middle of the Sahara. If he is to pass safely between the Scylla of scholarly decay and the Charybdis of the brain drain, then we—the international scientific community—must try to help him. We owe this not merely to our own charitable instincts but to our allegiance to the pursuit of learning, for he is not merely a fellow man but a scientific colleague. What can we do to keep his mind alive?

Here again, the only cure is to provide some means of personal contact with other scientists in the same field of study. The admirable practice of sending experts out on lecture tours from the major centres to the provinces, or from advanced countries to less developed regions, helps a little; but it has necessarily only a transitory effect that cannot be repeated often enough, or for long enough on each occasion. It is important to remember that we are not so much concerned with particular pieces of knowledge—the latest formula for the elixir of life, or where to look for quarks, or what would have been found at the bottom of a Mohole—but with recreating an intellectual environment. The visiting expert may convey a great deal of technical information in a brief visit and by criticism or encouragement may decisively alter the direction of research in particular cases. But he cannot transport the atmosphere of a whole institution.

In one of his sympathetic and realistic articles on this whole subject, Moravcsik (1964) has made much of what I should describe as the missionary approach to the problem—the sending out of young but competent scientists from the advanced countries to spend several years or more in the newer institutions to help them set high standards of teaching and research. This too, is an admirable activity, but it is not really cheap in money and manpower. It demands, indeed, a missionary spirit of charity and humility which is not always to be found in psychological combination with scientific ability, and it carries with it just a hint of the anxious condescension that is so damaging to the whole missionary enterprise. As I have already remarked, the exchange of scholars for substantial periods is far and away the best means of linking the institutions for which they work, but it needs to be justified on its own merits, and as between near equals.

Another valuable development is the organization of local, national and regional conferences, summer schools, and other meetings. One of the lamentable consequences of intellectual isolation is *isolationism*, the unconscious fear of exposing the inadequacy of one's achievement to one's fellow scientists. In the struggle to get any work done at all, the effort to create time, funds, and organization for such meetings begins to seem not worth the possible benefits.

This is a mistake that we have long been making even in Britain, and in Europe generally. The importance of a scientific meeting does not lie solely in the formal communication of the results of research; it is an opportunity for a scientific community to become conscious of itself, and for individual scientists to become aware of one another's difficulties and common needs. Meetings cost money, and they demand public-spirited initiative from those who arrange them, but they could be the saving of many sound scholars.

The above remedies for the chronic malady of intellectual isolation attack the cause, but the disease is so deep-rooted that more drastic cures are needed. A day's visit from a travelling expert, a week at a conference, a fortnight at a summer school —these are only occasional episodes that do not really change the pattern of one's life. Somehow we must make it possible for young scientists in developing countries to maintain prolonged contact with their fellows, over periods of months and years.

The most ambitious project with this goal is Abdus Salam's International Centre for Theoretical Physics, at Trieste. This was created in 1962 as a place to which active scientists from developing countries might go from time to time, for periods ranging up to a year, to bring themselves into contact with current research and with their contemporaries from both developing and advanced nations. One might say that it was conceived as a permanent meeting place for the members of the Invisible College of this particular scientific subject, where they might return again and again to recharge their intellectual batteries. By arranging lengthy advanced courses of study, and a programme of seminars by distinguished visiting scientists, Salam has made it an institution of the very highest standing, permeated with and radiating the best scientific traditions.

The work of the Trieste Centre is so very relevant to the whole theme of this lecture that I wish I had time to say more about it. I should have liked, for example to speak of the experience of directing one of these advanced courses, which was attended by young research workers from about the most diversified list of countries that you could imagine, and which nevertheless demonstrated the complete universality of the scientific attitude in this particular field. This experience was, indeed, so moving and revealing that it is the real basis for my temerity in giving this lecture; and much of what I am saying here is merely a reflexion of their own feelings about their situation.

The International Centre for Theoretical Physics can deal only with a part of the problems on a very small sector of the whole front line of modern science. I am by no means sure that it is in all respects an ideal solution, nor that the same thing could be done in all other fields. But it has produced one entirely new social device —the scheme of *Associateships* (Salam 1966; Editorial in *Science*, 1968)

Briefly, this is a scheme by which an individual scientist acquires the means to travel to Trieste and to spend up to three months there, once a year, over a period of several years. It is like a travelling scholarship, or post-doctoral fellowship, but broken up into small pieces, and spread over a much longer time. Travel and subsistence costs for each visit are paid for, but the associate must, for his part, continue his ordinary duties as a lecturer, professor, or scientific officer in his home country in the intervening periods.

The effectiveness of such an arrangement rests upon its being so well geared to the true needs and time scale of the isolated scientist. A few months at a time is as much as he can be spared from his official and domestic responsibilities, and yet is long enough to make significant progress on a new scientific problem. The annual repetition of this privilege over, say, 5 years, allows his work to continue and grow, with opportunities for critical evaluation each year. It stabilizes him in his native environment and yet gives him a seat at the international 'high table'. By assigning an appropriate interval of time to each of those conflicting responsibilities that tend to pull him apart, he can largely reconcile them, and serve each with honour.

But there is no reason at all why such a scheme should be arranged solely in connexion with an international centre of the Trieste type. Every good university is an international centre of learning. The cure for isolation is to come back, from time to time, to any active research centre in one's particular field. The very strength of the scientific community is that it is cross-linked in all directions, and does not depend upon just a few key 'centres'. It is still true, however, that ideas tend to flow outwards, from the most successful and active institutions (not always in advanced countries) to the isolated workers on the periphery. Let the means be found by which the latter can travel regularly to the places whence their ideas come, so that they may share a little in the making of them, and carry them home inside their own heads. The associate thus becomes, so to speak, a foreign member of an active research group, familiar as a person through his regular visits, and a party to its collective wisdom. On the scale on which money is lavished on the support of science, the cost of such a scheme would be negligible by comparison with its benefits. Here is the most practical means by which the scientists of advanced countries can give aid to their colleagues in developing countries—aid that is more precious than shiploads of books and apparatus.

As I come to the end of this lecture, I realize how much more there is to say. I am aware, too, that I have made an elementary error of scientific strategy; I have tackled an insoluble problem, and merely recapitulated all the familiar, trite, non-answers to it. But the problems of the growth of pure science in developing countries are not scientific problems, which by definition always have 'answers'. They are by no means always technological or economic problems—although these are large factors in the equations. They are historical, political, cultural, and psychological problems—that is, they belong in the realms where there are never 'solutions' but only influences. They demand more than knowledge; they demand wisdom, charity, and heroic strength. I bow to those men and women, in all quarters of the globe, who have shown by their scientific achievements under the most adverse circumstances, that these virtues are still attainable.

References

Amar Kumar Singh 1962 *Minerva* **1**, 43–53

Blackett, P. M. S. 1964 *British universities and the developing countries*, Convocation address to University of Leeds, 4 July 1964.

Blackett, P. M. S. 1967 *Science, N.Y.* **155**, 959–964.

Basalla, G. 1967 *Science, N.Y.* **156**, 611–622.

Crawford, M. 1966 *Minerva* **4**, 170–185.

Dedijer, S. 1963 *Minerva* **2**, 61–81.

Editorial in *Science, N.Y.* 1968 **161**, 421 'Foreign associates'

Feuer, L. S. 1963 *The scientific intellectual*, New York: Basic Books.

Hagstrom, W. O. 1965 *The scientific community*, pp. 177–184. New York: Basic Books.

Johnson, H. G. 1965 *Minerva* **3**, 299–311.

Johnson, H. G. 1966 *Minerva* **4**, 273–274.

Johnson, H. G. 1967 *Minerva* **6**, 105–112.

Karve, D. D. 1963 *Minerva* **1**, 263–284.

Kidd 1968 in *The politics of science* (ed. W. R. Nelson) New York: Oxford University Press.

Maheshwari, P. 1964 *Minerva* **3**, 99–113.

Merton, R. K. 1968 *Science, N.Y.* **159**, 56–63.

Moravcsik, M. J. 1966 *Minerva* **4**, 381–390.

Moravcsik, M. J. 1964 *Minerva*, **2**, 196–209.

Odhiambo, T. R. 1967 *Science, N.Y.* **158**, 876–881.

Orleans, L. A. 1967 *Science, N.Y.* **157**, 392–400.

Reagan, M. D. 1967 *Science, N.Y.* **155**, 1383–1386.

Salam, A. 1966 *Minerva* **4**, 461–465.

Toulmin, S. 1966 *Minerva* **4**, 155–169.

Vucinich, A. 1963 *Science in Russian culture: a history to 1860*, p. 384. Palo Alto: Stanford University Press.

Ziman, J. M. 1968 *Public knowledge: the social dimension of science*, Cambridge University Press.